Silver Recovery from Assorted Spent Sources

Toxicology of Silver Ions

Silver Recovery from Assorted Spent Sources

Toxicology of Silver Ions

Editor **Syed Sabir** *King Saud University, Saudi Arabia*

World Scientific

NEW JERSEY • LONDON • SINGAPORE • BEIJING • SHANGHAI • HONG KONG • TAIPEI • CHENNAI • TOKYO

Published by

World Scientific Publishing Europe Ltd.
57 Shelton Street, Covent Garden, London WC2H 9HE
Head office: 5 Toh Tuck Link, Singapore 596224
USA office: 27 Warren Street, Suite 401-402, Hackensack, NJ 07601

Library of Congress Cataloging-in-Publication Data
Names: Sabir, Syed, 1970– editor.
Title: Silver recovery from assorted spent sources : toxicology of silver ions /
edited by Syed Sabir (King Saud University, Saudi Arabia).
Description: New Jersey : World Scientific, 2018. | Includes bibliographical references.
Identifiers: LCCN 2017043642 | ISBN 9781786344571 (hc : alk. paper)
Subjects: LCSH: Silver--Recycling. | Silver ions. | Silver--Environmental aspects. |
Silver--Toxicology. | Silver mines and mining--Environmental aspects.
Classification: LCC TP245.S6 S55 2018 | DDC 622/.3423--dc23
LC record available at https://lccn.loc.gov/2017043642

British Library Cataloguing-in-Publication Data
A catalogue record for this book is available from the British Library.

For any available supplementary material, please visit
http://www.worldscientific.com/worldscibooks/10.1142/Q0134#t=suppl

Desk Editors: Herbert Moses/Jennifer Brough/Koe Shi Ying

Typeset by Stallion Press
Email: enquiries@stallionpress.com

Printed in Singapore

About the Editor

Dr. Syed Sabir is a Distinguished Research Professor at the King Saud University, Riyadh, Saudi Arabia, and Visiting Professor at Riyadh College of Technology, Saudi Arabia. His research interests include hydrometallurgy (Au & Ag), hydrophobic nanomaterials, hydrophobic biomass, carbon dioxide capturing, hydrogen energy, water pollution, and atmospheric degradation of materials. He has published over 60 peer-reviewed research papers in international journals, including three invited reviews, as well as five invited books. One of his articles titled, "Recovery of Gold from Secondary Sources — A Review," was ranked first on the Top 25 Hottest Articles for Hydrometallurgy in the years 2012–2017 Full Year (January to December). He has over 1,500 citations, 190,514 publications views, and has received the "Hydrometallurgy Top Cited Author" from 2012 to 2017, awarded by *Hydrometallurgy* journal (Elsevier Ltd., New York, USA). He was also awarded the "2016 — Highly Cited Researcher" (Thomson Reuters IP and Science, Boston, USA). He is a reviewer for more than 65 international journals and is a research proposals reviewer of various international research institutes. He is a member of many international scientific societies and has several international collaborations to his credit. His books and research papers are being used globally as standard reference sources.

Acknowledgments

The authorship of this book was supported by the Deanship of Scientific Research (DSR), and College of Engineering Research Center, King Saud University. The author would like to thank them for their support.

Contents

Chapter 1
Introduction

Syed Sabir

Chemical Engineering Department, King Saud University
P.O. Box 800, Riyadh-11421, Saudi Arabia
sabirsyed2k@yahoo.com; ssabir@ksu.edu.sa

1.1 Silver: Historical Perspective

Silver is a precious and durable transition metal, which occurs as a minor constituent, is 67th in abundance among the metals by weight in the earth's crust, with an average concentration of approximately 0.1 mg/kg [1], is found worldwide, and is a member of group 1B of the Periodic Table. The name silver is derived from different sources — the ancient name of silver refers to its bright white color; the Hebrew term originates from the verb "to be white", and the Roman and Greek terms come from the terms "argentum" and "argyros", respectively, which mean "to be shining". It seems that the word "argenos" is of Aryan origin and "arg" is of Sanskrit origin (i.e., "clear"). However, in Latin, "argentum" stands for the element silver [2]. This metal was known to ancient civilizations along with other metals, viz. gold, copper, iron, tin, lead, and mercury [3]. The physical beauty of silver has made this precious metal highly desirable to many civilizations, past and present. The metal's mechanical properties of ductility and malleability, and its high electrical and thermal conductivities, have also made it suitable for a wide range of industrial applications. The singular beauty of silver was well known, and it has been used by humans for over 6,000 years and has played significant roles in various aspects of human life [4, 5]. The use of silver in ornaments, utensils, daggers, pottery, and trade in the ancient Aegean period, and as the basis of monetary systems by ancient people living in the Indus Valley (i.e., in Harappa and Mohenjo-Daro cities), is supported by archaeological evidence [6–10]. Interestingly, many ancient cultures used silver and silver compounds as natural anti-microbial agents in numerous traditional medicinal preparations. Several studies have revealed that nanometer-scale silver particles have unique and potent anti-bacterial and anti-inflammatory properties that can promote faster wound healing. Importantly, in recent years, several bacterial and fungal species have developed immunity against many routinely used antibiotics. Therefore, there is an imperative need to develop new anti-microbial agents to fight these antibiotic-resistant strains. Thus, nanoforms of silver have been extensively studied and incorporated into a wide range of wound dressings and pharmaceutical preparations to prevent bacterial growth. Furthermore, due to the efficacy of nanosilver, it has also been used as a disinfectant agent in a variety of food/beverages containers and in water purification systems.

In pre-historic metallurgy, attention was focused mainly on metals such as lead, silver, copper, and iron, which may have witnessed social changes, the development of the civilization, and mastery of new weapons and articles during the ancient Aegean period. Native silver is relatively rare in the Middle East, and it is believed that the bulk of ancient silver was separated from lead by the process known as cupellation, which may have been discovered in the pre-history period of metallurgy [8]. However, the earlier pre-history in the subcontinent supported the use of copper, gold, silver, and iron. For centuries, the British, French, and Portuguese have developed trade with India and the Far East in silver or gold for their aesthetic value as well as their rarity [6]. The global demand of silver has been escalating progressively with the increasing consumption of electronic and electrical equipment (EEE) [11, 12]. Global silver mine production declined in 2016 for first time in 14 years by 0.6% to a total of 885.8 million ounces [13].

1.2 Silver Occurs Naturally

Silver is a naturally occurring element in the earth's crust, generally present in fairly low concentrations [14]. However, silver is present as minor and trace constituents of a great variety of minerals, which are listed by Boyle [2]. The most common silver-containing minerals are native silver (Ag), acanthite (Ag_2S), pyrargyrite (Ag_3SbS_3), proustite (Ag_3AsS_3), tetrahedrite tennanite [$(Cu,Fe,Ag)_{12}$ $(Sb,As)_4S_{13}$], chlorargyrite (AgCl), argentite, and argentojarosite [$AgFe_3(SO)_2$ $(OH)_6$] [15]. Most silver is produced as a by-product of copper, gold, lead, and zinc refining, whereas some amount of silver is extracted by Parkes and cyanidation processes. There are a number of places named after silver, for example, Argentina or the Rio de la Plata in South America. Canada (9%), Australia (9%), Soviet Union (7%), and Chile (7%) were traditionally the main silver-producing countries in the world [16], although today's main silver-producing countries are Peru, México, and China [7]. Silver production worldwide reached more than 2,700 tons in 2009 — there is now an estimated reserve of 550,000 tons in the earth's crust, but in 2014, the global production of silver reached 877.5 million ounces [17]. About 700 tons are consumed annually in heterogeneous catalysis (e.g., production of ethylene oxide and formaldehyde, purification of diesel emission gases), more than that used for jewelry [7]. Literature study indicates that the acidity of an indigenous rock has a pronounced effect on the silver concentration in rocks. Boyle [2] suggests that basic rocks contain a maximum amount of silver, with the concentration decreasing with increasing acidity. This makes perfect sense from a chemical standpoint since silver is most soluble in acidic condition. In the recent years, while the natural resources of silver are decreasing, the cost of silver productions has risen rapidly, and the market price of silver has undergone a spectacular rise in spite of its increased applications in numerous fields such as photography, radiography, electronics, photonics, electrical, catalysis, batteries, jewelry, silverware, dental material, biomedicine, medicines, disinfectants in wastewater treatment, food/beverages processes, etc. [18–22]. Among its wide applications, its anti-microbial activity is of great interest.

With the growth of all these industries, the recovery and removal of silver have been intensely studied because its present market demand poses an acute problem in the recovery of silver from spent sources through new economically efficient and environmentally clean technologies [23]. The recent interest in using eco-friendly and cost-effective recovery methods to reduce pollution levels offers an important secondary source of silver. Using plants to extract and recover silver from mine tailings, contaminated soils, and wastes from other sources (i.e., radiographic film and electronics) is an alternative and different method from conventional mining and extraction processes. Meanwhile, there has been a phenomenal increase in the regulations against environmental pollution caused by silver ion toxicity in the marine, microbial, invertebrate, and vertebrate community (including humans) through chains, which has caused numerous diseases and disorders [24–27]. Therefore, there is great concern in the recovery and removal of monovalent silver ion (Ag^+) from industrial effluents and spent sources, up to ppm level [28]. Moreover, there is a great interest in silver recovery from both environmental and economic points of view [29, 30].

1.3 Natural Sources

Generally, silver metal is obtained from natural sources as a co-product of gold and by-product of zinc, copper, antimony, lead, etc. [19, 31] through conventional open pit as well as underground methods. Silver is most commonly associated with the ores of Au, Cu, Pb, and Zn [32]. The silver-containing base metal ores are crushed and concentrated by flotation methods. The recovery methods are, however, different for ores of lead, zinc, copper, and gold.

Lead concentrates with silver content ranging between 0.08% and 0.12% are sintered and then smelted with coke in a low shaft blast furnace. The hard lead produced from the blast furnace, containing impurities such as copper, antimony, zinc, bismuth, nickel, etc., is pyro-refined in refining kettles in stages. The silver present in hard lead (0.15–0.25%) is separated from the melt by Parkes process involving the addition of zinc into the melt. After thorough mixing of zinc into the molten mixture of lead and silver, the bath is allowed to cool down. The top crust of silver–zinc–lead alloy containing 4–5% silver is removed using a Howard press, which helps in squeezing out the mechanically entrained lead. The silver–zinc–lead alloy is charged into a retort furnace, where zinc is condensed in the adjacent condenser and the retort bullion, containing 7% silver and lead, is cast out. In the next stage, lead is recovered by oxidation to litharge (lead oxide) lying behind enriched silver (996 fineness) in the cupellation furnace. In the year 2010, about 30.72 tons (0.72 M oz) of silver was recovered from lead/zinc concentrates.

Silver content in copper concentrates is in the range of 100–150 g/t" of concentrate. Almost all the silver present in the copper concentrates return to the anode copper in the smelter, since the silver loss through discard slag in the smelter is comparatively small. Silver is also recovered from gold ores by the cyanidation method in which ground gold ores are leached by a dilute cyanide solution. Zinc dust is used to precipitate the gold and silver from the cyanide solution. The precipitate is melted in a furnace to produce a doré metal.

In recent years, the extraction and recovery of silver was carried out either by hydro- or bio-metallurgical routes [33]. Leaching represents the first step of the hydrometallurgical route, and the recovery of silver by different techniques, including cementation, chemical precipitation, adsorption, bio-sorption, electrocoagulation, electrowinning, ion exchange, and solvent extraction, are very important cost-effective methods of extraction from various silver-containing sources.

1.4 Secondary Silver-bearing Raw Materials

Silver-containing materials originate from diverse sources, such as manganese silver ore, refractory antimony ore, silver sulfide concentrates, lead–zinc sulfide concentrates, mining wastewater, spent photography/radiography films, electronic and electrical materials, discarded jewellery, silverwares, brazing alloys, catalysts, batteries, dental amalgam, orthopedic materials, spent bleach-fixing photographic/radiographic solutions, electroplating solutions, and metallurgical processing solutions. Processing of such sources is, therefore, complicated. However, the basic processing technology is shown in Fig. 1.

1.5 Scope of the Book

The aim of this book is to provide a summary and comprehensive overview of the silver-containing sources and technical approaches in silver recovery. In addition, processing techniques for the recovery of silver from spent sources are described in the entire chapter.

In Chapter 2, some comparative studies are presented in order to be able to propose non-toxic leaching reagents that can, at the same time, give good results in the recovery of silver contained in mining wastes

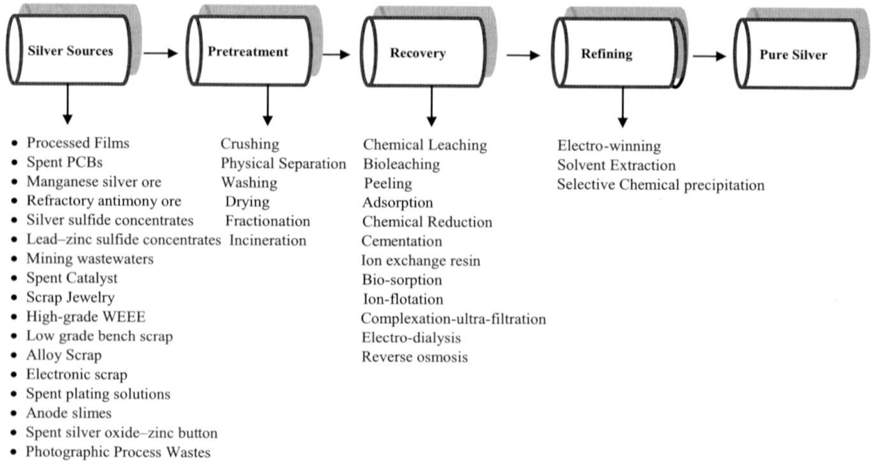

Fig. 1. Silver recovery and recycling basic process flow sheet.

using froth flotation, which is the cyaniding process of massive sulfide minerals. Initially, cyaniding tests were carried out with minerals considered as refractories to the process itself, but when the mineral was treated to a previous stage of pre-oxidation using ozone, a silver extraction rate of 94% was achieved, which was a really significant value that proves the efficiency of the process that is albeit still toxic and dangerous due to the use of cyanide. The substitution of cyanide for thiosulfate in the leaching process of mining tailings with silver content of the order of 60 g/ton showed that an increased oxidizing environment improved the silver dissolution in the system, from 31.7% to 61.07% at partial pressures of O_2 of 0.2 and 1 atm, respectively. On the other hand, the addition of Cu^{2+} greatly improved silver dissolution, reaching 96.4% extraction with the addition of 0.25 g/L of Cu^{2+} in the form of hydrated sulfate. Finally, the use of thiosemicarbazide in the leaching of silver contained in the same mining tailings was studied, with the result that at a concentration of 0.3 M of (CH_5N_3S) and a temperature of 328 K, silver dissolution of 91% was achieved. According to the obtained results, it was possible to confirm that the use of alternative reagents to cyanide (e.g., thiosulfate) can improve silver recovery and provide an alternative, non-toxic, and easy-to-use method for the alkaline leaching of minerals, mining tailings, and liquid wastes with relatively low silver contents.

In Chapter 3, adsorption is described as a feasible process to remove and recover silver ions from aqueous solutions. The choice of the most suitable adsorbent and the adsorption/desorption process system are some of the topics included. In addition, the final section of the chapter presents some experimental results obtained for silver adsorption processes using non-conventional adsorbents.

Chapter 4 discusses bio-mining using indigenous Australian plant species as an attractive alternative that uses green-chemistry–based techniques and processes that are eco-friendly and have the potential to remove silver from a variety of sources.

Chapter 5 outlines the results of electrochemical silver recovery from waste solutions. The use of units process (electrochemical reactors) of different configurations (Batch and Filter press type) is assessed through the determination of merit figures such as current efficiency (ϕ^e), energy consumption (E_s), and space–time yield (ρ_{ST}). Values obtained allow considering the electrochemical recovery of silver from waste solutions possible using electrochemical reactors, due to high current efficiencies, low energy consumption, and high space–time yields. To achieve good results, it is important to take into account the chemical composition of the solutions, purity of the electrodes used, and disposition of the flow through reactors. Studies on these types of items are of main interest, since they result in high energy savings, need low consumption of different reactants, and can be environmental friendly.

Chapter 6 focuses on the bio-hydrometallurgical route for the recovery of silver from various industrial wastes. Various micro-organisms involved in the process and their mechanisms of bio-leaching are discussed in detail.

Chapter 7 presents a brief history of photography, fundamental information about the photographic process, importance of silver recovery, and chemistry of the most used methods for silver recovery from photographic processing solutions. The third section in the chapter mainly discusses the methods used for stripping the silver layer from waste photographic films and recovering the silver. The fourth section describes the methods for the determination of silver and determination of impurities in the recovered silver.

Chapter 8 summarizes the recovery of silver present in trace concentrations in plasma TV monitor scraps. Leaching tests were carried out using lixiviants such as thiosulfate and acidothiourea, a mixture of thiourea in sulfuric acid solution, to screen the most suitable lixiviant. Among the tested lixiviants, acidothiourea was found to be the most suitable. The effect of different concentrations of thiourea and sulfuric acid, temperature, shaking time, and addition of ferric ions were investigated to find the optimum leaching conditions. From the pregnant leach liquor of acidothiourea, silver was quantitatively recovered by means of cementation using zinc powder. Following these results, recovery of silver was attempted from printed circuit boards of spent mobile phones, another e-waste consisting of much more complicated constituents, by means of leaching using acidothiourea followed by adsorption using persimmon tannin gel prepared by treating persimmon extract powder using boiling concentrated sulfuric acid for cross-linking. The optimum conditions were investigated to achieve the quantitative recovery of silver.

Chapter 9 presents the syntheses and silver extraction with calixarene derivatives as macrocyclic extraction reagents and alkyltrimethylol derivatives as linear tripodal extraction reagents. Most of the extraction reagents with a certain size of metal co-ordination sites showed high selectivity for silver based on structural effects, such as the size-fitting effect and converging effect of functional groups.

Chapter 10 outlines the most common sources of nanosilver from spent sources and explains the properties of silver nanoparticles that can influence the bio-accumulation, bio-availability, and environmental risk and demonstrate the impact that still-unexplored nanowaste can have on humans and the living environment in general.

Extensive work has been reported on the processing techniques for the recovery of silver from various spent sources, therefore, this book would serve as a new vista for scientists, researchers, engineers, and postgraduate students.

References

[1] Etris S (1997) Silver and silver alloys. In *Kirk–Othmer Encyclopedia of Chemical Technology*, 4th Edn, Vol 22, Howe-Grant M (ed.) (John Wiley & Sons, New York, NY, USA), pp. 163–179.

[2] Boyle RW (1966) Geological survey of canada. *Bulletin*, **160**: 264–285.

[3] *Encyclopedia Americana* (1972), New York, Americana Corporations, USA.

[4] Syed S (2005) A critical review on the recovery of silver from photographic processing waste. *Journal of Saudi Chemical Society*, **9**: 243–252.

[5] Martin F, Peter B, Jaroslav B (2009) Study of the silver ions cementation after mechanical activation of cementator. *Hydrometallurgy*, **97**: 15–20.

[6] Mohide TP (1985) Silver. Ontario Ministry of Natural Resources, Mineral Policy Background Paper No. 20, Toronto.

[7] Katharina MF (2011) Give silver a shine. *Nature Chemistry*, **3**: 178.

[8] Gale NH, Stos-Gale ZA (1981) Lead and silver in the ancient Aegean. *Science America*, **244**(6): 176–192.

[9] Clark G (1977) *World Prehistory — In New Perspective* (Cambridge University Press, Cambridge), p. 250.

[10] Renfrew C (1972) *Asia Pacific Mineral Year Book* (Methuen & Co. Ltd., London), p. 308.

[11] Vats MC, Singh SK (2015) Assessment of gold and silver in assorted mobile phone printed circuit boards (PCBs): Original article. *Waste Manage.*, **45**: 280–288.

[12] Cayumil R, Khanna R, Rajarao R, Mukherjee PS, Sahajwalla V (2015) Concentration of precious metals during their recovery from electronic waste. *Waste Manage.*, http://dx.doi.org/10.1016/j.wasman.2015.12.004.

[13] World Silver Survey (2016) GFMS Limited/the Silver Institute. Available at: http://www.silverinstitute.org.

[14] Janes N, Playle RC (1995) Modeling silver binding to gills of rainbow trout (*Oncorhynchus mykiss*). *Environmental Toxicology and Chemistry*, **14**: 1847–1858.

[15] Palache *et al.* (1944).

[16] Lee JD (1996) *Concise Inorganic Chemistry*, 5th Ed. (Blackwell Science Pvt. Ltd., Oxford, UK).

[17] World Silver Survey, 2014. GFMS Limited/the Silver Institute (http://www.silverinstitute.org).

[18] Rivera I, Roca A, Cruells M, Patiño F, Salinas E (2007) Study of silver precipitation in thiosulphate solutions using sodium dithionite. Application to an industrial effluent. *Hydrometallurgy*, **89**: 89–98.

[19] Lanzano T, Bertram M, Palo MD, Wagner C, Zyla K, Graedel TE (2006) The contemporary European silver cycle. *Resources, Conservation and Recycling*, **46**: 27–43.

[20] Huanzhen L, Qing T, Kening Y, Shaohua L, Jiajun K (2007) Preparation of metallic silver from Ag_2S slurry by direct hydrogen reduction under hydrothermal conditions. *Materials Letters*, **61**: 1020–1022.

[21] Hu-Chun T, Zhu-You G, Hui D, Nan X, Wei-Min W (2012). Recovery of silver from silver (I)-containing solutions in bio-electrochemical reactors. *Bioresource Technology*, **111**: 92–97.

[22] ATSDR (1990).

[23] Gromov OG, Kuz'min AP, Kunshina GB, Lokshin EP, Kalinnikov VT (2004). Electrochemical recovery of silver from secondary raw materials. *Russian Journal of Applied Chemistry*, **77**(1): 62–66.

[24] Richards RME (1981) Antimicrobial action of silver-nitrate. *Microbios*, **31**: 83–91.

[25] Tappin AD, Barriada JL, Braungardt CB, Evans H, Patey MD, Achterberg EP (2010) Dissolved silver in European estuarine and coastal waters. *Water Research*, **44**: 4204–4216.

[26] Chen CK, Lung TN, Wan CC, (1980) A study of the leaching of gold and silver by acidothioureation. *Hydrometallurgy*, **5**: 207–212.

[27] Yang W, Xiaojie M, Yanfeng L, Xiaoli L, Liuqing Y, Lei J, Yin H (2012) Preparation of a novel chelating resin containing amidoxime–guanidine group and its recovery properties for silver ions in aqueous solution. *Chemical Engineering Journal*, **209**: 394–400.

[28] De RQ, Santoro R, Proost J (2010). Kinetic transitions during Ag and Cu electrorecovery on reticulated vitreous carbon electrodes in flow-by mode. *Chemical Engineering Journal*, **162**: 273–277.

[29] Lee DY, Fortin C, Campbell PGC (2005). Contrasting effects of chloride on the toxicity of silver to two green algae, Pseudokirchneriella subcapitata and Chlamydomonas reinhardtii. *Aquatic Toxicology*, **75**: 127–135.

[30] Park YJ, Fray DJ (2009) Recovery of high purity precious metals from printed circuit boards. *Journal of Hazardous Materials*, **164**: 1152–1158.

[31] Salim MO, Mahir I, Mahmut B (2005) Leaching of silver from solid waste using ultrasound assisted thiourea method. *Ultrasonics Sonochemistry*, **12**: 237–242.

[32] Chansoo C, Yufeng C (2012) Recovery of silver from wastewater coupled with power generation using a microbial fuel cell. *Bioresource Technology*, **107**: 522–525.

[33] Kunda W, Etsell TH (1985) Recovery of silver from x-ray film. In: *Precious Metals, Metchem Research*, ED Zysk, JA Bonucci (eds.) (Boulder, CO), pp. 289–304.

Chapter 2
Leaching of Silver Contained in Mining Tailings: A Comparative Study of Several Leaching Reagents

Eleazar Salinas-Rodríguez*, Juan Hernández-Ávila, Eduardo Cerecedo-Sáenz, Alberto Arenas-Flores, Ma. Isabel Reyes-Valderrama, Edmundo Roldán-Contreras and Ventura Rodríguez-Lugo

Academic Area of Earth Sciences and Materials
Autonomous University of Hidalgo State, México
Carretera Pachuca-Tulancingo Km 4.5 s/n
Mineral de la Reforma, Hidalgo, C.P. 42184, México
**salinasr@uaeh.edu.mx*

2.1 Introduction

Countries involved in the mining and processing of massive sulfides for the recovery of precious metals have developed several processes in order to extract the minerals; however, in doing so, they create millions of tons of tailings as leftovers. This chapter focuses only on tailings that have been produced for over 400 years in a traditional Mexican mining site located in the state of Hidalgo in the mining district of Pachuca–Real del Monte. This region has a rich silver-mining tradition that has supported development since the pre-Hispanic era.

The Pachuca–Real del Monte mining district has traditionally had rich deposits of metallic minerals [1] with valuable deposits of silver (around 250 g/MT) [2] associated with Pb, Zn, Au, and Cu, which are of hydrothermal type [3] and correspond to mineralization dating to the tertiary geological period. Likewise, the types of mineral deposits found are distinguished based on the districts in which they occur. The first district comprises the municipalities of Pachuca and Mineral del Monte y Mineral del Chico, containing principally Ag–Pb–Zn–Au deposits with an approximate extension of 130 km^2 [2], where there are several mining works that have produced thousands of tons of Ag, Pb, and Zn. This zone also has three more important tailings dams as a result of the industrial activity carried out during the processing of these types of mineral deposits. The second district comprises the municipalities of Arenal and Actopan, which basically contain Au–Ag mineral deposits.

On the other hand, the zone of Zimapán, a town located 95 km northwest of Pachuca, bears hydrothermal mineral deposits of the Skarn type [2, 3], comprising calcium-bearing rocks with significant contents of Zn–Pb–Ag [3] and epithermal mineral deposits, containing Ag–Pb [Zn–Cu], in addition to polymetallic mineral deposits of Pb–Zn and Ag, such as sulfurs and sulfosalts, spread in an area of 45 km^2. This mining region is also where the fourth tailing dam is located, which is more important in the state of Hidalgo due to its significant Au and Ag content.

In this way, and because large quantities of Ag and Au ores have been extracted and processed in this region for a long period, this zone has contributed almost 6% of Ag produced worldwide, in addition to extracting a significant quantity of gold through several processes over the last 453 years of mining [1]. The mining activity, which has continued during all these years, has contributed to the generation of approximately 112 million tons of mining tailings, which can be classified into different types such as mine drainage waters, water transport solids [4], piles of abandoned mineral untreated (dumps), and mineral processed waste known as tailings, which are most abundantly present in the state of Hidalgo [5]. This type of tailings occupy a surface area of 1,500 hectares and have Ag and Au contents of 35–71 g/ton and 0.2–0.6 g/ton, respectively [6–8].

The principal characteristic of these tailings is that they were produced through three different technological periods:

(a) the first 350 years of mining was rudimentary and involved the use of grinding and amalgamation processes for extraction, which was known as "Patio process";

(b) the second extraction period, which covered the next 48 years, involved the use of grinding and cyaniding processes; and
(c) the last and current extraction period, continuing for 52 years, include the use of grinding, froth flotation, and cyaniding processes.

Obviously, the older residues are those that are at the base of the dump and have more Ag and Au contents [6].

In a major proportion of mining tailings having Ag and Au content in the state of Hidalgo, the two precious metals are present in different sulfide ores such as galena, pyrite, and blende. First, the occurrence of these minerals was determined by identifying suitable conditions such as chemical and mineralogical affinities. The presence of tons of minerals in the tailings containing valuable metals, especially silver, make these residues attractive from an economical point of view.

In the same way, many of the minerals from the state of Hidalgo are refractory to the cyaniding process, which hinders the recovery of silver that ultimately ends up in the tailings dam. A common reason for the refractoriness in these minerals is the distribution of Ag and Au in iron sulfides, mainly pyrites, arsenopyrites, or both [6–9]. To extract the precious metals from these types of minerals, it is necessary to perform a pre-oxidation treatment of the sulfurous matrix to allow direct contact of the valuable metals with the leaching reagent used [10].

Nowadays, many producers have directed their efforts at obtaining a high rate of silver recovery from several secondary sources. Based on the origin, these sources can be classified as dross, sludge, waste dumps, dust, ashes, effluents, etc., which can be treated by several methods to recover the silver present, depending on the state in which it is present [8]. Generally, the residues containing silver have been classified into two particular types:

(a) secondary metals in solids, consisting of low-grade ores and tailings of processed minerals, industrial sludge, industrial dusts, scrap metal, and shavings.
(b) secondary metals in solutions, comprising photographic and radiographic effluents, electroplating effluents, wastewater industry, brines, and washing water.

The treatment and use of these wastes can also contribute to environmental remediation caused by their disposal [8], since in many cases, the tailings may also contain harmful or toxic elements [9] and at the same time possess significant quantities of zinc, lead, and copper and minimum quantities of gold [2].

In the state of Hidalgo, traditionally, the cyaniding process has been most widely used to recover silver and gold. The minerals treated by this process usually have elements or mineral species considered as consumers of cyanide and do not allow an adequate extraction of metallic content. As result, in almost all the tailings dams in the region, there are appreciable quantities of Ag and Au (principally Ag), because these metals have been caught in the quartz particles, or they were not properly leached due to the presence of elements or cyanide-consuming species. This is the reason why these tailings are economically important, principally as a source of silver, and to a minor extent, gold. Therefore, there is the need to study the different leaching reagents to improve the recovery of these precious metals, including cyaniding, cyaniding with pre-oxidation treatment, and using thiosulfates [8]. It is also important to include the grinding process to obtain a complete characterization of these tailings.

However, according to previous studies [8], the valuable metals are present in particles of reduced size (mostly $-75\,\mu$m in diameter), which comprise 60–65% of the total amount of tailings in dams. This represents a technical problem because, from the metallurgical point of view, it is necessary to conduct an extensive study of the characterization and leaching [10] to obtain the valuable metals contained in this type of wastes. It is also important to note that, in addition to the above-mentioned metals, the wastes may contain other elements such as oxides of silica, iron, aluminum, sodium, and potassium [11]. The above was observed in its mineralogy for the distribution of the elements in the primary and secondary phases [8]. Therefore, grinding is used as the principal technique for determining the characterization of tailings, as it allows the optimal release of the metal content, thus obtaining a suitable configuration of the grinding circuits to achieve:

(a) above 47% milling with particles in the -400 to $+500$ mesh (-37 to $+25\,\mu$m);

(b) optimization of the use of water in the pulp;
(c) dimension the correct size for minimum classification; and
(d) tight control of particle size.

Based on the above criteria, the application of conventional leaching processes can establish the best conditions for the recovery of silver contained in these dumps [8].

In addition to the recovery of silver present in these tailings, the residues obtained after the leaching process could be used to improve products in several industries, such as in the manufacture of construction blocks, in the glass and ceramic industries, and as fillers used in roads or construction [11]. The residues could also be used as an alternative in the cement industry due to their quartz content [12] and also as a substitute of mixed feldspars [11].

With respect to the leaching studies made on this type of tailings, the most widely studied leaching systems to date are based on the use of cyanide as a leaching reagent, while very few studies deal with the use of less toxic agents such as thiosulfates [13–15]. Cyaniding has been in use for over a century, and it is well known that the efficacy of the process is based on the ability of CN^- to form stable complexes with silver. The oxidizing agent used in this process is oxygen or air. However, the process has two principal disadvantages: the first is the negative environmental impact due to the toxicity of the CN^- ion, and the second is that some minerals can contain species that are refractory to the cyaniding process [13].

Another recently studied leaching reagent is thiosulfate, which seems to be a good alternative for the extraction of silver in sulfide and refractory minerals. This method was first developed in ammoniacal systems to maintain the solubility of the copper ions, which is an important step in the dissolution of silver. One advantage of this method compared to the cyaniding process is the selectivity in the extraction of silver contained in refractory phases, but its major disadvantage could be the low solubility of the thiosulfate ion [14].

Besides, recent studies using thiosulfates have not clearly established the optimal conditions for leaching gold and silver contained in the ore concentrates and tailings obtained through mining and metallurgical processes. Therefore, the authors have studied this method to try and determine the optimal concentrations for thiosulfate leaching in order to obtain the highest recoveries of silver from mineral ores and tailings, which will be presented in the following sections.

2.2 Characterization of Tailings in Pachuca–Real del Monte Mining District

In order to properly understand the main characteristics that were used to test the different methods of leaching, the tailings were completely characterized. The results presented in this chapter reflect years of research conducted by the authors and provide an in-depth analysis of the advantages and disadvantages of all leaching reagents used, as well as variations in some processes. This is why the characterization study performed represents one of the first stages of tailings preparation, along with the grinding step, which was conducted to determine the concentration of precious metals in these tailings, and in this case, the important metal being silver.

The characterization of the tailings was carried out by X-ray fluorescence (XRF), X-ray diffraction (XRD), scanning electron microscopy (SEM), atomic absorption spectrometry (AAS), inductively coupled plasma spectrometry (ICP), and sieve analysis (SA). The final step involved a conditioning process that used grinding to promote the release of silver present in the tailings obtained from the Dos Carlos Dam located in the Pachuca–Real del Monte mining district, México.

2.2.1 *Chemical composition*

Results obtained from XRF reveal the presence of the major, minor, and trace mineral species, which are shown in Table 1. Based on the results observed, the presence of silica, alumina, hematite, and potassium oxide as the major species; magnesium oxide, calcium oxide, sodium oxide, P_2O_5, and SO_3 as the minor species; and zinc, copper, lead, silver, and gold in traces can be confirmed.

In this case, the presence of iron oxides, sodium, potassium, and sulfur was due to their primary concentration in the massive sulfur minerals treated with metallic zinc using the froth flotation–cyaniding–precipitation process; in addition, the presence of jarosite-type minerals was detected, which can incorporate silver into their structures [6].

Table 1. Chemical analysis performed by XRF.

Element	Wt.%
SiO_2	70.430
Al_2O_3	7.320
Fe_2O_3	2.800
MgO	0.548
MnO	0.731
CaO	0.690
Na_2O	0.080
K_2O	0.080
P_2O_5	0.120
SO_3	0.940
Zn	0.045
Cu	0.010
Pb	0.031
Ag	55.0 g/ton
Au	0.58 g/ton

On the other hand, Table 2 shows the results obtained by AAS and ICP on the same sample. Similar concentrations of Ag (56 g/ton) and Au (0.6 g/ton) were present, as seen in Table 1.

2.2.2 *Mineralogical and morphological description*

A qualitative description of the complex crystalline compounds contained in the sample was obtained by XRD, while the morphology and texture of particles of the same sample were determine by SEM. According to the results, in general, the sample principally contained non-metallic minerals, together with metal ores such as oxides and sulfurs, with silver and gold content found in massive sulfur minerals that were treated by cyanide. Table 3 shows the minerals species found in the sample by XRD.

In the same way, jarosite-type minerals were detected in these tailings, which could have originated from the mines of Paricutin and El Cristo [6]. These minerals may not have been adequately leached, thereby remaining in the wastes; in addition, silver sulfide was also detected, which was also possibly not well leached. Table 4 shows these species detected by XRD.

Table 2. Chemical analysis performed by AAS and ICP.

Element	Content wt.%	Technique used	Element	Content wt.%	Technique used
Au	0.6 g/ton	ICP	Ca	0.2000	ICP
Ag	56 g/ton	AAS	Si	56.000	AAS
Cu	0.0110	AAS	K	2.3200	AAS
Bi	0.0002	AAS	Na	0.3000	AAS
Ni	0.0050	AAS	Al	6.0950	AAS
Mn	0.0460	AAS	Ti	0.2780	AAS
Fe	2.6900	AAS	P	0.1400	AAS
Sn	0.0002	AAS	Mg	0.0050	AAS
Cd	0.0170	AAS	B	0.3300	AAS
Se	0.0065	AAS	Sc	0.0004	AAS
Te	0.0051	AAS	Co	0.0080	AAS
Zn	0.0400	AAS	Cr	0.3700	AAS
Pb	0.0260	AAS	Be	0.0001	AAS
As	0.0430	AAS	W	0.0053	AAS
Sr	0.4632	AAS	Sb	0.0007	AAS
S	11.450	ICP	Hg	0.0003	AAS

Table 3. Principal mineral species found by XRD.

Non-metallic	Metallic
Quartz	Hematite
Orthoclase	Magnetite
Berlinite	Pirolusite
Anorthoclase	Pyrite
Calcite	Sphalerite
Gypsum	Galena
Albite	Chalcopyrite
	Argentite

Figure 1 is an XRD spectrum showing the presence of jarosite-type compounds of three samples taken from different sites in the Dos Carlos tailings dam, confirming the presence of this mineral that could incorporate silver into its crystalline structure.

To determine the average particle size and describe the texture of minerals contained in samples, an SEM analysis was performed. Figure 2 shows images obtained by Scanning Electron Microscopy [Secondary

Table 4. XRD analysis showing the presence of jarosite-type minerals.

PDF	Mineral species	Formula
10-0443	Potassium jarosite	$KFe_3(SO_4)_2(OH)_6$
30-1230	Sodium jarosite	$NaFe_3(SO_4)_2(OH)_6$
10-0443	Argentite	Ag_2S
31-0650	Hydronium jarosite	$(H_3O)Fe_3(SO_4)_2(OH)_6$

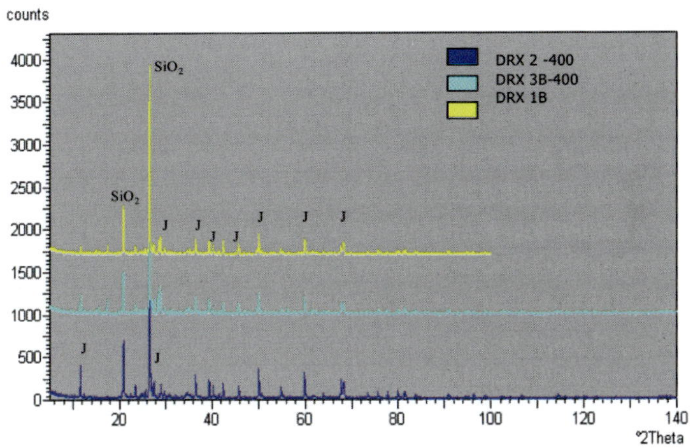

Fig. 1. XRD spectrum showing jarosite-type compounds.

Electrons (SE)] of samples sieved at different meshes (80, 140, 270, and 400), which show that the samples comprised irregular particles in shape of constant size, composed principally of quartz. Also, the presence of silver at the smaller particle sizes was determined, which is why SA assumes importance for the subsequent leaching study.

On the other hand, Fig. 3 shows an EDS spectrum obtained from a characteristic particle observed by SEM, where the presence of elements already determined by XRF, XRF, AAS, and ICP coincide with the silver content determined previously by the above-mentioned techniques.

2.2.3 *Sieving analysis*

Based on the results from SA, it can be assumed that minerals obtained from tailings were treated with conventional metallurgical processes, but

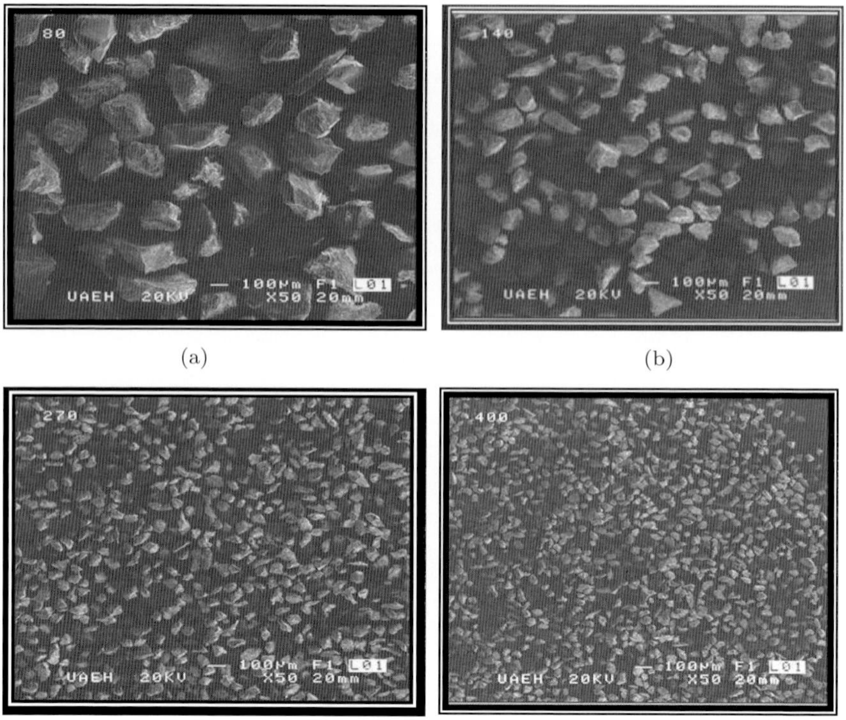

(a) (b) (c) (d)

Fig. 2. Scanning electron microscopy [secondary electrons (SE)] images of samples sieved at (a) 80 mesh, (b) 140 mesh, (c) 270 mesh, and (d) 400 mesh.

without being subjected to an adequate milling process, because there are particles sizes relatively bigger, where silver and gold is still present after the mentioned processes (from mesh 80, 177 μm to mesh 200, 74 μm). Table 5 shows the SA results according to particle size distribution, percentage retained, and percentage accumulated in relation to the percentage of Ag and Au contained.

Also, it can be observed from Table 5 that the major retained weight is at mesh 100, with 19.81% retention, showing that the particle size is too big. Next a sharp drop to mesh 400 is noted, with a minimum retained weight of 4.50%.

In the same way, it can be observed that there is an increase in the silver and gold content as particle size decreases, which is indicative and confirms

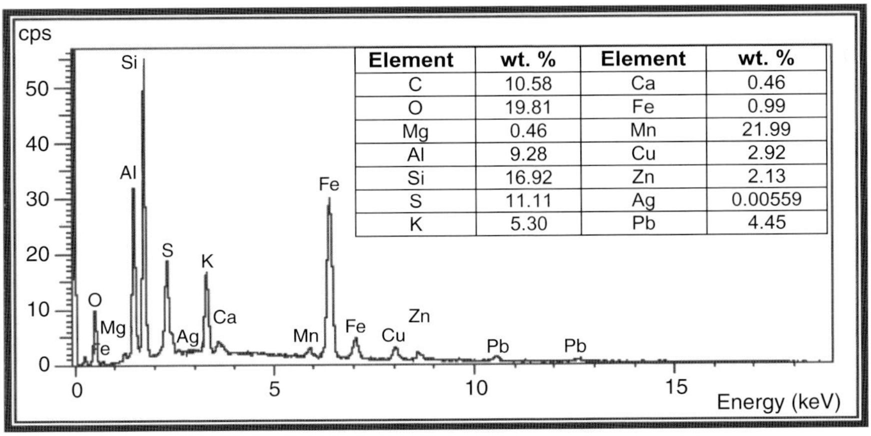

Fig. 3. EDS spectrum showing a punctual and semi-quantitative composition of a characteristic particle of tailings (SEM-EDS).

Table 5. Particle size and metallic values distribution (SA).

Mesh	Aperture (mμ)	Weight (g)	Retained (%)	Accum. (−)	Accum. (+)	Ag (g/ton)	Au (g/ton)
+80	+177	16.80	16.80	83.20	16.80	18	0.20
−80	−177						
+100	+149	19.18	19.18	64.02	35.98	34	0.35
−100	−149						
+140	+105	17.99	17.99	46.03	53.97	39	0.40
−140	−105						
+200	+74	15.58	15.58	30.45	69.55	59	0.60
−200	−74						
+270	+53	10.12	10.12	20.33	79.67	51	0.55
−270	−53						
+400	+37	4.50	4.50	15.83	84.17	60	0.65
−400	−37						
		15.85	15.85	—	100.00	147	1.20

that these elements were not adequately leached. Therefore, another way to carry out a modified process to obtain better silver recoveries using different leaching reagents must be found.

In relation to the above-mentioned results, it can be determined that the Dos Carlos tailings effectively have silver and gold content of around

56 g/ton and 0.6 g/ton, respectively, which could be trapped in the quartz particles and distributed in additional minerals species such as jarosite-type minerals present in residues [7]. For the above reason, and as there is an increase in the metal content as particle sizes decrease, it is necessary to carry out a milling process prior to leaching to promote additional liberation of the metal.

2.3 Leaching of Silver Contained in Mining Tailings

2.3.1 *Background*

The processing of ores containing silver and gold involves the use of varied technologies, ranging from the patio technique and the Pachuca tanks, to the use of circuits involving current techniques, such as grinding, froth flotation, and cyaniding. These processes have generated a huge volume of waste mounds of approximately 112 million tons distributed into four deposits, which have gained importance due to their Au and Ag content [15–18]. The above deposits are the result of the mining and metallurgical activity carried out during the past 455 years in Hidalgo, México [17]. Among the problems associated with these waste mounds, we can cite the presence of pyritic and quartziferous ores, with gold and silver encapsulated principally in the quartz particles, whose predominant size is within the order of $-75\ \mu$m (60–65%) [15], which has an important effect in reducing the extraction efficiency for these types of elements. On the other hand, the presence of some cyanide-consuming elements and compounds does not allow, in many cases, the achievement of an adequate extraction of the metallic content. Therefore, the recovery of silver contained in refractory and jarositic ores constitutes a metallurgical problem for conventional processes [19, 20].

Until now, several research studies on silver leaching have been carried out. While cyanide-based extraction systems have been extensively studied, there has been very few studies on the thiosulfate medium ($S_2O_3^{2-}$) for the recovery of the silver contained in the waste mounds, particularly in the state of Hidalgo. The cyaniding technique has been widely used for more than 100 years due to the ability of the CN^- ion to complex with the silver. The oxidant agent used in this process has been oxygen or air. The disadvantages of this process are the negative environmental impact and the occurrence of ores that are inert to the cyanide solution (known as refractories). On the other hand, the toxicity and high levels of pollution caused by cyanide, as shown by the alarming figures regarding its impact on human health, has now led to the worldwide prohibition of its use as a reagent [9]. Another interesting reason for the search of alternatives to cyanide leaching is the need to increase the dissolution rate of the precious metals contained in the refractory ores, for which cyanide is not very selective, as low levels of metal recoveries are obtained. Additionally, the systems based on thiosulfate leaching have been considered as a non-toxic alternative to the conventional cyanide process [21, 22] and as an effective dissolution media for refractory ores [21]. One of the advantages of thiosulfate in comparison to cyanide is the higher selectivity for the extraction of silver contained in refractory phases, whereas one of the main disadvantages is the low stability of the thiosulfate ion with respect to temperature. However, nowadays, the chemistry and kinetics of this process represent a challenge in the metallurgical sector because little research has been published on the subject. Therefore, the dissolution kinetics of the silver contained in the waste mounds was studied using an alternative technique to the conventional cyaniding process, that is, silver dissolution by the O_2–$S_2O_3^{2-}$ system. In addition, other leaching reagents were also probed, such as thiosemicarbazide, for which very significant results have been found by the authors during several studies, and the results are presented here.

2.3.2 *Use of cyanide as a leaching reagent*

The cyaniding technology of minerals containing silver and gold is well known and has the advantage that, in many cases, it avoids the use of mercury in the amalgamation process. Besides, this technique has been useful because it can be used even in minerals with low metal content.

Generally, the cyaniding process is the most widely used for the extraction of gold and silver from their minerals using oxygen or air as oxidant and CN^- as the complexing reagent [10]. This process is based on the fact that gold and silver are easily dissolved in a dilute aqueous solution of sodium or potassium cyanide, maintaining favorable oxidizing system conditions [23]. However, due to the natural conditions of some minerals that are associated with different sulfide minerals such as pyrite, pyrrhotite, and marcasite, among others, the cyaniding process faces difficulties in the treatment of these types of minerals. Therefore, these minerals have been considered as refractories to the conventional cyaniding process, which is a problem because if the silver is not leached well, it accumulates in the tailings, thus constituting a double problem: the excessive consumption of cyanide and the loss of metal content, which deposits in the tailings dam [24].

For small mining processes, the alternative of leaching is vat leaching together with heap leaching, but the selection of a more convenient method of leaching depends on several factors, such as the type of mineral, value of the metal content, volume of minerals present, accessibility, and water resources at hand.

Actually, many mining companies use alkaline cyanide solutions to recover precious metals such as gold and silver, and although the cyaniding process using stirring can reach recovery rates beyond 90% for gold and silver, similar rates of recovery can be attained using vat leaching and heap leaching as well, but with longer reaction times. In all cases, the success of the process is related to the liberation of the metal content.

Generally, from an operative point of view, the stages carried out in this process can be classified into dynamic and static cyaniding, depending on whether the solid mineral is subjected to movement. In almost all cases, the extraction of dissolved metals was carried out by the precipitation of the metal using an ionic reduction process through the replacement of the valuable metal by other metals such as zinc, under the basic atmosphere of de-oxygenation by the use of vacuum. Another alternative process that is more widely used during the last few years is the application of activated carbon to recover the dissolved metals by adsorption. This is possible by allowing the solution to come in contact with the carbon in the columns, but another possibility is the addition of carbon into the pulp by stirring: both leaching and adsorption simultaneously occur, which is more convenient as it eliminates the need to pull apart the solution as the carbon acts without posing any major inconvenience.

In the mining district of Pachuca–Real del Monte, silver extraction began in the pre-Hispanic era by using rudimentary methods to extract the native metals. The industrial or large-scale extraction of these metals dates back to 1557, when Bartolomé de Medina developed the process known as "Patio process", in which minerals rich in silver were treated using mercury form an amalgamation with NaCl, and then the mercury was evaporated to recover the silver. However, due to the adverse effects of mercury on the health of workers, the process was abandoned and transformed into a leaching process (cyaniding), which is still used [24].

The process of silver cyaniding consists of dissolving the metal in an aerated sodium cyanide solution, resulting in a silver complex $[Ag(CN_2)]^-$ from which silver is obtained by adding zinc dust through the cementation process. Finally, pure metal is obtained by electrolysis from a solution where the silver is in the form of nitrate into a solution of nitric acid, so impure silver is the anode and a lamella of pure silver is the cathode [11, 25].

The use of cyaniding to recover precious metals began when Elsner in 1844 determined that it could be possible to dissolve silver and gold from ores using cyanide. This process required the presence of oxygen in the solution, and this modification saw its first industrial application in 1887 with the cyaniding of gold using oxygen, which was performed by MacArthur and Forrest. Due to the excellent results obtained and the simplicity, better selectivity, and economics, this process became very popular [26].

In [25], Habashi described the proposal of an electrochemical model in which gold was completely dissolved in the anode sites, while oxygen was reduced in the cathode sites. Metallic silver shows a similar behavior like gold in cyanide solutions, dissolving in the anodic area formed in the surface of metal, according to the following reaction:

$$Ag(CN)_2^- + e^- \rightarrow Ag + 2CN^- \quad E^0 = -0.31\,V \tag{1}$$

On the other hand, in the cathodic area, the reduction of oxygen was carried out according to the following reactions:

$$O_2 + 2H_2O + 2e^- \rightarrow H_2O_2 + 2OH^- \quad E^0 = -\,0.15\,V \tag{2}$$

$$H_2O_2 + 2e^- \rightarrow 2OH^- \quad E^0 = +\,0.88\,V \tag{3}$$

The overall reaction for silver dissolution is given by:

$$4Ag + 8CN^- + O_2 + 2H_2O \rightarrow 4Ag(CN)_2^- + 4OH^- \tag{4}$$

Although this process works effectively with most of the minerals containing silver and gold in adequate contents, there are some factors affecting the dissolution of metallic values. The principal factors influencing gold and silver dissolution are cyanide and oxygen concentrations, temperature, pH, surface area, stirring rate, and the presence of other ions in solution, which may compete for the consumption of cyanide [27, 28].

On the other hand, while cyanide is the most employed reagent used to recover precious metals, it is toxic and dangerous to the environment and public health, so its use worldwide has been increasingly restricted [29, 30]. Finally, the other reason to search for alternatives to cyanide is due to the existence of minerals considered as refractories to the cyaniding process, where cyanide is less selective, thereby resulting in very low recoveries of precious metals [31].

2.3.3 *Use of thiosulfate as a leaching reagent*

Traditionally, it has been difficult to find practical information on the leaching process using thiosulfate, such as the thiosulfate concentrations used, chemical fundamentals, leaching times, reagent consumption, metallurgical balances, temperatures used, etc. However, there is much recent information on leaching using thiosulfate in papers, patents, academic works, and theses, but with very different criteria or points of view that pose a barrier to obtaining practical information on how to apply the process correctly. Replacement of cyanide by thiosulfate is very interesting but requires expertise to understand all the chemical mechanisms.

The leaching systems based on thiosulfates ($S_2O_3^{2-}$) are considered as a non-toxic alternative to the conventional cyaniding process [20, 32] and as a medium of dissolution for refractory minerals [9], because one of the advantages of the system with respect to the use of cyanide is the selective silver extraction contained in the refractories phases. However, the system offers some disadvantages, the principal one being that the thiosulfate ion shows low stability with respect to temperature. That is the reason why the chemistry and kinetics of the leaching system using thiosulfate is a challenge in the metallurgical sector; also, few investigations have been published so far on this issue. For this reason, it is important to extensively study the kinetics of dissolution of silver contained in concentrates obtained from the froth flotation–cyaniding process using the O_2–$S_2O_3^{2-}$ system.

2.3.4 *The $AgS_2O_3^{2-}$ system*

Thiosulfate ($S_2O_3^{2-}$) is widely used to solubilize silver halides present in emulsions of photographic processes. At low concentrations of thiosulfate, the principal species of silver is $Ag(S_2O_3)_2^{3-}$, while at high concentrations of thiosulfate, other species are formed such as $Ag_2(S_2O_3)_6^{10-}$; both compounds are highly soluble in water and are the main base of the waste fixative [14].

However, in the case of silver sulfide minerals, and particularly in waste dumps, there have been few studies on the use of thiosulfate to dissolve silver and may be due initially to the mineralogy and chemical complexity of the solutions [33–35].

It is evident that silver recovery from mining tailings has a significant impact on the economy of a country or a region, and for this reason, it is important to study these kinds of issues. The economic impact is very important because the tailings in the mining district of Pachuca–Real del Monte contain around 180.377 million ounces of silver [15], and that is the reason to probe the use of thiosulfate to recover them effectively without the use of toxic cyanide; certainly, an important amount of silver, which could not be recovered by conventional cyaniding processes, is discarded daily.

On the other hand, with regard to the environmental impact caused by the presence of these tailings dams, it is not necessary to mention that these wastes are a risk to environmental and public health since they are considered as hazardous; according to environmental regulations, any waste material containing more than 4 g/ton of silver is considered harmful.

For the above-mentioned reasons and due to the fact that cyanide is extremely toxic, there is a need to search for a new alternative to this reagent in the leaching process. Therefore, a new system is proposed that is based on using thiosulfates as the leaching reagent to describe the entire kinetics and determine the effect of temperature, partial pressure of oxygen, thiosulfate concentration, concentration of Cu^{2+} as a catalyst, OH^- concentration, and the settings involved in the study.

The proposed system has been studied before [14, 32, 33], and important results were found regarding its use in refractory minerals; the obtained results could be directed to the recovery of silver from mining tailings, where studies are scarce. Berezowsky and Sefton [36] were the first to conduct research on the leaching of silver, achieving recovery rates of up to 83%, using ammonium thiosulfate in copper concentrates with gold and silver content.

On the other hand, other research works [37] studied the leaching of a concentrate of silver sulfur (argentite) using thiosulfate solution in presence of copper, observing that the copper ion reacted with thiosulfate to form tetrathionate and cupric thiosulfate ions according to Eq. (5), then inside, the mineral silver content was substituted by copper according to Eq. (6).

$$5(NH_4)_2S_2O_3 + 2CuSO_4 \rightarrow Cu_2S_2O_3 \cdot 2(NH_4)_2S_2O_3 + 2(NH_4)2S_2O4 + (NH_4)_2S_4O_6 \quad (5)$$

$$Cu_2S_2O_3 \cdot 2(NH_4)_2S_2O_3 + Ag_2S \rightarrow Cu_2S(\downarrow) + Ag_2S_2O_3 \cdot 2(NH_4)_2S_2O_3 \quad (6)$$

2.3.5 *Use of thiosemicarbazide as leaching reagent*

It has been determined that silver leaching with organic compounds containing the "thio" group, such as thiourea [$CS(NH_2)_2$], thiosulfate ($S_2O_3^2$), thiocyanate (SCN), and thiosemicarbazide (CH_5N_3S), have some advantages over cyaniding, such as the ease of carrying out the silver dissolution under acidic conditions and the use of strong oxidizing reagents. Thus the valuable metals can be recovered by adsorption or precipitation and, finally, the leaching reagent showed less toxicity, thereby reducing environmental problems.

The leaching reagent that has been studied extensively is thiourea, but in the case of thiosemicarbazide, some studies on the kinetics and thermodynamics of this system have been recently performed using the rotating disc with pure silver, obtaining favorable results to conduct further research applied to refractory ores, concentrates, and tailing from them [38]. Therefore, in this chapter, the results of the kinetics study of silver leaching using these two reagents under acidic and oxidative conditions to determine the optimum conditions for silver recovery in both systems are presented. In addition to the results obtained, the authors propose new alternative systems to cyanide leaching.

2.4 Results

This chapter presents the most important and recent results found by authors related to the leaching of concentrates, refractory minerals, and tailings containing silver using different leaching reagents. In the first part of the work, the cyaniding process is presented as a conventional process after oxidation of a refractory concentrate with ozone, which resulted in major silver recovery rates and showed that conventional cyaniding was less effective on refractory minerals. After this step, some results of the cyaniding process performed to dissolve the silver contained in the mining tailings of the Dos Carlos tailings dam in the mining district of Pachuca–Real del Monte are detailed.

The second part of the work is related to the kinetics study of silver leaching of the same mining tailings using an oxygen–copper–thiosulfate system to determine the advantages and disadvantages of the process and to compare the results with those obtained in the conventional cyaniding system. Finally, a kinetics study of silver leaching using thiosemicarbazide as the leaching reagent is performed to try and explain the best leaching conditions for this system and is compared with cyaniding and thiosulfate leaching.

Practically, the study of different leaching systems apart from cyaniding represents opportunities for the dissolution of silver in the mining tailings without using toxic leaching reagents like cyanide. This could be an alternative for the marginal minerals and tailings that undergo traditional leaching processes. The results presented here are of great interest and part of a wide-ranging study of kinetics to determine the best experimental conditions for silver dissolution and the subsequent recovery of the valuable metal.

2.4.1 *Leaching of tailing using cyanide*

2.4.1.1 *Ozone pre-oxidation (for a concentrate)*

As was mentioned earlier, the cyaniding process has shown some disadvantages when the mineral is considered refractory, independent of the silver content, so the following case is studied where the mineral treated is refractory to the cyaniding process. For this reason, the mineral was pre-oxidized using ozone and then cyaniding was performed using the traditional method to try and obtain the best result and achieve higher silver recovery rates. So in the following paragraphs [39], the authors will show the preliminary results of this method of oxidizing the mineral with ozone, which was generated by high voltage and then injected to the mineral in suspension in the aqueous phase, to try and modify the structure of mineral and alter its matrix, allowing the reagent to have direct contact with the desired metal to promote complexing with cyanide and obtaining a major extraction of the metal.

According to the preliminary results, it was determined that during pre-oxidation of the refractory sulfide ore, a significant alteration of the structure of the mineral was achieved, thereby obtaining a greater dissolution of the gold and silver contained in the ore.

The chemical composition of the mineral used is as follows: S (24.31%), SiO_2 (9.96%), Ca (0.12%), Fe (38.75%), Co (97 g/ton), Cu (0.41%), Zn (6.27%), Pb (2.6%), Ag (551.3 g/ton), and Au (1.05 g/ton). Significant quantities of Cu, Fe, and Zn were present; in addition, a high level of pyrite was detected due to the S content. Therefore, the cyaniding process could present difficulties due to the occurrence of these elements, which are consumers of cyanide. Traditionally, to avoid the harmful effects of the cyanide-consuming elements, caustic or lime have been used to neutralize acidity, but in some cases, this corrective step delayed silver and gold dissolution, so it is necessary to strictly control the alkalinity level during the process [23].

On the other hand, the same results can be observed when there is a significant presence of pyrite; the desired metal could be encapsulated in the pyritic matrix, which also hinders silver and gold dissolution in cyanide systems.

The study was carried out by performing leaching proofs with the mineral without oxidation and by oxidizing the mineral with ozone by leaching it previously with a solution of distilled water saturated with ozone for 5 min. The mineral thus oxidized was subjected to conventional cyaniding, and the results obtained for the two kinds of proofs were compared.

Figure 4 shows the results obtained for silver leaching, in which the results obtained for the mineral oxidized with ozone in comparison with the mineral subjected to the conventional cyaniding process can be observed. It

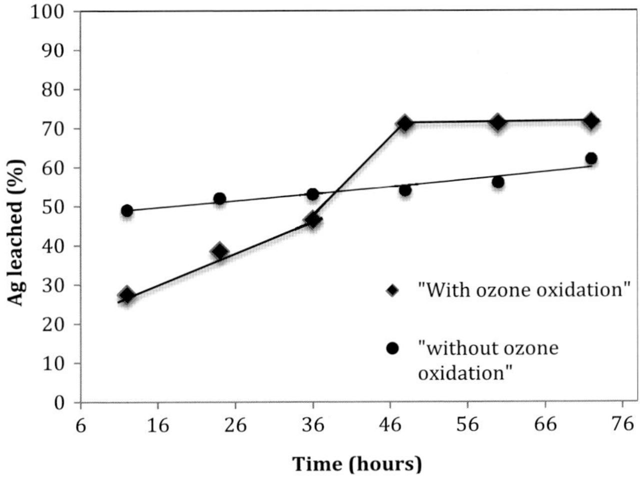

Fig. 4. Silver dissolution during cyaniding of mineral with and without ozone oxidation.

can also be observed from Fig. 4 that there appears to be a high rate of silver dissolution at the beginning of the process for the mineral that is not treated with oxidation, but this could be attributed to the absence of refractory species such as galena and sphalerite in this concentrate, and to the high concentrations of silver in these mineral species, which increases the rate of the cyaniding process. However, as the process continues, there is a decline in the rate of dissolution of silver for the unoxidized ore, reaching only 61% of dissolution in approximately 72 h, while for the oxidizing mineral, 71% of silver dissolution was achieved at only 48 h. Approximately 29% of silver remains in the leached mineral, which is probably composed of very small-sized particles, so the mineral must be milled before treatment to achieve a good yield of silver; also, some silver content could be trapped in quartz particles and due to the major affinity of cyanide for gold, as seen in Fig. 4.

On the other hand, Fig. 5 shows the results obtained for gold leaching obtained using the same conditions that were used to treat the mineral for silver dissolution.

In this case, there is a higher and faster dissolution of gold in the case of the mineral pre-oxidized with ozone, due to the alteration of its structure, thereby resulting in 93% of gold extraction in 48 h of reaction; in

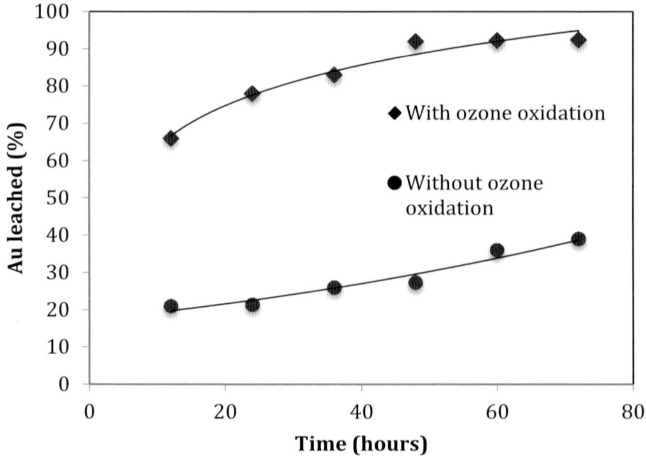

Fig. 5. Gold dissolution for the mineral with and without ozone oxidation.

(a) (b)

Fig. 6. Particles of mineral (a) without and (b) with ozone pre-oxidation.

comparison, only 40% of gold extraction is reached at 72 h of reaction for the mineral that is not subjected to ozone oxidation.

Figure 6 shows the mineral particles with and without ozone oxidation, and it can be seen that matrix structure is altered during oxidation, principally breaking the pyritic structure, removing sulfur, and making the particles more porous, thus allowing the entry of the reagent with greater ease.

The above results indicate that pre-oxidation of this kind of minerals (i.e., refractories) considerably improves their treatment by cyaniding and, as a consequence, the dissolution rates of silver and gold particles are greater and involve shorter reaction times. This phenomenon has also been detected by other researchers [40], who have carried out comparative studies of cyaniding on refractory and non-refractory sulfur ores, finding good results for the dissolution of gold and silver contained in minerals originating northern México. This confirms that the use of ozone as an oxidizing reagent is a viable method to obtain higher Au and Ag dissolution rates with shorter reaction times and less consumption of cyanide.

Thus the obtained results for the ozone-oxidized refractory minerals from the state of Hidalgo in México showed good rates of silver and gold dissolution. In summary, this pre-treatment achieved 70.78% silver dissolution within 48 h of reaction for the oxidizing ore. The dissolution obtained during cyaniding of the unoxidized mineral was 61.7% at 72 h of reaction, which indicates a difference of 9.08% but with a reaction time that was 24 h lesser than that of the pre-oxidation method. A higher silver dissolution rate could not be achieved due to the presence of silver encapsulated in quartz particles, the very small size of the particles, and the major affinity of cyanide for gold.

On the other hand, for gold extraction, a 93% gold dissolution rate can be achieved from the mineral oxidized with ozone at 24 h of reaction, while the unoxidized mineral elicited 40% gold dissolution at 72 h of reaction. This shows that the pre-oxidized mineral showed an increase of 53% gold dissolution rate at less than 48 h of reaction compared to the unoxidized mineral. This data indicates that pre-oxidizing of this kind of minerals considered as refractories to the conventional cyaniding process can lead to a higher rate of silver and gold leaching but with minor leaching reaction times. Similarly, it is important to mention that pre-oxidation with ozone is a cheap and easy method.

2.4.1.2 *Cyaniding leaching of mining tailings*

The results obtained for the leaching of mining tailings from the Dos Carlos dam, whose characterization is presented in Sec. 2.2, show that cyaniding carried out in the NaOH medium does not show an induction period, resulting in an instant cyaniding of the silver contained in the tailings which was released from quartz particles, the silver contained in argentite mineral, and the silver associated with jarosite-type minerals (both sodium and potassium jarosites).

Due to the complex nature of the tailings, it is very difficult to determine the stoichiometry of the system based on the experimental techniques. Therefore, only the steps of leaching of silver in its four forms present in the tailings are presented, as native silver, silver sulfide, jarosite potassium bearing siver and jarosite sodium bearing siver.

For the case of native silver the reaction is shown in Eq. (7) [10]:

$$4Ag_{(s)} + 8CN^{-} + O_2 + H_2O \rightarrow 4Ag(CN)_2 + 4OH \tag{7}$$

The dissolution of silver sulfur in a cyanide solution is given by the following sequence of equations [41].

$$Ag_2S \rightarrow 2Ag^{+} + S^{2-} \tag{8}$$

$$Ag_2S + 4CN^{-} \rightarrow 2Ag(CN)_2^{-} + S^{2-} \tag{9}$$

$$2S^{2-} + 2H_2O + O_2 \rightarrow 4OH^{-} + 2S\downarrow \tag{10}$$

According to the above equations, the sulfur ion was oxidized to free sulfur, and for this reason, silver sulfide was totally dissolved [41].

Cyaniding of potassium and sodium jarositeargentiferous can be described through a series of two stages: a first stage of decomposition controlling the global process followed by a rapid step of complexation of products from reaction, as is shown in the following reactions [42, 43]:

Potassium jarositeargentiferous:

$$\begin{aligned}(K)Fe_3(SO_4)_2(OH)_{6(s)} + 3OH_{(aq)} &\rightarrow 3Fe(OH)_{3(s)} \\ &+ 2SO^{2-}_{4(aq)} + K^{+}_{(aq)} \text{ slow}\end{aligned} \tag{11}$$

$$Ag(OH)_{(s)} + 2CN_{(aq)} \rightarrow Ag(CN)_{2(aq)} + OH_{(aq)} \text{ fast} \tag{12}$$

Sodium jarositeargentiferous:

$$\begin{aligned}NaFe_2(SO_4)_2(OH)_{6(s)} + 3OH_{(aq)} &\rightarrow 3Fe(OH)_{3(s)} \\ &+ 2SO^{2-}_{4(aq)} + Na^{-}_{(aq)} \text{ [slow]}\end{aligned} \tag{13}$$

$$Ag(OH)_{(s)} + 2CN_{(aq)} \rightarrow Ag(CN)_{2(aq)} + OH_{(aq)} \text{ [fast]} \tag{14}$$

With the above stoichiometry already defined, the kinetics study was executed evaluating the effects of cyanide concentration, temperature, sample amount, leaching time, stirring rate, and OH^- concentration, taking samples at determined times and evaluating for silver in solution by AAS at pre-determined sampling times, maintaining pH by adding 0.2 M of NaOH solution constantly.

For the study of cyanide concentration (from 5×10^{-2} to 5.1×10^{-3} M), the other parameters were kept constant: stirring rate $= 750\,\text{min}^{-1}$, temperature (298 K), mineral sample amount (40 g/L), [NaOH] $= 1 \times 10^{-2}$ M, volume of solution = 500 mL, and particle size = uniform. Figure 7 shows the effect of cyanide concentration on the cyaniding of mining tailings from the Dos Carlos dam. These results of these variables were mapped to the shrinking core model for chemical control, $K\text{t} = [1-(1-x)^{1/3}]$, and it was found that they adhered well to the proposed model, where the chemical reaction was the controlling stage of cyaniding process.

It can be seen that, as the cyanide concentration increases, k_{exp} increases slightly as well, but only up to a concentration of 2.04×10^{-2}M; the overall effect of cyanide concentration is negligible, as seen in Fig. 7, where the dependence of k_{exp} against $[\text{NaCN}]^-$ provided an order of the reaction approximately equal to 0 ($n \approx 0$).

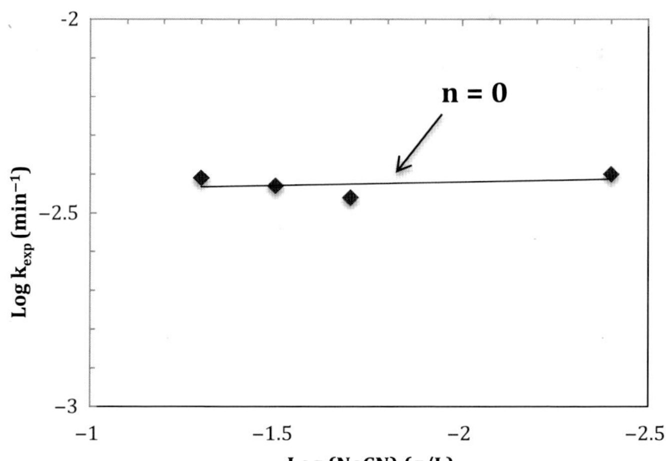

Fig. 7. Dependence of [NaCN] with respect to the k_{exp} values obtained. Order of cyaniding reaction $n = 0$.

For the study of the temperature effect, a series of experiments were carried out by varying the temperature in the range of 288–333 K and keeping the other parameters constant; stirring rate $= 750\,\text{min}^{-1}$, [NaCN] $= 2.4 \times 10^{-2}$ M, sample $= 40$ g/L, [NaOH] $= 1 \times 10^{-2}$ M, volume of solution $= 500$ mL, and particle size = uniform.

Figure 8 shows the obtained results for the effect of temperature, which shows that at temperatures higher that 323 K, silver dissolution was above 90%, reaching maximum dissolution at a temperature of 333 K, while at temperatures between 288 and 318 K, there was no significant increase in silver dissolution. Also shown in Fig. 8 are the values of experimental rate constants (Ln k_{exp}) versus temperature (1,000/T) from which the activation energy was calculated through Arrhenius equation taking in account the different temperatures tested, which was $E_{\text{a}} = 42$ kJ/mol, corresponding to a reaction controlled chemically.

To determine the effect of particle size, all parameters were kept constant as shown above, and only particle size (average) was changed in each experiment (180, 150, 105, 75, 53, and 38 μm). Figure 9 shows the obtained results for this effect, where it can be observed that as the particle size decreases, the experimental rate constant increases, simultaneously resulting in higher silver dissolutions. It can also be seen that the k_{exp} values

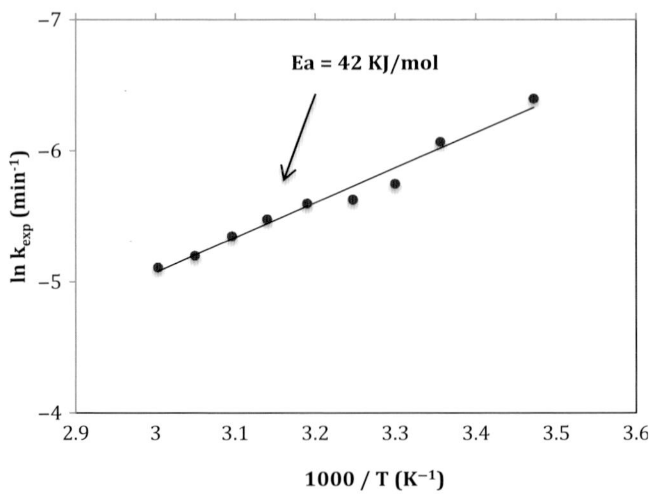

Fig. 8. Dependence of k_{exp} with temperature, $E_a = 42$ kJ/mol.

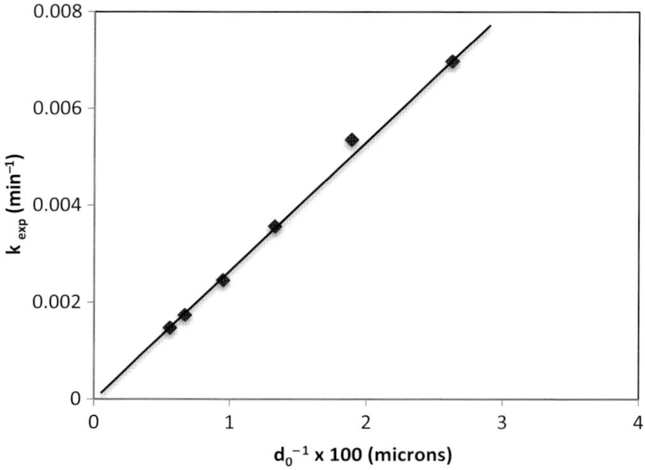

Fig. 9. Dependence of k_{exp} with respect to particle sizes.

were directly proportional to the inverse of the average particle diameter, confirming (joint with E_a value found), that the cyaniding process was controlled by the chemical reaction.

Finally to determine the effect of [NaOH], all parameters were kept constant as shown earlier, and this parameter was changed in each experiment (5×10^{-2}, 3×10^{-2}, 1×10^{-2}, 3.3×10^{-3}, and 1.1×10^{-3} M). Figure 10 shows the effect of Log $[OH^-]$ with respect to Log k_{exp}, and the order of reaction for the effect of NaOH can be observed, which was of $n \approx 0.214$.

2.4.2 *Leaching of tailing using thiosulfate*

A kinetics study on the leaching of silver contained in the mining tailings was carried out to compare the results with respect to those obtained through cyaniding kinetics. The kinetics study includes the study of the effect of milling time of tailings to promote a higher rate of liberation of the target metals and the effect of thiosulfate concentration to determine the order of reaction with respect to the concentration level. The effect of the stirring rate to observe the influence of liquid film during the reaction and the effect of temperature were also studied to determine the energy of activation (E_a). Other variables studied include the effects of partial pressure of oxygen (P_{O2}), pH, and finally Cu^{2+} concentration.

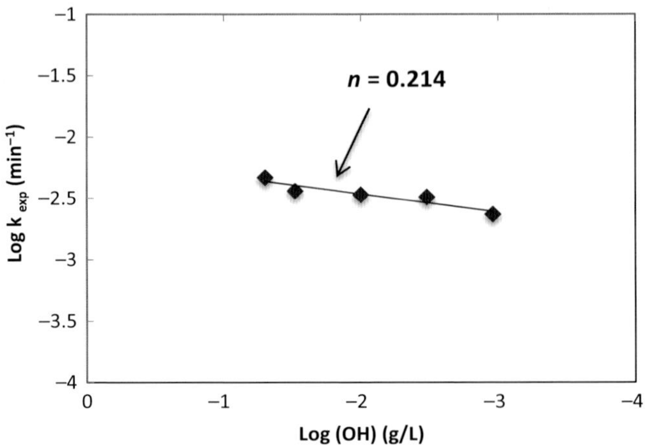

Fig. 10. Dependence of k_{exp} with respect to $[\text{OH}^-]$, $n \approx 0.214$.

To study the effect of milling time on the amount of silver liberated during this stage, some experiments were carried out using samples that were milled at different times and then were leached based on the following values of the process variables: the variable parameter was time of milling (0, 2, 4, 8, 16, 32, and 46 min) and the constant parameters were $[S_2O_3] =$ 79 g/L, temperature $=$ 298 K, sample $=$ 40 g/L (uniform), pH $=$ 9, volume of solution $=$ 500 mL, stirring rate $=$ 500 min^{-1}, and partial pressure of $O_2 =$ 1 atm. In all the experiments, pH was kept constant by adding 0.2 M NaOH solution, and the samples were taken at pre-determined times to determine the silver dissolved using AAS.

According to the obtained results for this effect, it can be seen that as the milling time increases, silver dissolution gradually increases up to a maximum dissolution rate of 54%. The results for the effect of milling time showed that there is no significant difference between 8 and 32 min of milling, therefore, 8 min of milling was deemed adequate for leaching using thiosulfate.

The study of the effect of thiosulfate concentration was performed using the following experimental conditions: the variable parameter was the thiosulfate concentration $[S_2O_3] =$ 15.8, 31.6, 79.0, 126.4, and 189.6 g/L. On the other hand, the constant variables were temperature $=$ 298 K, sample $=$ 40 g/L, pH $=$ 9, volume of solution $=$ 500 mL, stirring rate $= 500\,\text{min}^{-1}$, partial pressure of $O_2 = 1$ atm, and time taken for milling $= 8$ min.

As mentioned earlier, and similar to the cyaniding study, the experimental results were analyzed by measuring the amount of dissolved silver in the solution by AAS. Likewise, the values obtained were treated by the decreasing core model and chemical control $k_t = [1 - (1 - x)^{1/3}]$.

Figure 11 shows the effect of thiosulfate with respect to the constant rates of silver dissolution, showing that as the thiosulfate concentration increases, the rate of silver leaching also increases, reaching a point at which there is a change in the reaction order, going from $n = 0.321$ for the range from 15.8 to 79.0 g/L of thiosulfate, after which the reaction order value nears 0, which indicates that above a threshold level (i.e., 79.0 g/L), higher concentrations of thiosulfate have no effect on the rate of silver dissolution.

The effect of the stirring rate was determined by keeping all other variables constant as in the above procedure and varying only the stirring rate (300, 500, 750, and 1,000 min^{-1}). The values of the constant parameters were: temperature $= 298$ K, $[S_2O_3^{2-}] = 79$ g/L, sample $= 40$ g/L, pH $=$ 9, volume of solution $= 500$ mL, and partial pressure of $O_2 = 1$ atm. The reaction was monitored by the dissolution of the silver in system and analyzed by AAS, and the treatment of the data was carried out using the same model already described.

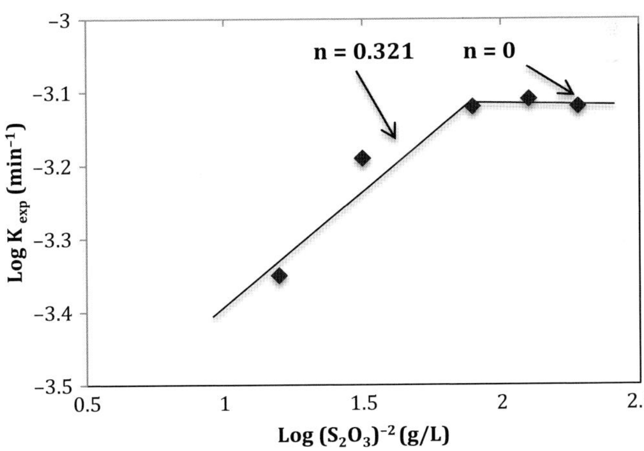

Fig. 11. Dependence of k_{exp} with respect to $[S_2O_3^{2-}]$: $n = 0.321$ (15.8–79 g/L) and $n = 0$ (79–189.6 g/L).

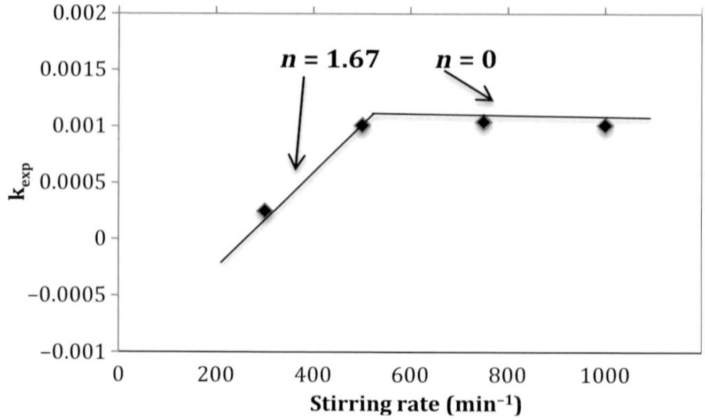

Fig. 12. Dependence of k_{exp} with respect to stirring rate.

Figure 12 shows the obtained results, where it can be seen that from 300 to 500 min^{-1}, there is a great influence on rate constants of silver dissolution ($n = 1.67$), while above 500 min^{-1}, there was no noticeable effect on silver dissolution ($n = 0$). This indicates that, at low stirring rates, the pulp is not sufficiently dispersed, thereby creating a layer of liquid film that restricts the diffusion of the reagent into the particles' surface. However, upon total dispersion of the pulp, irrespective of the stirring rate, the particles are in constant motion and in contact with the reagent, which promotes the reaction.

The study of the effect of temperature was carried out similar to the two above-mentioned experiments. In this case, the temperatures analyzed were 288, 298, 308, 323, and 333 K. The following parameters were kept constant: stirring rate = 750 min^{-1}, $[S_2O_3^{2-}] = 79$ g/L, sample = 40 g/L, pH = 9, volume of solution = 500 mL, and partial pressure of O_2 = 1 atm. As mentioned above, the reaction and treatment of the results were similarly monitored.

Figure 13 shows the dependence of temperature with respect to rate constants of silver dissolution, and it can be seen that, as the temperature increases, the rate constant also increases. Figure 13 also provides the activation energy which, in this case, was 38 KJ/mol, indicating a chemically controlled reaction.

The study of the effect of partial pressure of O_2 was carried out with only two experiments at different pressures, 0.2 and 1 atm, using air and

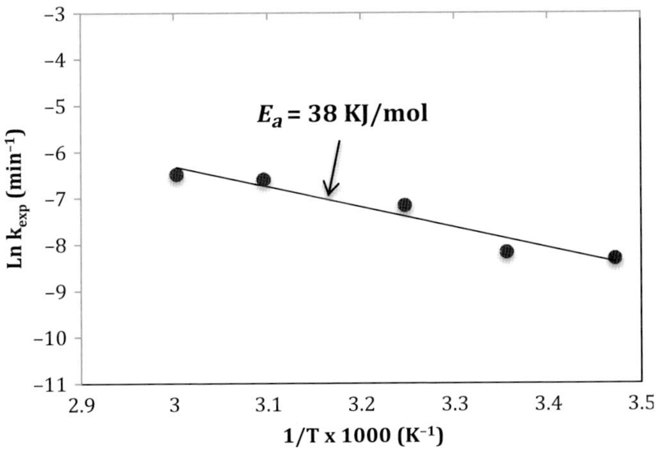

Fig. 13. Effect of temperature ($E_a = 38$ kJ/mol).

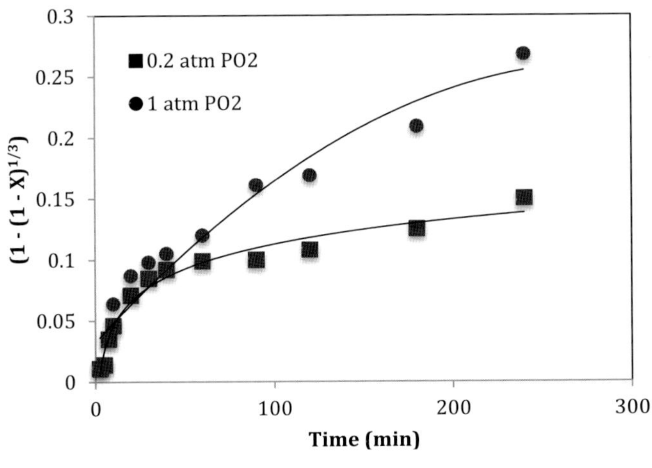

Fig. 14. Effect of partial pressure of O_2 in leaching of silver.

oxygen, respectively. The following were the experimental conditions; temperature $= 298$ K, $[S_2O_3^{2-}] = 79$ g/L, sample $= 40$ g/L, pH $= 0$, volume of solution $= 500$ mL, and stirring rate $= 750\ \text{min}^{-1}$.

For this case, Fig. 14 shows the obtained results, where it can be seen that leaching using pure oxygen was more successful than that carried out using only air.

The study of the effect of pH was executed by varying the pH values: 8, 9 10, and 11, and keeping constant the following parameters; temperature = 298 K, $[S_2O_3^{2-}]$ = 79 g/L, sample = 40 g/L, stirring rate = 750 min^{-1}, volume of solution = 500 mL and partial pressure of O_2. The monitoring of the reaction and interpretation of the results was conducted as for the previous experiments. Figure 15 shows the effect of pH on rate constants for silver dissolution, and it can be seen that the pH had no practically significant effect on the rate constant, which shows an apparent order of reaction of $n \approx 0$.

Finally a complementary study of the effect of Cu^{2+} concentration to determine the catalytic effect of copper ion was carried out. The copper was added as $[CuSO_4 \cdot 5H_2O]$ in the following concentrations for each experiment: 0.125, 0.25, 0.50, 1.0, 2.0, and 4.0 g/L. The constant parameters were temperature = 298 K, $[S_2O_3^{2-}]$ = 79 g/L, sample = 40 g/L, stirring rate = 750 min^{-1}, volume of solution = 500 mL, pH = 9, and partial pressure of O_2 = 1 atm. Similar to the above-mentioned experiments, the monitoring of the reaction and the interpretation of the results were conducted.

According to the results, addition of $[CuSO_4 \cdot 5H_2O]$ increases silver dissolution, because without Cu^{2+} the maximum silver dissolved was of 61% at 298 K and with the addition of only 0.125 g/L of Cu^{2+} at the same

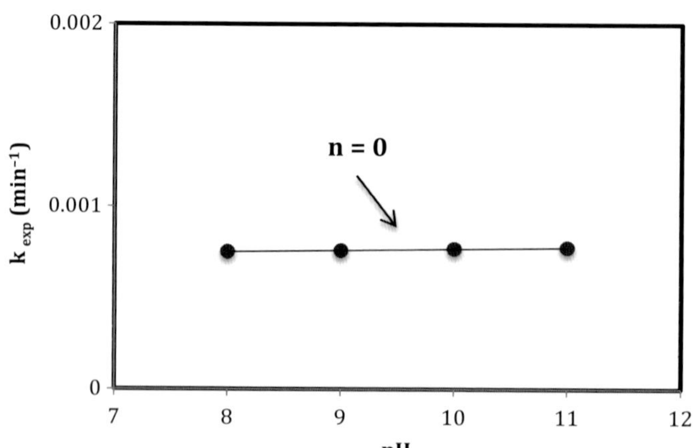

Fig. 15. Dependence of k_{exp} with respect to pH.

temperature, it was possible to obtain a silver dissolution rate of 84%, which reached 96.43% on increasing the Cu^{2+} concentration up to 0.5 g/L. It means that the leaching of silver using thiosulfate in the system O_2–$S_2O_3^{2-}$ without a catalyst basically depends on the dissolved oxygen in system. Thus the reaction rate was principally controlled by the transport of oxygen to the solid–liquid interface. However, in the presence of the copper cation, a reduction reaction from Cu^{2+} to Cu^+ occurs, promoting the oxidation of Ag^0 to Ag^+, according to Eq. (15) [14]:

$$Cu^{2+} + Ag^0 + 5S_2O3^{2-} \rightarrow Ag(S_2O_3)_2^{3-} + Cu(S_2O_3)_3^{5-} \qquad (15)$$

Figure 16 shows the dependence of k_{exp} with respect to Cu^{2+} concentration.

2.4.3 *Leaching of tailings using thiosemicarbazide*

The study of the leaching of silver contained in mining tailings using thiosemicarbazide as the leaching reagent was performed according to similar conditions and methodology as the previous leaching processes reported in this chapter. The experimental results show the effect of molar concentrations between 0.1 and 0.5 g/L, obtaining a maximum silver dissolution of 33.9% and 76.9%, respectively. The other factors also studied were temperature, partial pressure of O_2, pH, and stirring rate.

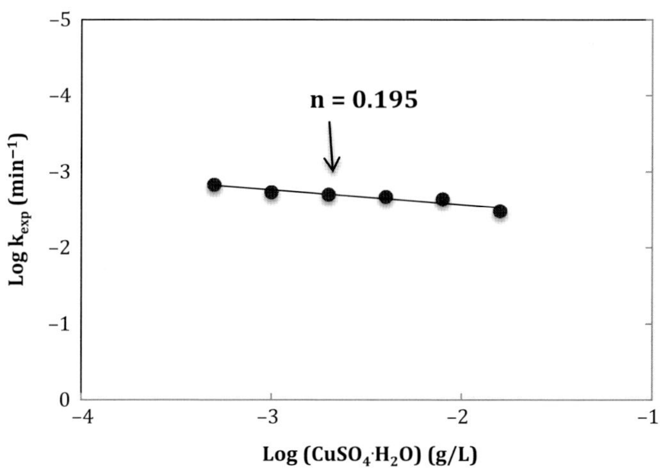

Fig. 16. Dependence of k_{exp} with respect to $[Cu^{2+}]$, $n = 0.195$.

The effect of thiosemicarbazide concentration was studied by varying the concentration values as follows: $[CH_5N_3S] = 0.1, 0.2, 0.3, 0.4$, and 0.5 M. The following parameters were kept constant: temperature = 298 K, stirring rate = 600 min^{-1}, partial pressure of O_2 = 1 atm, sample = 40 g/L, volume of solution = 500 mL, and milling time = 8 min for tailings that were not sieved. Silver dissolved in the solution was tested using AAS for samples of aliquots taken at pre-fixed times, and the obtained results were also mapped with the decreasing core model $[1-(1-x)^{1/3}]$ to compare with the other studies done previously.

Figure 17 shows the effect of CH_5N_3S concentration with respect to the rate constants of silver dissolution. It can be observed that, as the reagent concentration increases, the reaction rate also increases, with a maximum silver dissolution of 76.9% obtained at a $[CH_5N_3S]$ of 0.3 M. Also in Fig. 17, the order of reaction obtained for this effect can be observed as $n = 0.475$, which was indicative of the reagent concentration having a significant effect over rate constants and silver dissolution.

The evaluation of the effect of the stirring rate was carried out by varying this parameter and keeping the others constant. The experimental conditions were as follows: the stirring rates evaluated were 200, 300, 400, 600, and

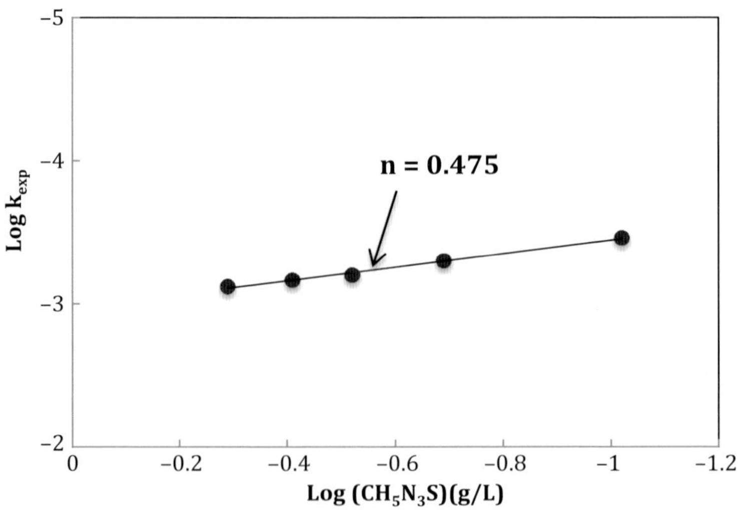

Fig. 17. Dependence of k_{exp} with respect to $[CH_5N_3S]$, $n = 0.475$.

800 min^{-1}, and the parameters kept constant were temperature = 298 K, $[CH_5N_3S]$ = 0.3 M, sample = 40 g/L, pH = 10, volume of solution = 500 mL, partial pressure of O_2 = 1 atm, and using tailings milled for 8 min (constant size). The reaction was monitored by the determination of the dissolved silver by AAS and using the shrinking core model $[1 - (1 - x)^{1/3}]$ for the treatment of results to compare with the previous studies.

Figure 18 shows the relation of k_{exp} with respect to the stirring rates studied. It can be seen that, as in the thiosulfate study, only the low stirring rates have a slight influence, which could be because the mineral is not suspended enough to allow a good interaction of the reagent with the material surface. At high stirring rates, the reaction is not influenced by this parameter, so we again have two orders of reaction, one at low stirring rates and another at high stirring rates.

Another parameter also studied was the temperature, and as in the other cases, the temperatures were varied while keeping constant the other parameters. In this case, the temperatures evaluated were: 288, 298, 308, 318, and 328 K. The parameters that were kept constant were the following: stirring rate = 600 min^{-1}, $[CH_5N_3S]$ = 0.3 M, sample = 40 g/L, pH = 10,

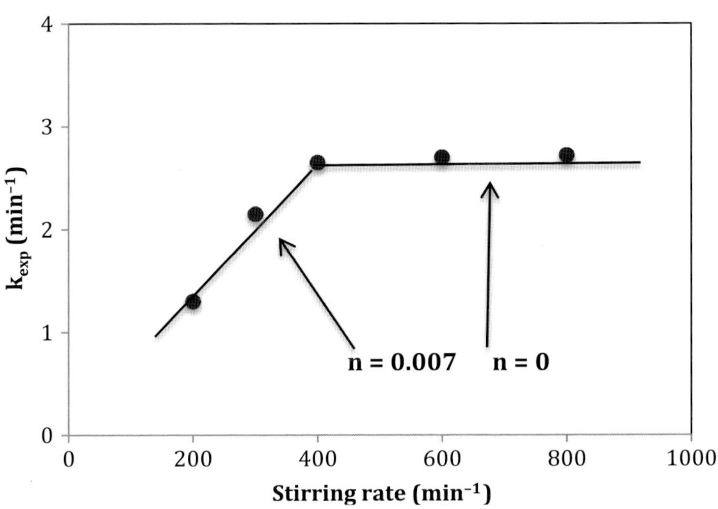

Fig. 18. Dependence of k_{exp} with respect to stirring rate, n = 0.007 (200–400 min^{-1}) and n = 0 (400–800 min^{-1}).

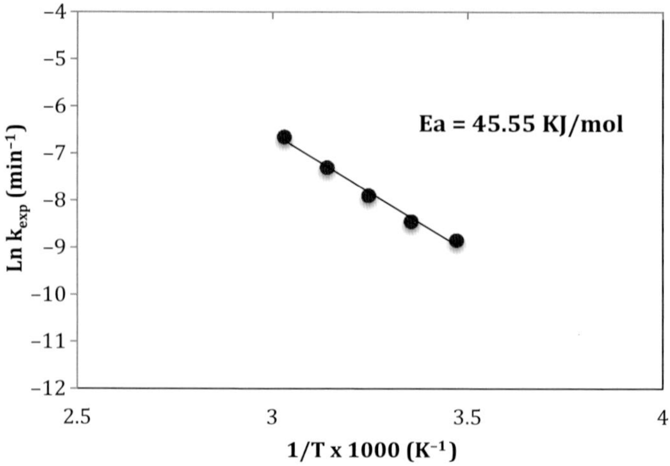

Fig. 19. Dependence of k_{exp} with respect to temperature, $E_a = 45.55$ KJ/mol.

volume of solution = 500 mL, partial pressure of O_2 = 1 atm, and using tailings milled for 8 min (constant size).

Figure 19 is the representation of the dependence of k_{exp} with respect to temperature, where it can be seen that, as the temperature increases, the rate constants of reaction also increase, showing a strong effect of this parameter on silver dissolution, reaching 91% for the highest temperature tested. Based on the data obtained, the activation energy was calculated, which gave a value of 45.55 kJ/mol that indicates a chemical control of the leaching process using thiosemicarbazide.

For the effect of partial pressure of O_2, the experiments were carried out similarly to those mentioned above, but here only two pressure values were probed (0.2 and 1 atm) while maintaining the following parameters: temperature = 298 K, pH = 10, $[CH_5N_3S]$ = 0.3 M, sample = 40 g/L, stirring rate = 600 min^{-1}, and using tailings milled for 8 min (all one).

Figure 20 shows the results obtained from the study of the effect of partial pressure of O_2, during silver dissolution using thiosemicarbazide. As can be seen, with increasing partial pressure of O_2, silver dissolution increases, from 33% at 0.2 atm to 60% at 1 atm.

For the study of the effect of pH, the experiment was performed by varying the pH values as follows: 4, 6, 8, 10, and 12, while keeping the following parameters constant: temperature = 298 K, $[CH_5N_3S]$ = 0.3 M,

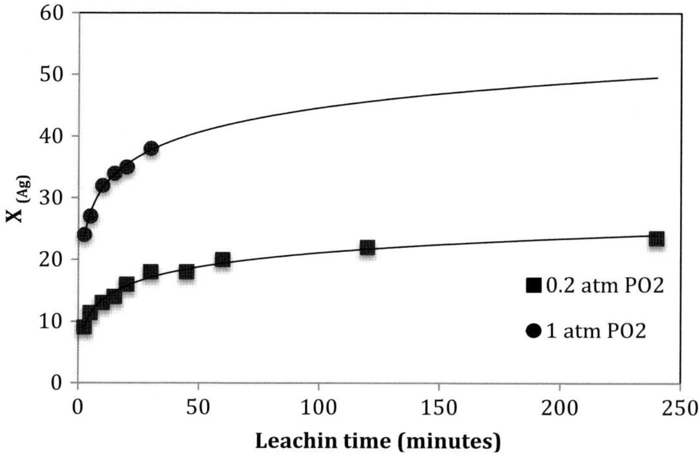

Fig. 20. Dependence of partial pressure of O_2 in the system.

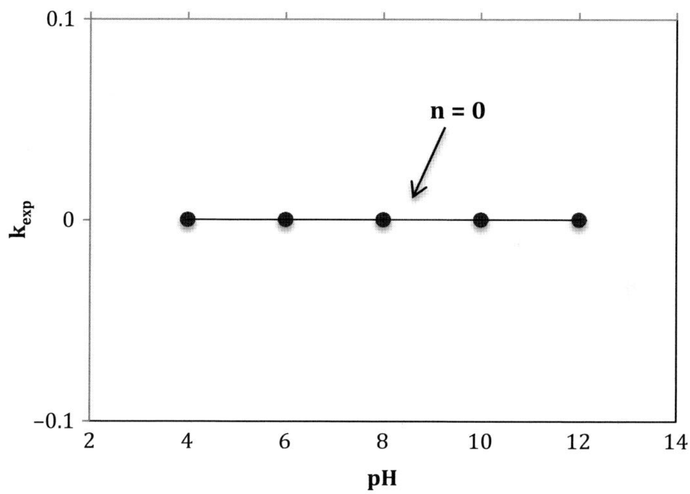

Fig. 21. Dependence of k_{exp} with respect to pH.

sample = 40 g/L, stirring rate = 600 min^{-1}, partial pressure of O_2 = 1 atm, and using tailings milled for 8 min (constant size).

Figure 21 shows the obtained results for the study of the effect of pH on silver dissolution, observing that this parameter has practically no significant effect on silver dissolution.

2.5 Conclusions

2.5.1 *Characterization of tailings*

The tailings used for the comparative study of leaching using different reagents showed that they had adequate values of silver to be used in this study (56–60 g/ton). In addition, a large amount of quartz was also detected, which was present as very fine particles, in which probably some of the silver was trapped. On the other hand, jarosite-type minerals were detected, in which silver could be present, and finally, argentite was also detected. The tailings were subjected to a preliminary milling process, and the obtained results at 8 min of milling were found to be the most favorable.

2.5.2 *Cyaniding*

The cyaniding process was adjusted to the shrinking core model $[1 - (1 - x)^{1/3}]$, and the results found for cyanide concentration showed that cyanide concentration has no significant effect on silver dissolution, resulting in an order of reaction of $n = 0$. On the other hand, temperature has a marked effect on the rate of dissolution, with an activation energy of 42 kJ/mol, which is indicative that the cyaniding process is chemically controlled. The study of the effect of particle size was also carried out, where the results showed a straight line passing through origin.

On the other hand, on increasing NaOH concentration, the rate constant of silver dissolution process also increased, obtaining an order of reaction of $n = 0.214$. For this study and with the optimal conditions for cyaniding, it was possible to get up to 94% of silver dissolution in the system.

2.5.3 *Thiosulfate*

The stoichiometry for silver dissolution in the system $Ag–S_2O_3^{2-}–O_2$ was carried out in three stages depending on what form the silver is present in tailings: as native silver, silver sulfur, or silver involved in the jarosite-type compounds present in tailings. For this system, the optimum conditions of leaching of silver contained in tailings were the following: stirring rate of 750 min^{-1}, partial pressure of O_2 of 1 atm, pH of 11, sodium thiosulfate concentration of 126.4 g/L, $[Cu(SO_4) \cdot H_2O]$ of 0.5 g/L, and a temperature of 323 K. Under these conditions, a maximum silver dissolution of 72.57% was obtained, with an order of reaction of $n = 0.21$ for the effect of thiosulfate concentration. The stirring rate had no influence on the reaction, whereas variation in temperature showed a very significant effect, with the obtained activation energy of 38 kJ/mol indicating that the leaching process using thiosulfate was also controlled by chemical reaction. Varying the partial pressure of O_2 showed a notable difference because silver dissolution rates of 31.07% and 61.07% were obtained at 0.2 atm and 1 atm, respectively. Finally, the effect of pH was found to be insignificant and independent of the process.

Finally, the addition of Cu^{2+} was found to significantly improve the silver dissolution in the system studied, increasing by almost 40% With the addition of 0.25 g/L of Cu^{2+}, silver dissolution rates can reach up to 96.4%, establishing an order of reaction of $n = 0.43$, indicating the oxidizing behavior of copper sulfur.

As a major conclusion, we can say that the leaching of silver contained in mining tailings by using thiosulfate as leaching reagent is a good alternative to conventional processes of cyaniding, because silver dissolution up to 96% can be achieved under optimal experimental conditions.

2.5.4 *Thiosemicarbazide*

In the process of silver dissolution proposed, it was observed that increasing thiosemicarbazide concentration also increases the silver dissolved in the system, increasing from 38% of silver dissolved at 0.1 M of [CH_5N_3S] to 76% when using 0.3 M of [CH_5N_3S] at room temperature, with an order of reaction of $n = 0.93$. Similarly, increasing the temperature had a marked influence on the amount of silver dissolved in the system as it was observed that dissolution increased from 32% (at 288 K) up to 91% (at 328 K) using 0.3 M of [CH_5N_3S].

On the other hand, the activation energy (E_a) found was 45.55 kJ/mol, which indicates that leaching process of mining tailings using thiosemicarbazide is controlled by a chemical reaction.

Based on the obtained results, the system proposed and studied can be considered as an alternative to the treatment of refractory minerals and tailings due to the high content of silver dissolved (up to 91%).

2.5.5 *Comparative study*

According to the obtained results, both thiosulfate and thiosemicarbazide are preferable alternatives for the leaching of silver contained in mining tailings instead of cyanide, which is toxic. The efficiency of the processes was clear, and they could operate at low temperatures as well. With regard to silver dissolution, the maximum silver dissolved in each system was 94% during cyaniding process, 96% for the leaching using thiosulfate, and 91% for leaching using thiosemicarbazide.

Acknowledgments

The authors would like to acknowledge the support of the UAEH – AACTy M (ICBI) for the development of this work and recognize the work of their colleague Isauro Rivera Landero for his support in the use and interpretation of thiosulfate systems applied to precious metals leaching.

References

[1] Carrera SM (1973) Bartolomé de Medina y el beneficio de patio. *Revista de la CámaraMinera de México*, **1**(12): 12–31.

[2] Geyne AR (1963) Geology and mineral deposits of the Pachuca-Real del Monte District, State of Hidalgo, México. *Consejo de RecursosNaturales no Renovables*. https://scholar.google.com.mx/scholar?hl=es&as_sdt=0%2C5&q=Geyne+AR+%281963%29+Geology+and+mineral+deposits+of+the+Pachuca-Real+del+Monte+District%2C+State+of+Hidalgo%2C+Me%CC%81xico.+Consejo+de+RecursosNaturales+no+Renovables.&btnG=.

[3] Comisión de FomentoMinero (1961) Experimentación metalúrgica preliminar corrida sobremuestras de colas de la planta de flotación de Santa Julia de la Comisión de Fomento Minero en Pachuca, Hgo., para la concentración de cuarzo, pp. 1–2.

[4] American Public Health Association, American Waterworks Association, Water Pollution Control Federation, & Water Environment Federation. (1915). Standard methods for the examination of water and wastewater (Vol. 2). American Public Health Association.

[5] Yta M, Moreno TR, Hernández AJ, Ezquivel VC (2003) Sistema de información geográfica de residuos mineros: Valorización, inventario y caracterización en el estado de Hidalgo. *COREMI Mineral Resources Council*, 14–16.

[6] Hernández AJ, Salinas RE, Patiño CF, Rivera LI, Yta M, Pérez LM, (2006) Caracterización de los jales del distrito Pachuca — Real del Monte, en el estado de Hidalgo y suposible aplicacióncomo material industrial alterno, *Boletín de Mineralogía*, **17**(1): 29–134.

[7] Patiño F, Ramírez J (1991) Grinding of the 'Jales de Santa Julia de la Compañía Real del Monte y Pachuca', SA de CV, Mexico. In *EMC'91: Non-FerrousMetallurgy — Present and Future*, J Vereecken (ed.) (Springer, Netherlands), pp. 19–24.

[8] Patino F, Hernandez J, Rivera I, Salinas E, Longoria GLC, Gonzalez JM (2007) Characterization and kinetics of the grinding of the Dos Carlos burrows in the state of Hidalgo. *International Precious Metals Institute*, 895–901.

[9] Schmitz PA, Duyvesteyn S, Johnson WP, Enloe L, McMullen J (2001) Ammoniacal thiosulfate and sodium cyanide leaching of preg-robbing Gold strike ore carbonaceous matter. *Hydrometallurgy*, **60**: 25–40.

[10] Habashi F (1986) *Principles of Extractive Metallurgy, Vol 3. Pyrometallurgy* (Gorgon and Breach Science Publisher, New York, USA), p. 486.

[11] Young RS (1971) *Chemical Analysis in Extractive Metallurgy* (Charles Griffin & Company, Ltd. London, England), p. 318.

[12] Salisbury, DE (1985) *Tratado de Mineralogia* (In Spanish), 4th Edition S.A. Edición (CECSA, Mexico), pp. 905–912.

[13] Dai X, Jeffrey MI, Breuer PL (2005) The development of floe injection analysis method for the quantification of free cyanide and cooper cyanide complexes in gold leaching solutions. *Hydrometallurgy*, **76**: 97–96.

[14] Rivera I (2003) Estudiocinético de la precipitación/lixiviación de plata en el sistema O_2- $S_2O_3^{2-}$ - $S_2O_4^{2-}$, PhD Thesis, Universitat de Barcelona, España.

[15] Salinas-Rodríguez E, Hernández-Ávila J, Rivera-Landero I, Cerecedo-Sáenz E, Reyes-Valderrama MI, Correa-Cruz M, Rubio-Mihi D (2016) Leaching of silver contained in mining tailings, using sodium thiosulfate: A kinetic study. *Hydrometallurgy*, **160**: 6–11.

[16] Patino F, Hernandez J, Rivera I, Salinas E, Longoria GLC, Gonzalez JM (2007) Characterization and kinetics of the grinding of the Dos Carlos burrows in the state of Hidalgo. *International Precious Metals Institute*, pp. 895–901.

[17] Badillo JF, Ávila JH, Cardona FP, Pineda NYT, Santos, JAO (2014) Developing alternative building material from mining waste. *Advanced Materials Research*, **976**: 202–206.

[18] Hernandez J, Salinas, E, Patino F, Rivera I, Flores J, Trapala N, Perez M, Reyes M, Flores U, Reyes I (2012) Tile production using wastes from mining

industry of the mining District Pachuca and Real del Monte. TMS (The Minerals, Metals & Materials Society), February 2012, USA, pp. 203–209.
[19] Hernández J, Rivera I, Patiño F, Juárez JC (2013) Estudio Cinético de la Lixiviación de Plata en el Sistema S2O3 (2-)-O2-Cu2+ Contenido en Residuos Minero-Metalúrgicos, *Información tecnológica*, **24**(1): 51–58.
[20] Jeffrey MI, Breuer PL, Choo WL (2001) A kinetic study that compares the leaching of gold in the cyanide, thiosulfate, and chloridesystems. *Metallurgical and Materials Transactions B*, **32**(6): 979–986.
[21] Hernandez J, Patino F, Rivera I, Reyes IA, Flores MU, Juarez JC, Reyes M (2014) Leaching kinetics in cyanide media of Ag contained in the industrial mining-metallurgical wastes in the state of Hidalgo, Mexico. *International Journal of Mining Science and Technology*, **24**(5): 689–694.
[22] Patiño F, Reyes IA, Flores MU, Pandiyan T, Roca A, Reyes M, Hernández J (2013) Kinetic modeling and experimental design of the sodium arseno-jarosite decomposition in alkaline media: Implications. *Hydrometallurgy*, **137**: 115–125.
[23] Sancho J, Verdeja LF, Ballester A (2000) *Metalurgia extractiva: Volumen II; Procesos de obtención* (Editorial Síntesis, Madrid, Spain).
[24] Salinas RE, Patiño CF, Hernández AJ, Hernández CLE, Rivera LI (2006) La problemática de los jales en el estado de Hidalgo. *Memoria del IV Congreso Nacional de Metalurgia y Materiales*, Monclova, Coahuila, México, pp. 327–328.
[25] Habashi F (1980) *Principles of Extractive Metallurgy, Vol. 2: Hydrometallurgy* (Gorgon and Breach Science Publisher, New York, USA).
[26] Gupta CK, Mukherjee TK (1990) *Hydrometallurgy in Extraction Processes*, Vol. 2 (CRC Press, Boca Raton, Florida USA), pp. 174–177.
[27] Marsden J, House I (1992) *The Chemistry of Gold Extraction* (Ellis Horwood, Allentown, PA), pp. 192–193, 259–287, 378–398.
[28] Senanayake G (2006) The cyanidation of silver metal: Review of kinetics and reaction mechanism. *Hydrometallurgy*, **81**(2): 75–85.
[29] Johnson NW (2006) Liberated 0–10 μm particle from sulphide ores, their production and separation — Recent developments and future needs. *Minerals Engineering*, **19**(6): 666–674.
[30] Luna RM, Lapidus GT (2000) Cyanidation kinetics of silver sulfide. *Hydrometallurgy*, **56**(2): 171–188.
[31] Senanayake G (2005) Kinetics and reaction mechanism of gold cyanidation: Surface reaction model via Au (I)–OH–CN complexes. *Hydrometallurgy*, **80**(1): 1–12.

[32] Muir DM, Aylmore MG (2004) Thiosulphate as an alternative to cyanide for gold processing — issues and impediments. *Mineral Processing and Extractive Metallurgy*, **113**(1): 2–12.

[33] Jeffrey, MI (2001) Kinetic aspects of gold and silver leaching in ammonia–thiosulfate solutions. *Hydrometallurgy*, **60**(1): 7–16.

[34] Breuer PL, Jeffrey MI (2000) Thiosulfate leaching kinetics of gold in the presence of copper and ammonia. *Minerals Engineering*, **13**(10): 1071–1081.

[35] Feng D, Van Deventer JSJ (2007) The role of oxygen in thiosulphate leaching of gold. *Hydrometallurgy*, **85**(2): 193–202.

[36] Berezowsky RMGS, Sefton VB (1979) Recovery of gold and silver from oxidation leach residues by ammoniacal thiosulphate leaching, *AIME Annual Meeting*, New Orleans, LA, pp. 102–105.

[37] Briones R, Lapidus GT (1998) The leaching of silver sulfide with the thiosulfate–ammonia–cupric ion system. *Hydrometallurgy*, **50**(3): 243–260.

[38] Oraby EA, Browner RE, Nikraz HR (2014) Effect of silver in thiosulfate leaching of gold–silver alloys in the presence of copper and ammonia relative to pure gold and silver. *Mineral Processing and Extractive Metallurgy Review*, **35**(2): 136–147.

[39] Salinas E, Rivera I, Carrillo FR, Patiño F, Hernández J, Hernández LE (2004) Mejora del proceso de cianuración de oro y plata, mediante la preoxidación de minerales sulfurosos con ozono. *Revista de la Sociedad Química de México* **48**: 315–320.

[40] Carrillo PFR, Soria AMJ, Rodríguez MJ, González AJA (2001) procesamiento de minerales con metales preciosos y minerales siderúrgicos empleando ozono, *AIMMGM Convención Internacional*, October 17–20 2001, Acapulco, Guerrero México.

[41] Xie F, Dreisinger DB (2007) Leaching of silver sulfide with ferricyanide–cyanide solution. *Hydrometallurgy*, **88**(1): 98–108.

[42] Patiño F, Salinas E, Cruells M, Roca A (1998) Alkaline decomposition–cyanidation kinetics of argentian natrojarosite. *Hydrometallurgy*, **49**(3): 323–336.

[43] Rodríguez ES, Sáenz EC, Ramírez M, Cardona FP, Labra MP (2012) Kinetics of alkaline decomposition and cyaniding of argentian rubidium jarosite in NaOH medium, *Metallurgical and Materials Transactions B*, **43**(5): 1027–1033.

Chapter 3
Adsorption and Recovery of Silver from Aqueous Solutions

Emanuelle Dantas de Freitas, Thiago Lopes da Silva, Meuris Gurgel Carlos da Silva and Melissa Gurgel Adeodato Vieira

Department of Processes and Products Design
School of Chemical Engineering, University of Campinas — UNICAMP
Albert Einstein Avenue, 500, Zip code 13083-852
Campinas, SP, Brazil

3.1 Introduction

Silver occurrence in the water environment has caused great concerns regarding its high toxicity to aquatic life forms. The concern is even greater when silver is in the free ion form (Ag^+), since this is the most toxic form of silver. As the most extensively used noble metal, silver-contaminated may originate from different kinds of sources, such as electroplating, photographic process, mining, and electronic components [1]. Historically, the photography industry has used more silver than any other industry; however, in the past decade, with the increase of virtual media, this situation has been reversed. Medical and non-medical applications of silver, such as in X-rays, paintings, clothes, and children's toys, have grown significantly [2]. Silver is generally bio-accumulated and can cause diseases and disorders through the food chain when discharged into the environment by anthropogenic applications. Due to its toxicity and high economic value, silver removal and recovery is now a great challenge [3].

Several techniques have been studied for silver removal, such as membrane separation, electrolysis, ion exchange, precipitation, and adsorption.

Among them, adsorption is highlighted as an effective and environmentally friendly technique when compared to other alternatives, especially in treating wastewaters with low silver concentrations, when other conventional treatments are not feasible.

In this chapter, different adsorption processes will be described as an efficient alternative for silver-contaminated wastewater treatment supported by several published results. The theory and fundamental aspects on silver contamination issues, its removal, and its recovery using adsorbents are also discussed.

3.2 Toxic Metals

Of the chemical elements, metals comprise the largest group. Metals are elements that conduct electricity, are malleable and ductile, form cations, and have metallic luster [4]. Among those elements, heavy metals are defined as metals or metalloids with an elevated toxic potential, depending on their current form. The term "heavy metal" is no longer used due to its difficulty in describing toxic properties. Thus, it has been replaced with the term "toxic metal" [5].

Metals are naturally occurring elements due to the phenomena of rocks weathering and soil leaching [6]. Nevertheless, metal contamination began to affect the environment during the Industrial Revolution, when a large amount of contaminated wastewater began to be discharged into natural systems. This issue has become one of the main environmental concerns in the past few decades. Human activities, including mining, electroplating, gas and energy production, fertilizer and pesticide use, among others, have increased pollution [7]. Zinc, magnesium, copper, and iron are required metals for human life (essential metals) and are needed for biological functions. However, humans may be affected in different ways by the long-term, low-level exposure to metals. Symptoms including gastrointestinal, renal, and neurological disorders; bronchitis; skin rashes; decalcification; and anemia may eventually occur. Mostly, poisoning occurs via contaminated drinking water and food [8].

Over the course of time, bio-accumulation leads to a high level of toxic metal accumulation in the organism compared to the environmental levels [9]. Furthermore, metals are not bio-degradable, which means

Table 1. Adverse human health effects related to toxic metals exposure.

Metal	Adverse human health effects
Cadmium	• Cardiovascular disease • Kidney effects • Lung and prostate cancers • Hepatic damage
Copper	• Accumulation in brain, skin, liver, pancreas, and myocardium
Iron	• Cardiomyopathy
Lead	• Neurotoxic effects • Cardiovascular disease • Kidney effects • Affect the process of hemoglobin synthesis
Manganese	• Neurotoxic effects
Mercury	• Neurotoxic effects • Cardiovascular disease • Kidney effects

Source: Igwe and Abia [10] and Nordberg *et al.* [8].

they cannot be biologically destroyed [7]. Table 1 presents some of the damages caused by the most common toxic metals found in the environment.

Due to its toxicity, metal uptake techniques such as complexation, ionic exchange, adsorption, and micro-precipitation are employed in order to remove these elements, especially from contaminated effluents that could reach aqueous environments [11].

3.2.1 *Noble metals*

Many industrial activities generate a large quantity of wastewater containing noble metals. The recovery of these metals has assumed importance due to their high commercial value and their adverse environmental impact, as several problems are related to the presence of these metals in waste aqueous solutions [12].

Noble metals are characterized by their high conductivity level and great resistance to oxidation and corrosion [13]. Because of their properties, noble metals have been extensively used as catalysts in chemical processes; in jewelry, electronic, electrical, and glass industries; and in medical applications, among others [14]. The list of these metals includes the platinum-group metals (PGMs) and gold and silver elements; due to their scarcity, durability, and beauty, they can be considered as precious metals [15].

Several noble metals may present high toxicity, and the increasing generation of wastes containing these elements demands their removal. Furthermore, these metals have a high economic value and a growing demand in many industrial applications, which emphasizes the need of these metals recovery and recycling [16].

PGMs include the elements platinum, rhodium, osmium, iridium, palladium, and ruthenium, all noble metals found together in nature [17]. Among them, the most used are palladium (Pd), found as +2 state, and platinum (Pt), which can be found as +2 and +4 states, all in complexes [18]. PGMs are commonly used as materials in electrodes, catalysts, and jewelry and as anti-cancer drugs in chemotherapy. However, the greater discharge of PGMs into the environment occurs from their use in automobile catalytic converters [19]. Toxicity data about PGMs involve specifically Pt^{+4} and Pd^{+2}, which may be allergens and can cause gastrointestinal irritation, and Pt^{+2}, which presents mutagenic and carcinogenic potential [20].

Gold (Au) is a noble and precious metal mined in 94 countries around the world, of which Republic of South Africa is the largest producer, followed by United States. Historically, gold has been mostly used as jewelry and money and in dentistry; in the twentieth century, applications such as in radioisotope medicine, electrical and electronic circuitry, radiation shielding, and scientific instrumentation have been explored more, leading to the increase in wastewater containing this metal [21]. Pure gold is non-toxic and non-irritating if ingested, although it presents an allergen potential that may cause dermatitis when it comes in direct contact with the skin. Chemically, it presents six oxidation states (+1, +2, +3, +4, +5, +7), although +1 and +3 states are the most common [15]. In water, Au^{+}

and Au^{+3} occur as complexes with ligands and may cause damage to living organisms, especially in the +3 state [22].

Silver (Ag) is another of the eight noble metals and is usually found with gold in nature. These two metals were probably the first known metals to man. Silver is considered as highly interesting due to its color, brilliant luster, and physical–chemical properties. Nowadays, other silver characteristics have transformed this decorative/monetary metal into an industrial metal [23]. This will be better described in the following sections.

3.3 Silver

Most silver is produced as a by-product of copper, gold, lead, and zinc mining [2]. Silver ion belongs to the highest toxicity class of metals. This metal class includes cadmium, chromium (VI), copper, and mercury, which is considered the most toxic metal [24]. In this chapter, silver ion refers to Ag^+, save when otherwise indicated.

A vast range of applications is possible due to the properties of silver. Pure silver presents great ductility and malleability, besides having the highest thermal and electric conductivities among all metals [25]. Silver is a naturally occurring white metal in the pure form or in ores. Its color, shine, non-corrodibility, and ease of working are what makes it more attractive [26]. Finally, silver presents anti-microbial properties, which enables its application in the medical field and purification processes [27].

Of the noble metals, Ag is the most reactive and may present +1, +2, and +3 oxidation states. Most Ag compounds are based on the +1 state, although complexes may be formed with Ag^+ and Ag^{+2} [26]. From the beginning of civilizations, silver has been considered quite valuable and used to produce small tokens representative of a person's wealth. Nowadays, silver is present in many applications around us, such as in batteries, computer keyboard, photographic paper, solar cells, water purifiers, and medical products [28]. The anthropogenic life cycle of silver involves the mining activities, manufacture of silver products, and disposal as waste material. The activities of mining carried out in the lithosphere, the production process of silver products, and the disposal of the silver products as waste material discharges silver into the environment [29].

Table 2. Physical–chemical properties of silver.

Silver		Ag^{+1}	
Chemical symbol	Ag	Pauling electronegativity	1.93
Atomic number	47	Ionic radius (Å)	1.26
Oxidation states	+1, +2, +3	Log of the first hydrolysis constant	−12.0
Melting point (°C)	961.93	Hardness index	0.074
Boiling point (°C)	2212	Hydrated radius (Å)	3.41 Å
Density (g cm^{-3})	10.5		
Ionization energies (kJ mol^{-1})			
1st	730.995		
2nd	2,072.26		
3rd	3,360.58		

Source: RSC [30], Lansdown [31], and Kinraide and Yermiyahu [32].

Table 2 presents some physical and chemical properties of silver and the soluble silver ion. The high value of ionization energy of the second and third electrons of silver explains why the oxidation state +1 is the most common to silver ion.

Silver compounds present extreme photosensitivity and stability in air and water. When exposed to sulfur compounds, silver compounds instantly tarnishes. Although silver salts, such as $AgNO_3$ (silver nitrate), are water soluble, the metallic form is rather insoluble. In nature, silver may be found as silver sulfide (Ag_2S) or associated with insoluble metal sulfides, as those of gold, iron, copper, and lead, among others [33].

In the natural environment, the monovalent form of silver (Ag^+) is quite rare or negligible. The most important component of aqueous silver is the colloidal phase, which is composed of organic and inorganic materials. Silver is capable of readily forming compounds with antimony, arsenic, selenium, and tellurium [33], besides complexing with organic and inorganic compounds, such as thiosulfate, sulfide, chloride, and organic matter [1]. The physicochemical environmental conditions dictate the occurrence of silver species. Thus, individual silver forms influence silver toxicity more than the total silver concentration. In surface waters and wastewaters, covalent complexing or colloidal binding of silver with several ligands is very likely to occur. Therefore, the dissolved silver toxicity is usually more evident in laboratory tests than in those environments [24].

Soluble silver compounds, mainly Ag^+ ions, are extremely toxic species, even at trace levels. These species are more readily absorbed than metallic or insoluble silver and thus may produce undesirable effects on the human body and environment, depending on the type and extent of exposure [34]. Silver absorption by any route may be determined by the nature of the product or source of silver contamination, its capacity to release biologically active silver ion, and the route and duration of exposure. The contamination is reduced in presence of anions, proteins, and materials that chelate or otherwise bind the free silver ion in the form of insoluble precipitates [31].

Silver ions are extremely toxic to bacteria, so bio-accumulation can only be investigated in silver-tolerant organisms or at low concentrations [24]. The anti-microbial action is related to many mechanisms. One possible mechanism considers that silver promotes an uncoupling of oxidative phosphorylation in the cell, and another considers the cell wall integrity is compromised by the interaction between Ag^+ and enzyme bounded to the membrane and the thiol groups from proteins [2].

Silver may be very critical to aquatic environments, considering its toxic potential [26]. In conditions of low concentration of dissolved silver ion, high water pH, hardness, presence of sulfides, and organic matter as dissolved or particulate species, lower toxicity of silver ions was observed. When the organisms were well fed, the toxicity was also less pronounced. Although silver ions are considered extremely toxic to micro-organisms, this effect is not observed in wastewater treatment plants. Silver ions may quickly suffer adsorption and complexation reactions in these environments, leading to reduced inhibitory action. Silver ions proved lethal to aqueous organisms at concentrations in the range of 1–5 $\mu g\ L^{-1}$ [33].

Silver is a non-essential metal, which means it is not required for any biological function in the human body [35]. However, it can be easily absorbed by humans through inhalation or after ingestion. Then, it may accumulate essentially in liver and spleen, although it can be found in muscles, skin, and brain [36]. Silver may produce other adverse effects on the human body, such as stomach irritation, decreased respiration, and changes in blood cells [25]. The prolonged over-exposure to silver nitrate may cause a decrease in blood pressure and respiration rates, stomach irritation, and diarrhea. The ingestion of small doses of silver salts for a long time may lead to chronic symptoms, such as compromised liver and kidneys and damages to blood cells [37]. Long-dated ingestion or inhalation of soluble silver compounds or colloidal silver possibly causes argyria and/or argyrosis, which is characterized by silver deposition on skin and hair, resulting in the skin exhibiting grey-blue pigmentation [25, 36]. The lethal dose for humans of the most common silver compound, silver nitrate, is at least 10 g as a total lifetime oral intake [38]. The US Environmental Protection Agency (EPA) has established the maximum contaminant level of silver in drinking water as 0.1 mg L^{-1} [39].

The widespread applications of this metal in industries lead to the generation of large amounts of silver-contaminated wastewater, making it more available to the environment and humans. Silver is present in photographic effluents as thiosulfate complex and easily transforms into silver sulfide in the sludge. Both compounds do not inhibit micro-organisms' activities as silver ions do, thus the biological treatment of wastewater is possible in this case [24].

In medicine, the most widely known use of silver is in combination with sulfadiazine, forming a topical anti-bacterial agent, available in liquid and cream forms, for burn treatment [25]. Nowadays, because of the biocide properties, the use of silver in material also includes those at the nanoscale (nanosilver). Nanoscale-level silver particles have been produced by several methods, such as electrochemical reduction, cryochemical synthesis, solution irradiation, and spark discharge (with plasma vapor deposition) [2]. The applications of engineered nanosilver are rapidly growing, especially in the healthcare sector. Nanosilver particles may be found in coating dressings, contraceptives, and surgical devices, besides prostheses. In the daily routine, silver nanoparticles are observed in soaps, water purifiers, paints, and washing machines [40]. However, despite the rapid growth of nanosilver applications, its toxic effects on humans and the environment are not well recognized and measured [2]. There is a lack of agreement on the silver toxicity in these products and an absence of absolute consensus on their use for long exposure periods, especially if there is a possibility of the silver being released in an ionic form. Benn *et al.* [41] studied some consumer products, by quantifying and characterizing silver use and potential release, based on whether the products contained nanosilver, colloidal silver, or ionic silver. One has investigated toothpaste, shampoo, medical cloth and mask, and laundry detergent. All products analyzed have demonstrated that both consumers and the environment are subjected to some exposure of silver by using or disposing these products.

Because of the high commercial value of silver and its toxicity, the removal and recovery from water and wastewater are studied through several techniques: electrochemical deposition [42], electrolysis [43], precipitation [44, 45], electrodialysis (ED) [46], ion exchange [47–49], membrane separation [50], and adsorption [51, 52]. Choosing the best removal mechanism takes account the silver concentration, costs, available technology, and the method's efficiency. The need for conjugated techniques in order to obtain the necessary removal is quite common.

3.3.1 *Silver in wastewater*

Among the chemical industries around the world, those that deal with heavy metals may be considered to be the most hazardous due to the large amount of discharged toxic metal-contaminated effluent. Usually, a vast range of toxic metals may be found conjugated in such media [53].

Silver may be found in several wastewater sources, in which it can be subjected to complexing or precipitating reaction with organic or inorganic compounds. The solubility products (K_{sp}) of silver chloride and silver sulfide are $10^{-9.75}$ and $10^{-48.97}$, respectively. Thus, the presence of solubilized silver (Ag^{+1}) is strongly reduced if chloride or sulfide is present [1]. Silver nitrate and silver chloride are the most used silver salts [54], and silver nitrate consumption has greatly increased in the past few years.

The silver-contaminated wastewater generation is related to industrial use in the processes such as electroplating, photography, coinage, mining, ink formulation, mirroring, porcelain, jewelry, silverware, catalysts, batteries, electronic components, superconductors, heat sink, and bactericide components, among others [25, 55, 56]. Mining includes the removal of the desired mineral from the surface, and during the process, dissolved pollutants as heavy metals may migrate to the local ground and surface water. The most common metals found near mine sites are arsenic and cadmium, although noble metals as gold and silver may also be present [57, 58].

An important source of silver-contaminated wastewater involves processes involving electroplating and metal surface treatment [53]. The contamination occurring from metal surface cleaning and rinsing may present silver concentration in the range of 1–3 g L^{-1}. Considering the spent electrolyte and silver stripping wastewaters, silver concentration may reach up to 20–60 g L^{-1} [59].

Silver-contaminated wastewater may also be originated from photographic processes, which also includes X-ray operations. Silver compounds may reach 30–40% of the total wastewater weight in those cases [42]. It is estimated that 18–20% of all silver used in the world is obtained from recycling photographic wastewaters [60]. Table 3 shows some silver concentrations found by different studies in different aqueous media.

The presence of silver in wastewater may result in the reduction of efficiency of wastewater treatment processes, mainly in biological treatment plants. Silver is extremely toxic for bacteria, and the effectiveness of silver as anti-bacterial agent is a problem for biological wastewater treatment, essentially in its soluble forms. In addition, for this reason, the microbial recovery of silver from wastewater is not usually applied or investigated by scientific researches. Only mercury is more toxic to bacteria than silver, and this is clearly a limiting factor in the extraction of silver by live biomass [69]. This application may be employed since the silver toxicity limit for bacteria is not achieved. Kim *et al.* [70] evaluated silver's minimal inhibitory

Table 3. Silver concentration in common aqueous media.

	Silver concentration (mg L^{-1})	References
Electronic waste wastewater	640	[61]
Municipal wastewater (Santiago de Compostela, Spain)	0.0014	[62]
San Francisco Bay Estuarine Waters	0.000027	[63]
Photographic effluent	7800	[64]
Electroplating company effluent	8.04	[65]
Photoprocessing wastewater	1.85	[66]
Radiographic wastewaters		
Rinsing water of X-ray films	8.0	[67]
Radiographic fixer waste	2,175	
Wastewater from a nanosilver-producing washing machine	0.011	[68]

concentration for yeast, *Escherichia coli*, and *Staphylococcus aureus*, and obtained values of 6.6, 3.3, and 33 nmol L^{-1}, respectively.

3.3.2 *Silver and other toxic metals*

Typically, a vast range of metals is present in the same wastewater. Industrial processes use different metals in the same operation to obtain their products, and the presence of many metals in their effluents is common. For example, the electroplating industry may present different types of plating processes, such as plating of copper, zinc, gold, nickel, silver and chromium, and electroplating wastes may contain all these kind of metals [71].

Multicomponent composition of wastewaters directly affects the removal process of these metals. The interaction between metals may affect the removal efficiency, and optimized treatment conditions cannot be achieved for all components simultaneously. Knowledge of the metal composition of the wastewaters and the properties of the involved metals is important to evaluate the treatment processes. The competitive effect is more remarkable when the number of metal ions increases and the adsorbent may present different behaviors for different ions [72]. It is very hard to predict the uptake performance of a material in a complex multicomponent environment [60].

Naeemullah *et al.* [73] characterized an aqueous environment that receives untreated effluents from an industrial area. The metals found were Ag, Cd, Ni, Co, and Pb, which presented the following concentrations when entering the aqueous media: 54.4, 25.6, 33.6, 16.5, and 18.5 $\mu g\ L^{-1}$, respectively. Chiu *et al.* [71] have established different plating operation wastes concentrations from an electroplating industry. For the plating of Cu, Ni, Ag, and Cr effluents, the concentrations were determined as follows: 3 mg L^{-1} Cr, 11 mg L^{-1} Ni, and 3 mg L^{-1} Ag.

The uptake of Pb^{2+}, Ag^{+}, Sr^{2+}, and Cs^{+} ions, by waste brewery biomass was evaluated by Chen and Wang [74]. The maximum bio-sorption amount ($q_{\max}$) for the metal ions were 0.413 mmol g^{-1}(Pb^{2+}), 0.396 mmol g^{-1} (Ag^{+}), 0.091 mmol g^{-1} (Sr^{2+}), and 0.076 mmol g^{-1} (Cs^{+}). The metal ion accumulation in equilibrium (on the molar basis) has presented the following sequence: $Pb^{2+} > Ag^{+} > Sr^{2+} > Cs^{+}$. The observed order was directly correlated with their electronegativity and covalent index, and inversely correlated with the dissociation constant.

Carbonaceous sorbent from date palm leaflets treated with sulfuric acid was applied to the adsorption process of Cu^{+2} and Ag^{+} [75]. The sorption capacity was much superior for Ag^{+} than for Cu^{2+}, especially when the temperature was increased (from 25°C to 45°C). The silver ion was reduced to elemental silver on the surface of the sorbent. In the case of copper ions, no reduction processes have been observed. The q_{max} obtained was 48.08 mg g^{-1} (0.757 mmol g^{-1}) for Cu^{2+} and 217 mg g^{-1} (2.012 mmol g^{-1}) for Ag^{+}.

Petrova *et al.* [60] also evaluated Ag^{+} and Cu^{2+} adsorption on a sulfoethyl derivative of chitosan. A complex composition solution was applied, involving different kinds of metal ions. They have observed that the adsorbent was able to selectively remove silver and copper from that solution. In addition, different degrees of sulfoethylation have been investigated, and the increase in the degree has caused an improvement in adsorbent selectivity for silver.

3.4 Methods of Removing Toxic and Noble Metals from Wastewater

Due to the high toxicity of some metals, it is commonly necessary to remove them from the wastewater before disposal into water bodies [76]. For this purpose, conventional treatments such as ion exchange and chemical precipitation have been employed. Nevertheless, regulations have become more limited with regard to toxic metals in recent years, demanding more effective technologies development, such as adsorption and membrane separation [53]. These processes are briefly described in the following sections.

3.4.1 *Chemical precipitation*

Chemical precipitation is the most used treatment process in order to remove toxic metals from wastewater, due to its simplicity and low-cost operation [77]. Metal chemical precipitation usually occurs in the form of hydroxide, and the mechanism may be represented by Eq. (1) [78].

$$M^{2+} + 2(OH)^{-} \leftrightarrow M(OH)_2 \downarrow \qquad (1)$$

where M^{2+} is the dissolved metal ion, OH^{-} is the precipitant, and $M(OH)_2$ is the precipitated metal hydroxide, which can be separated from water by filtration or sedimentation. The main disadvantage of chemical precipitation is the large amount of sludge generation, which must be properly disposed. Also, the ideal precipitation pH for each metal may be different. So, the removal of one metal may cause the other metal to disperse back into solution [77]. The other disadvantage involving chemical precipitation is its low efficiency in reducing toxic metal concentration to acceptable levels. Thus, a further stage is usually required [79]. Other precipitation mechanisms involve sulfide precipitation or chelating ligands. The use of sulfide precipitation is effective in toxic metal removal, although it may result in the generation of H_2S fumes, which are toxic [77]. The use of chelating precipitants may reach low levels of toxic metals after treatment, although the production of toxic gases may occur [80].

3.4.2 *Ion exchange*

Ion exchange involves a reversible interchange of ions between two different phases: a solid phase that will attract the soluble ions, and a liquid phase, where the ions are primarily present. A synthetic resin is usually employed as the solid phase [81]. This technology has been extensively employed due to its high efficiency and treatment capacity, even from low concentration solutions [77]. When compared to precipitation, ion exchange is beneficial due to the reduced amount of sludge generated and the selectivity and possibility of valuable metal recovery [82]. A usual mechanism for ion exchange is shown in Eq. (2) [78].

$$n\mathrm{RSO_3^-} - \mathrm{H^+} + \mathrm{M}^{n+} \leftrightarrow n\mathrm{RSO_3^-} - \mathrm{M}^{n+} + n\mathrm{H^+} \qquad (2)$$

where ($-RSO_3^-$) is equivalent to the ion-exchange resin, M is the metal ion, and n depends on the metal oxidation state. The disadvantage of ion exchange is the need for resin regeneration using chemical reagents. Also, the process may be very expensive due to the resin, making it unattractive for large-scale processes [77].

3.4.3 *Solvent extraction*

Solvent extraction is a separation process involving an organic phase and an aqueous phase, containing the metal. In this case, by mixing both phases, the metal is removed by transferring into the organic phase [83]. The solvent extraction process does not require high energy for operation and allows obtaining the metal of expected purity, besides complete regeneration of the organic phase. The disadvantage of the process is the high salinity in treated water [84].

3.4.4 *Electrochemical treatments: Electrolysis*

Electrolysis is a metal-removal technique based on the passage of electricity by an aqueous solution containing a metal. The solution contains an insoluble anode and a cathode plate [81]. Electrolysis is the ideal technique to remove and recover metals back to their original form. Some of the advantages of this technique are the absence of sludge generation, low cost of operation, and no need for reagents [85]. However, it is a process with low space–time yield, with regard to the removal of low-concentration solutions, besides the concern over contaminated electrodes disposal [86].

3.4.5 *Cementation*

Another promising process to remove toxic metals from wastewater involves cementation. In this process, the toxic metal ion is replaced by a more active non-toxic metal ion [87]. The toxic metal is precipitated to its elemental metallic form by another electropositive metal. Equation (3) represents the general reaction for cementation.

$$m\mathrm{N}^{n+} + n\mathrm{M} \rightarrow n\mathrm{M}^{m+} + m\mathrm{N} \tag{3}$$

where m and n are the stoichiometric coefficients, and m^+ and n^+ refers to the oxidation state of the element, N is the toxic/noble metal to be removed and M is the reductant metal. Cementation allows metal recovery in relatively pure form, is a low-cost and simple control process, and requires low energy consumption. However, a large consumption of sacrificial metal in the process is a disadvantage [88].

3.4.6 *Membrane filtration*

The use of membrane filtration has extensively increased, especially due to its capacity of removing several types of contaminants, such as organic compounds, suspended solids, and inorganic contaminants, what includes toxic metals. Different types of membranes may be employed depending on the size of the retained particle [81]. The most common membrane filtration types are ultrafiltration (UF), nanofiltration (NF), and reverse osmosis (RO). UF (pore size 5–20 nm) removes inorganic and organic compounds and suspended solids based on the size of the particle. NF (pore size 0.5–2 nm) presents a membrane charged with anions that develop an electric potential and helps to remove the positive metal ions during filtration. The membrane surface charge associated to the little pores let small and charged solutes passing through it (steric and electrical effects). In RO (pore size 4–10 μm), a pressure-driven membrane is used in order to obtain a concentration gradient. The effluent is pumped through the membrane, helping in toxic metal removal [78, 80].

3.4.7 *Electrodialysis*

Electrodialysis, which is another membrane process, is based on the driving force of an electric field, promoting metal ion removal [77]. The main advantage of ED is the capacity to produce a highly concentrated metal solution for recovery. As a membrane process, periodic maintenance and careful operation are required [53]. Usually, ED and RO are competitive processes. When using ED, it is possible to employ high temperatures and extreme pH values solutions, which is not possible for RO. However, in case of ED, only charged components may be removed [89].

3.4.8 *Evaluation of silver removal by different treatments*

Chemical precipitation is one of the most used methods for silver removal and recovery from wastewater [90], mainly when there is a high concentration of metal in solution. Bas *et al.* [44] evaluated silver chemical precipitation using hydrogen peroxide (Ag initial concentration: 1,100 mg L^{-1}). High silver recovery rates, greater than 95%, have been obtained at high levels of hydrogen peroxide ($\geq$37.6 g L^{-1}). Silver removal and recovery from highly concentrated solutions (5,500 mg L^{-1}) have been investigated by Ratanamskul *et al.* [91]. At pH = 12.0, about 95% silver recovery was achieved.

Gao *et al.* [92] evaluated silver removal by electrolysis from a solution containing 200 mg L^{-1} of silver (industrial plating wastewater). A silver removal rate of 99.8% was obtained under appropriate conditions. Goshima *et al.* [93] have compared electrolysis and ion exchange as silver recovery methods from radiographic fixer containing an average 11.25 mg of silver.

Using electrolysis, 11% silver recovery was achieved, while using ion exchange, silver recovery was more than 99.9%. Figueroa *et al.* [94] have studied electrocoagulation with iron electrodes to recover silver from a 283 mg L^{-1} initial concentration solution. A silver recovery rate of 95% was achieved by this method.

Shimojo and Goto [95] used solvent extraction to remove silver from a synthetic solution. A 90% silver recovery rate was achieved, in addition to demonstrating high selectivity for Ag^{+} among other metal ions in solution (Cu^{2+}, Zn^{2+}, Co^{2+}, Ni^{2+}). Lee *et al.* [96] have also evaluated solvent extraction for silver removal (initial concentration: 32.3 mg L^{-1}) and obtained 95.2% silver extraction efficiency.

Bortoletto *et al.* [67] have evaluated the precipitation of Ag from radiographic fixer waste used in X-ray film printing. Precipitating agents were hydrogen peroxide, sodium hydroxide, and sodium chloride, and the initial concentration of silver in the wastewater was 2,175 mg Ag L^{-1}. The precipitation rate of silver was 83.9% when the hydrogen peroxide concentration was 0.17 mol L^{-1}. Hydrogen peroxide promotes silver precipitation to a metallic silver form, but this reaction is non-selective, and many other reactions can occur with all components of this complex wastewater, including reactions of hydroxide with ammonium thiosulfate and sodium sulfite yielding sodium sulfate. The removal of silver through changes in pH by reaction with sodium hydroxide yields AgOH and achieves 40.1% silver recovery with a hydroxide concentration of 1.8 mol L^{-1}, which leads to a high final pH of 13.0 in the effluent. Silver precipitation by NaCl addition has not been observed due to the stability of silver thiosulfate. In photographic and X-ray effluents, silver is mostly found as a thiosulfate complex [24]. After precipitation processes, the final concentrations of silver in the aqueous phase of these real effluents were about 350 mg Ag L^{-1} and 1,302 mg Ag L^{-1} in hydrogen peroxide and sodium hydroxide, respectively. These results show the difficulty in metal removal, mainly silver, in complex wastewaters. The same researches have applied the adsorption process in the water used to rinse X-ray films, during film printing. The initial concentration was 8.0 mg L^{-1} of Ag, and the adsorbents evaluated were zeolites, mordenite, and activated carbon (AC) from *babassu* coconut. The researches have shown zeolites and mordenite were not efficient in removal of silver complexes. Zeolites have high ionic exchange capacity for silver, but in the form of Ag^+. AC produced from *babassu* coconut is capable of uptake of 0.42 mg g^{-1}. The presence of the Ag^+ is almost negligible, so the adsorbents must have the ability to adsorb silver ions and its compounds.

3.5 Adsorption

Adsorption is a separation process based on the mass transfer operation between a fluid phase and a solid phase (adsorbent). The contaminant (adsorbate) in the fluid phase will be retained in the interface of the solid phase due to their contact during adsorption process [97], as shown in Fig. 1.

The contaminant may be retained by physical or chemical interactions with the solid phase. Table 4 presents the main differences between these two types of adsorption.

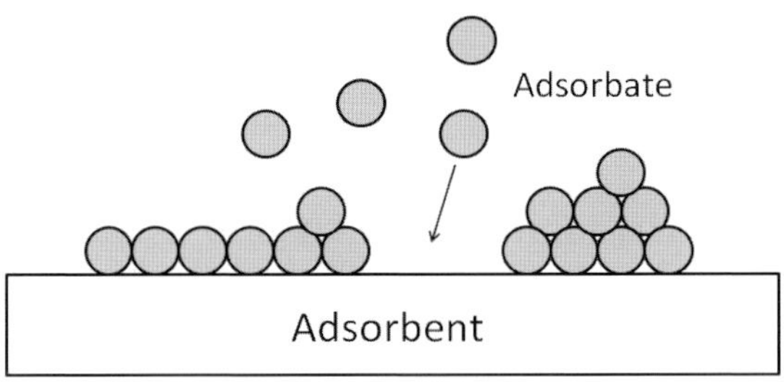

Fig. 1. Adsorption mass transfer mechanism.

Table 4. Physical and chemical adsorption main differences.

	Physical adsorption	Chemical adsorption
Heat of adsorption	Low (5–40 kJ mol^{-1})	High (40–800 kJ mol^{-1})
surface coverage	Monolayer or multilayer	Monolayer
temperature	Significant at low temperatures	Occurs at a wide temperature range
Reversibility	Reversible	Irreversible
cause	No electron transfer. Relative weak intermolecular force (van der Waal's force, hydrogen-bonding and ion-exchange processes)	Electron transfer. Formation of chemical bonds (electrostatic, covalent, and co-ordinate displacements)

Source: Atkins and de Paula [99], Bajpai and Rajpoot [98], and Ruthven [100].

The interest on adsorption as an environmental alternative has increased due to the need of better-quality treated wastewater [97]. The advantages of adsorption include design and operation flexibility, besides the great efficiency in wastewater treatment. Also, some adsorption processes are reversible, allowing adsorbent regeneration and the metal recovery. This is an important step to improve the economic benefit of the process [77]. However, to make adsorption interesting for large-scale process, it is necessary to choose the right adsorbent. It must be inexpensive, non-toxic, and of known adsorption characteristics [101].

3.5.1 *Silver adsorption and recovery*

Adsorption has been the subject of many works involving silver removal and recovery, especially when in low concentrations. Since silver is a quite common metal found in wastewaters and is considered a noble metal, its removal and recovery from aqueous media is frequently necessary. Among the many methods applied for this purpose, adsorption has gained attention in the last few decades. High removal efficiency and the possibility of adsorbent regeneration and adsorbate elution make adsorption a good option with respect to silver recovery through wastewater treatment.

One of the most important steps for adsorption is to choose an appropriate adsorbent. That adsorbent must present selectivity for the adsorbate, must be abundant, enable regeneration, and be of low cost. In order to reduce adsorption process costs, alternative adsorbents ought to be evaluated, replacing conventional adsorbents, such as ACand zeolite. A low-cost adsorbent may be a material found abundantly in nature, or a by-product, or a waste material from another industrial process [102] or agricultural waste. Several adsorbents have been studied for Ag^+ adsorption, such as AC nanospheres [103], silica gel encapsulated by amino-functionalized polystyrene [104], chelating resins [105], and concrete particles [106].

Adsorbents with complexing groups, produced on synthetic polymer matrices or inorganic substances bases, can be applied for silver adsorption. Nevertheless, in terms of environmental safety, biodegradable and non-toxic, natural, polymer-based sorbents are of highest interest [60]. The presence of the amino group in bio-sorbents establishes its complexation properties for the metal ion. Also, selective uptake of the silver ion by adsorbents with sulfo groups can occur, like chitosan [60] and sericin [107].

Chelating resins, which are able to remove contaminants by the adsorption process, have been the focus of studies due to their high selectivity and strong adsorption capacity. Frequently, these resins present as chelating ligands: amino, carboxy, oxo, thio, phosphoryl, with N, O, S, and P as their donor atoms, respectively. The complexing agent added into the polymeric chain influences metal ion selectivity. Regarding the Hard–Soft Acid–Base theory (HSAB) by Pearson, metals ions will preferably complex with ligands that have more or less electronegative donor atoms. The agents responsible for the chelating process that contain nitrogen and sulfur groups present high efficiency for precious metals. Chelating resins with amine, mercaptan, and thiourea groups are extensively applied to extract, separate, and recover precious and noble metals in hydrometallurgical field. Ag^+ is a soft Lewis acid with a small effective ionic radius, what leads to a strong affinity with ligands, including nitrogen atoms. Thus, the chelating ligands with more nitrogen atoms can be used to remove silver ions. Several chelating ligands with nitrogen atoms have been adopted for separation and adsorption of silver ions. Among them, the amidoxime group has been frequently reported [105, 108].

3.5.1.1 *Conventional adsorbents*

One of the most used adsorbents nowadays is AC, which presents high porosity and large surface area for the adsorption process. AC can be considered a versatile material, acting as an adsorbent, catalyst, or catalyst support. The major disadvantage of AC involves its elevated production cost, despite their high efficiency in removing a large amount of pollutants. ACs can be found in different physical forms, such as granular, powdered, fiber, and cloth. They may also be prepared from different raw materials like coconut shells, wood sawdust, polymeric materials, phenolic or viscose rayon, and more frequently, coal. ACs can also be manufactured from agricultural and wood industry wastes, thereby reducing environmental problems caused by their disposal [109].

Gold extraction from low-grade ore is performed by a metallurgical technique called gold cyanidation or MacArthur–Dingus process. Gold, and others metals including silver, is transformed into a water-soluble cyanide species and can be recovered later. It has been observed that, in multicomponent mixtures, free cyanide ions cause a decrease in the adsorption of silver cyanide species on charcoal. Na^+ and Ca^{2+} ions improve silver cyanide adsorption on charcoal, whereas free cyanide ions reduce its adsorption. Cho and Pitt [110] have evaluated the adsorption process of gold and silver cyanide on commercial coconut shell charcoal and observed the order of adsorption capacity as follows: gold cyanide > silver cyanide > cyanide, which indicates that the larger the ions present, the greater the adsorption potential. Jia *et al.* [111], in a similar study, observed that the presence of excess free cyanide ions in solution has inhibited silver cyanide ionic species adsorption on the active carbon, while gold cyanide adsorption was not significantly affected.

Other usual adsorbents are zeolites, which are naturally occurring hydrated porous aluminosilicate minerals. Their high ion-exchange capacity makes them an appropriate material for heavy metal removal. A wide range of species may be considered as zeolites, such as clinoptilolite and chabazite [112]. The main disadvantages of zeolites are their variable composition and low purity, leading to unpredictable adsorption levels [113].

Activated alumina may be also employed in toxic metal removal due to its large surface area (200–300 m^2 g^{-1}). It is produced by thermal treatment of hydrous alumina granules, in which a porous solid structure is formed by the elimination of hydroxyl groups. The last conventional adsorbent is silica gel, which presents a porous structure and a large surface area contributing for adsorption process. It can be found as regular-, intermediate- and low-density gels, prepared from $Si(OH)_4$ concentration on siloxane chains [114].

3.5.1.2 *Alternative adsorbents*

When the cost issue plays an important role in the adsorption process, conventional adsorbents might have to be replaced by low-cost alternative adsorbents. Therefore, different kinds of materials may be chosen, such as natural materials and agricultural or industrial wastes. Compared to conventional adsorbents, these materials may cost little to nothing.

Some natural materials used as adsorbents are mentioned here. Chitin is a substance found in shellfish and crustacean exoskeletons and can reach a surface area about 350 m^2 g^{-1}. Another extensively used adsorbent is chitosan, a bio-degradable cationic polyelectrolyte derived from chitin [115]. Cellulose is another natural material usually found in the cell wall of woody plants and is one of the most abundant polymers found worldwide. Although nature cellulose has low capacity for heavy metal adsorption, this can be improved by chemical modification, such as esterification or halogenation [116].

Clay is also a naturally occurring material mainly composed of silica, alumina, and water. In recent years, it has been considered an efficient adsorbent for the removal of trace amounts of toxic metal. Based on their variable composition, clays may be classified into different groups, such as smectites (montmorillonites), micas (illites), kaolinites, vermiculites, and sepiolites. Among them, bentonite, which is mainly composed of the clay mineral montmorillonite, is fairly abundant and has been quite extensively studied with regard to heavy metal removal. The advantages of using clays as adsorbent rely on their abundance, low cost, ion exchange potential, and elevated specific surface area [117].

Bio-sorbents can be also cited as natural materials. They are of biological origin and possess metal-sequestering property, which includes some bacteria, fungi, and algae. In this classification, some agricultural and industrial wastes, and some aquatic plants, can be included. In order to physico-chemically bind metal ions, these materials are employed as inactive or dead biomass [118]. The most abundant micro-organisms are bacteria, which are interesting due to their small size and resilience to different environmental situations. Concerning fungi, molds, yeasts, and mushrooms are the most attractive groups regarding the adsorption process. Fungi can be easily and cheaply produced in large quantities, which makes them a good alternative. Algae are materials with elevated adsorption capacity and are available in the seas and oceans, which is of high interest as a bio-sorbent [119].

Agricultural wastes have been studied in order to uptake toxic metal compounds and ions from effluents because of their abundance, low cost, unique chemical composition, and renewability. A vast range of materials can be included in this group of adsorbents, such as sugarcane bagasse, some fruits peels (orange, apple, coconut, and banana), rice and wheat (bran and husk), oilseeds shells (almond, hazelnut, walnut, and groundnut), seeds, sawdust, and leaves, which may be employed in their natural form or after being physically or chemically modified [120]. Agricultural wastes may present some disadvantages, such as low exchange capacity and low physical stability, besides the possibility of color release [121].

Industrial wastes are also potential adsorbent materials for toxic metal removal. They are usually produced as by-products or leftover materials and are found in large amounts, which reduces their cost. This group of adsorbents includes waste slurry, red mud, ashes, protein waste, and other industries sludge [122, 123].

Many studies may be found involving alternative adsorbents for silver removal. Jeon [124] has used immobilized crab shell beads to adsorb silver ions from industrial wastewater (100 mg L^{-1} Ag). The highest efficiency obtained for silver removal was 83%. Luo *et al.* [125] evaluated silver adsorption on rice husk and obtained a maximum removal efficiency of 99.76%. Table 5 presents the silver adsorption results for different types of adsorbent.

3.5.2 *Adsorption processes*

Another important parameter in the adsorption processes involves the system under study, which can be static (batch) or dynamic (continuous). In batch systems, the adsorbent contacts a fixed volume of the fluid solution containing the adsorbate. Environmental conditions as temperature, pH, adsorbent dosage, stirring, and solute initial concentration are specific, and any change may affect the adsorption process. Usual batch assays include mechanical (magnetic stirring) or orbital shaker (individual flasks). In magnetic stirring, only one solution is employed during the entire adsorption cycle, and aliquots are collected at pre-determined times. In orbital shaker, various solution flasks are placed for simultaneous shaking, and each solution is equivalent to one aliquot of the pre-determined time. Orbital shaker has the advantage of better control of temperature and stirring. Nevertheless, when pH is a significant factor, this mechanism is not appropriate, due to the difficult in maintaining the pH in all solutions. This parameter is better controlled in magnetic stirring.

Important information is obtained from the static systems, such as the adsorbent capacity for several adsorbates and the adsorption kinetics. When the batch assay is performed in the form of an isotherm, only

Table 5. Silver adsorption capacity of conventional and alternative adsorbents.

Adsorbent	q_{max} (mmol g^{-1})	References
AC nanospheres	1.409	[103]
Clinoptilolite (zeolite)	0.308	[126]
Thiol-functionalized silica gel	0.703	[127]
Silica gel microspheres encapsulated with 5-sulfosalicylic acid functionalized polystyrene	0.822	[128]
Chitin nanocrystals	0.184	[129]
Cellulose nanocrystals	0.318	
Chitosan	0.389	[130]
Raw vermiculite	0.428	[52]
Manganese oxide–modified vermiculite	0.642	
Montmorillonite	0.390	[131]
Natural bentonite	0.158	[132]
Expanded rice husk	0.172	
Aspergillus niger	0.915	[133]
Neurospora crassa	0.633	
Fusarium oxysporium	0.533	
Saccharomyces cerevisiae	0.387	[134]
Azolla filiculoides	1.600	[135]
Coconut fiber — pure	0.502	[136]
Coconut fiber — NaOH	0.612	
Rice husk	0.271	[137]
Waste yeast	0.397	[138]
Fly ash	0.060	[139]
Phosphogypsum	0.025	
Paper mill sludge	0.078	[140]

equilibrium data are collected, which means it is not possible to analyze the concentration variation as a function of time. Also, batch systems are not feasible at a large scale, where large amounts of wastewater must be treated. Thus, continuous systems are a reasonable option.

Continuous systems can be employed in several designs, such as stirred tank, fixed bed, moving bed, or air lift chamber. However, as regards adsorption processes, fixed bed has presented some advantages, such as ease of handling, cost-effectiveness, flexibility, and high level of purification. Figure 2 presents the experimental design for a continuous system, in which the solution may flow in upward or downward direction. From that, an S-shape curve can be obtained by the ratio of metal concentration in effluent and influent (C/C_0), versus time. The time at which the concentration

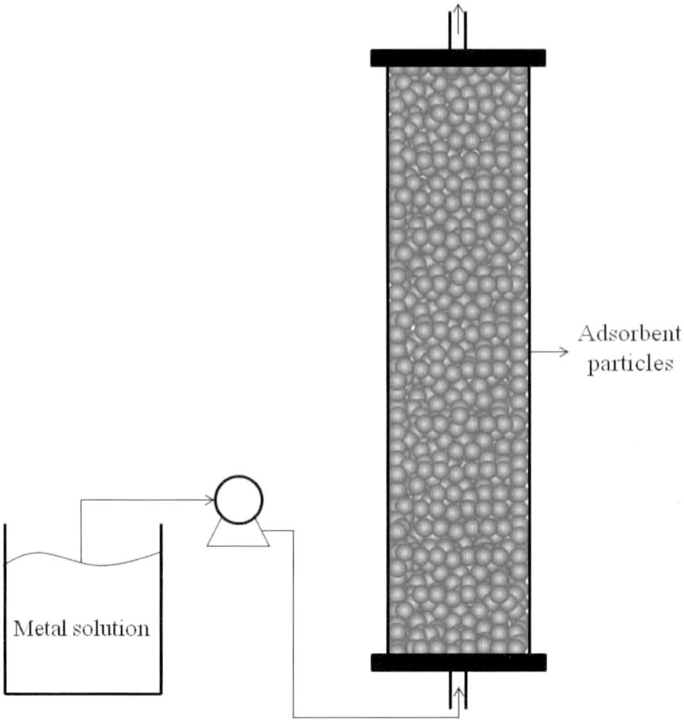

Fig. 2. Experimental design for continuous system.

abruptly rises is called the breakthrough point, which depends on the adsorbent capacity for the adsorbate. That point means the adsorbent is getting to saturation [141].

3.5.3 *Models applied for silver adsorption*

Kinetic models have been employed to investigate adsorption mechanisms and their rate-controlling steps. Adsorption occurs quickly in the beginning and slows down until the equilibrium stage, and the time to equilibrium varies with the adsorbent, adsorbate, and initial concentrations. The most usual kinetic models for silver adsorption are pseudo-first-order [142] and pseudo-second-order [143], which help quantify the adsorbed amount in adsorption kinetics [138].

The pseudo-first-order model admits that the mechanism controlling the adsorption process rate is the solute transfer from the fluid to the adsorbed phase, besides assuming that one adsorbate molecule occupies only one active site. The pseudo-second-order model assumes the possibility of one adsorbate molecule interacting with two or more adsorption sites [144].

Thermodynamically, adsorption processes may occur endothermically (adsorption increases when the temperature increases) or exothermically (negatively affected by the temperature increase). Parameters such as variations in Gibbs energy (ΔG), enthalpy (ΔH), and entropy (ΔS) can be estimated in order to evaluate the thermodynamics of an adsorption process. Negative ΔG values indicate the adsorption is of spontaneous nature; positive ΔH values are equivalent to endothermal processes, while negative values mean an exothermal process; and positive ΔS values state the affinity of the adsorbent with the adsorbate.

Several equilibrium models have been used to evaluate isothermal adsorption. Both Langmuir and Freundlich models are the most widely used to describe silver adsorption. The Langmuir model is based on monolayer adsorption, equivalent surface sites with the ability to adsorb only one adsorbate molecule, and active sites occupancy regardless of their neighboring sites. This model enables maximum adsorption capacity ($q_{\max}$) to be determined. The Freundlich model considers a highly heterogeneous surface and is based on an empirical equation [138].

A common isotherm presents one of the four main shapes shown in Fig. 3.

The "C" isotherm presents a curve as a line of zero-origin in which the ratio between the equilibrium concentration in solution and the adsorbed amount is the same, regardless of the concentration. This ratio is called partition coefficient. The "L" isotherm is a concave curve which may or may not reach a plateau. The adsorbed amount decreases when the concentration increases. The "H" isotherm is a specific case of "L" isotherm in which the adsorbate presents a high affinity with the adsorbent. The "S" isotherm is sigmoidal and results from two different mechanisms: the adsorbate has low affinity with the adsorbent; however, when it adsorbs, other adsorbate molecules are adsorbed more easily. The "L" and "H" isotherms are the most common and can be well described by Langmuir or Freundlich models [145].

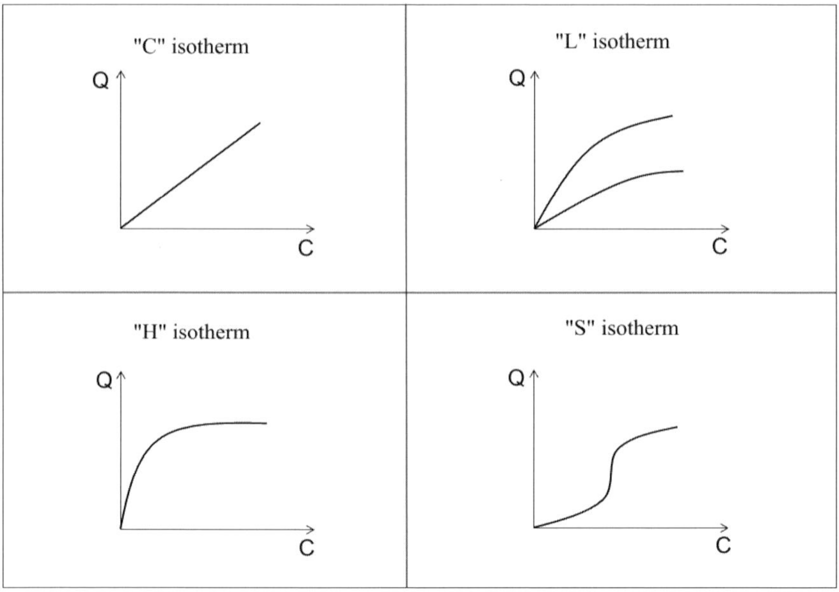

Fig. 3. Common isotherm shapes.

Silva-Medeiros *et al.* [146] evaluated silver adsorption on coconut shell AC through kinetic and equilibrium assays. In order to study the adsorption kinetics, the models of pseudo-first-order and pseudo-second-order have been applied, with the latter showing the best adjustment. Adsorption thermodynamics presented a non-spontaneous (positive ΔG values) exothermic process (negative ΔH value) and a decrease in the state of order (negative ΔS value). Langmuir and Freundlich equilibrium models have been used, and the former presented the best adjustments, i.e., monolayer adsorption. Silver adsorption thermodynamics was also evaluated by Zafar *et al.* [137] using rice husk as the adsorbent. In that case, the process was spontaneous (negative ΔG values) and endothermic (positive ΔH value), and the randomness showed an increase (positive ΔS value). Jeon [124] studied silver adsorption on immobilized crab shell beads, and the equilibrium data were adjusted by Langmuir and Freundlich models. Again, Langmuir provided the best-adjusted data, confirming monolayer adsorption. A similar result was observed by Liang *et al.* [132], who studied silver adsorption on expanded rice husk, natural diatomite, and natural bentonite. All adsorbents have presented experimental data best adjusted by Langmuir model.

Fixed bed assays provide the so-called breakthrough curves. These curves present the concentration profile as a function of time, usually as an S-shape. From that, it is possible to evaluate removal efficiency, based on the Mass Transfer Zone (MTZ). MTZ refers to the length of unused bed until the breakthrough point. Ideally, MTZ should be close to zero, and the breakthrough curve looks like a step function, meaning the adsorbent maximum use, but industrially, the usual breakpoint is near the legal limit tolerance of discharged streams.

Song *et al.* [103] evaluated silver adsorption on carbon nanospheres under batch and fixed bed conditions. The pseudo-second-order kinetic model showed the best adjustment to the experimental data when compared to pseudo-first-order. Continuous assays have shown the total silver removal in the initial 35 h. Cantuaria *et al.* [118] studied batch and continuous silver adsorption onto the macrophyte *Salvinia cucullata* and noticed that the removal efficiency depended on the system employed. Langmuir and Freundlich models have been applied to the equilibrium experimental data, and Langmuir has shown the most suitable adjustment, endorsing monolayer adsorption.

Al-Khawaja and Al-Fatlawi [147] evaluated cadmium and silver binary adsorption on fixed bed on nanoactivated alumina. It was verified that the silver adsorption was negatively affected by the presence of cadmium, showing how multicomponent removal presents differences compared to single ion removal.

Vakula *et al.* [148] investigated the adsorption of the silver atom and silver ion on pure α-quartz. Ag atom and its ion Ag^+ adsorption occurred on equivalent adsorbent sites. However, the energetic characteristics of both processes were different because of the different type of interactions in these situations. For the neutral silver atom case, only electrostatic interactions were involved, whereas the silver ion was chemically bound to the adsorbent surface. For the silver atom, the sorption energy was 1.84 and 2.07 kcal mol^{-1} in periodic and cluster models. Silver ion also bonded in the same site with sorption energy of 53.59 kcal mol^{-1}. The higher energy value of silver ion adsorption shows the highest force involved in its sorption on adsorbent.

3.6 Desorption and Recovery of Silver Ions

The adsorption process has been extensively employed in studies on wastewater contaminants removal. The adsorbent regeneration and adsorbate recovery is still poorly explored, although it means economic and environmental gain to the process. The current need of regeneration depends on the costs of adsorbate, adsorbent, and recovery, and on the adsorbate's environmental effect.

Regeneration may occur under chemical, thermal, solvent, and biological methods. Except thermal regeneration, the other methods do not allow the complete removal of the adsorbed material, leading to a decrease in the adsorbent performance after a sequence of regenerations, and thus its discard [149]. In general, toxic metals are desorbed by chemical regeneration. In order to obtain a successful desorption, the eluent must be low-cost, effective, and environmentally friendly and not cause damage to the adsorbent.

Beigzadeh and Moeinpour [150] evaluated silver adsorption and desorption using aloe vera shell ash supported $Ni_{0.5}Zn_{0.5}Fe_2O_4$ magnetic nanoparticles. In the desorption studies, Na_4–EDTA was effectively used as an eluent. Five adsorption/desorption cycles have been performed, indicating the adsorbent could be used repeatedly with no significant loss of adsorption capacity. Song *et al.* [103] used nitric acid (HNO_3) to recover silver from carbon nanospheres. Almost 100% of Ag^{+1} was recovered regardless of eluent concentration. Yurtsever and Şengil [151] have evaluated different eluents for silver desorption from valonia tannin resin. H_2SO_4, HNO_3, HCl, thiourea (NH_2CSNH_2), and thiourea–HCl solutions have been applied as eluents. In the presence of thiourea, a high level of desorption was achieved (99.6% efficiency), but the different acids have not varied very much in efficiency.

Wang *et al.* [105] produced beads of chelating chitosan containing amidoxime–guanidine functional groups. The adsorption performance of the resin for the silver ion was evaluated by batch method, and the results have shown that almost 100% of the silver ions were adsorbed in a large range of pH. The Langmuir model has described the experimental data more accurately than the Freudlich model. The q_{max} calculated by the Langmuir model was 212 mg g^{-1}, consistent with the saturated adsorption of resin. It is also related to the monolayer adsorption proposed by Langmuir. The adsorption multicomponent system, with Zn^{2+}, Cd^{2+}, Mg^{2+}, and Ni^{2+} was investigated in this study, and the adsorption capacity of silver was found to be much larger than the other metals. The uptake percentage of silver reached 99.95 %. The high selectivity of silver is important when the interest is to separate silver among others metal. The desorption process was evaluated using HCl, thiourea, and thiourea–HCl in various concentrations. The best recovery results have been obtained using HCl (6 mol L^{-1}) as an eluent. Ag^+ could be quantitatively desorbed with recovery above 87%. Thiourea with concentrations of 1% and 3% as eluents recovered about 77% and 81% of silver, respectively. The introduction of the amidoxime–guanidine group in the resin provides N ligands for the resin and, thus, it is supposed that the adsorbent has a high adsorption capacity for Ag^+.

Elwakeel *et al.* [12] studied the adsorption of silver from aqueous solution in modified chitosan resins, in batch and column techniques The resins have been produced using ethylene glycol diglycidyl, as crosslinking agent, and ethylenediamine and 3-amino-1,2,4-triazole-5-thiol, which contains four N and one S in its structure, to produce chitosan/amine and chitosan/azole resins. The addition of S and N groups improves the adsorption capacity of silver, as discussed previously. The chitosan/amine resin presented a removal capacity of 1.13 mmol g^{-1} in 20 min at 25°C. The pseudo-second-order model has predicted the adsorption mechanism. For desorption and recovery of Ag^+, thiourea (0.5 mol L^{-1}), acidified with H_2SO_4 (0.01 mol L^{-1}), was used as an eluent. The desorption of Ag^+ from the resin was performed using column technique, and the process was repeated for three cycles. The column maximum adsorption capacities in the three successive cycles were 1.373, 1.320, and 1.276 for chitosan/amine resin, and these values were 1.130, 1.081, and 1.088, respectively, for chitosan/azole resin.

3.7 Experimental Results for Silver Adsorption

3.7.1 *Case study 1: Silver adsorption on bentonite clay*

Verde-lodo bentonite clay from Boa Vista (Paraíba Brazil) was employed as the adsorbent for the adsorption of silver single ion in static and dynamic systems. Kinetic, equilibrium, thermodynamic and fixed bed analyses were performed with the purpose of studying silver adsorption [51, 152].

Adsorbent preparation

The adsorbent was milled and sieved to obtain an average diameter of 0.855 mm. Then, it was calcined at 500°C for 24 h to improve its stability and fluid dynamic resistance.

Silver single ion static and dynamic assays

Kinetic assays have been performed at room temperature by mixing Verde-lodo clay (10 g) with silver solution (1 L) in different concentrations (50, 100, and 200 mg L^{-1}). It was shaken for 16 h, and the metal concentration measured by absorption spectrophotometer AA-7000 (Shimadzu, Japan). Equilibrium and thermodynamic assays were carried out by mixing Verde-lodo clay (0.5 g) and silver solution (50 mL) in different concentrations (5–600 mg L^{-1}) at temperatures of 283, 293, 313, and 333 K. The assays were performed in a shaker SI-600R (Lab Companion, Korea) at 200 rpm for 24 h. Dynamic assays were carried out in a 15 cm glass column (1.5 cm, inner diameter) by evaluating the flow rate (3, 4, and 5 mL min^{-1}) and adsorption/desorption cycles using HNO_3 as the eluent.

Adsorption results

Kinetic studies have shown the amount of adsorbed silver increased with the increase in the initial silver concentration. Equilibrium isotherms presented favorable behaviors, with the greatest adsorbed amount at the lowest temperature (283 K). The maximum adsorption capacity was checked through Langmuir model adjustment, which provided 0.570 mmol g^{-1} at 283 K. Thermodynamically, the process was found to be spontaneous, exothermic, and controlled by physisorption.

Dynamic assays have presented curves similar to the ideal, which is a step curve. Although the 3 mL min^{-1} assays have shown a large amount of total and useful removal, the 5 mL min^{-1} assays have presented a high total removal percentage and low mass transfer zone (MTZ), indicating its great efficiency. Adsorption/desorption cycles have shown the subsequent removal steps increased the time between rupture and saturation. Also, it was observed that a maximum two cycles should be carried out in order to obtain an efficient removal. Desorption curves have shown the expected behavior of the exponential decay. Also, the amount of silver

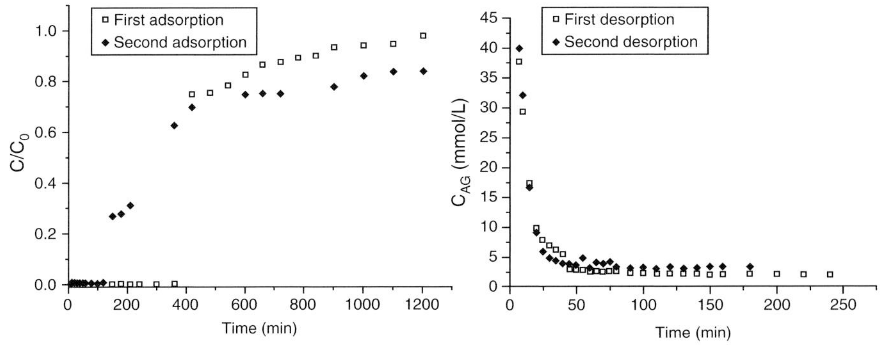

Fig. 4. First and second adsorption–desorption cycles.

eluted was higher in the first desorption cycle. Figure 4 shows adsorption and desorption curves for the two cycles.

3.7.2 *Case study 2: Silver adsorption on sericin/alginate particles*

Particles produced from a blend of sericin, alginate, and poly (ethylene glycol) diglycidyl ether (PEG-DE) have been employed for silver adsorption in static systems. Sericin is a glue protein from silkworm cocoons and it is discharged in silk manufacturing wastewater. Alginate is a polysaccharide extracted from brown algae. Both are bio-polymers that have been used to produce particles in order to remove toxic metals from the aqueous solution.

Adsorbent preparation

The cocoons were manually cleaned and cut into small pieces (about 1 cm^2). A degumming process was carried out in an autoclave (40 min, 120°C) to solubilize sericin, which was precipitated by freezing/thawing procedure. The sericin solution was frozen and then thawed at room temperature, which promotes protein agglomerate formation that can be recovery by filtration. The concentrated sericin solution was heated again in autoclave (10 min) to solubilize the sericin in order to obtain a homogeneous protein solution. The concentration was adjusted by dilution to 25 g L^{-1}, and sodium alginate was added at a concentration of 20 g L^{-1}. The mixture was stirred (IKA, Ultraturrax) to prepare the blend and the PEG added to the blend at a ratio of 0.2 g of PEG/g of sericin. The particles were prepared by the ionic gelation process, where the blend is dripped into aqueous Ca $(NO_3)_2$ (4,1% w/V) using a peristaltic pump and stirred magnetically for 24 h in its respective solution. After this period, the particles were washed, rinsed in deionized water, and dried first at 40°C for 24 h and then at 100°C for 24 h. The gelation process crosslinks the alginate in the blend, producing hydrogels, and the addition of PEG and drying the particles decrease their solubility in aqueous media. Silva *et al.* [107, 122, 153] described the procedure of particle production.

Silver adsorption

Kinetic studies in batch were performed at 25°C at three different concentrations: 0.33 mol L^{-1}, 0.90 mmol L^{-1}, and 1.65 mmol L^{-1}. The experiments lasted for 24 h. The fitting of pseudo-first-order, pseudo-second-order, intra-particle diffusion, and Boyd kinetic models to the data were investigated. Table 6 shows the results.

Except Boyd model, kinetic models have shown satisfactory fits to the experimental data obtained in the adsorption of silver by particles of

Table 6. Parameters obtained from kinetics models to experimental data.

		Ag^+ concentration		
Model	Parameter	0.33 mmol L^{-1}	0.90 mmol L^{-1}	1.65 mmol L^{-1}
Experimental	q_e (mmol g^{-1})	0.012	0.060	0.115
Pseudo-first-order	q_e (mmol g^{-1})	0.012	0.067	0.110
	Error	1.7 E-04	1.4 E-03	1.6 E-03
	R^2	0.952	0.884	0.966
Pseudo-second-order	q_e (mmol g^{-1})	0.012	0.069	0.115
	Error	1.4 E-04	1.0 E-03	1.8 E-03
	R^2	0.979	0.949	0.971
Intra-particle diffusion	k_{in}	3.9 E-04	2.5 E-03	8.3 E-03
	Error	2.7 E-05	3.5 E-04	6.6 E-04
	R^2	0.976	0.849	0.963
Boyd Model	D_i ($cm^2 \cdot s - 1$)	1.5 E-04	1.8 E-03	5.4 E-05
	R^2	0.753	0.887	0.444
	Error	1.2 E-03	1.5 E-02	8.8 E-04

sericin/alginate/PEG-DE. The pseudo-second-order model has shown high values of R^2, with a slightly smaller error in the model parameters. The pseudo-first-order model cannot properly represent the number of active sites from the adsorbent, so the pseudo-second-order model has shown a suitable fit to the results.

The high values of R^2 for the intra-particle diffusion model have shown that this mechanism is also related to this adsorption process. This model provides information about the controlling steps of the speed of adsorption on the solid particle. The graphics are divided into three stages, and each one is linearized. In the first stage (first linear section), the diffusion and silver adsorption on the outer surface of the particle occur, and it is followed by the second stage, in which the intra-particle diffusion is the controlling speed stage, and then the third-stage dynamic equilibrium is reached. The lowest value of R^2 for the concentration of 0.90 mmol L^{-1} is due to the multistage process that occurs in the second step of linearization of the results.

The adsorption of silver by particles of sericin/alginate/PEG occurs primarily due to diffusion in the boundary layer of the liquid–solid interface and adsorption on the surface of the adsorbate. Thereafter, the intra-particle adsorption stage also controls the speed of adsorption of silver, and this step is slower, since ion diffusion occurs in the inside the particle.

3.7.3 *Case study 3: Binary silver and copper adsorption on bentonite clay*

Verde-lodo bentonite clay (Boa Vista, Paraíba, Brazil) has been employed for silver and copper binary adsorption on static systems. This study aimed to evaluate the effect of another metal ion on silver adsorption by varying each metal molar fraction [154]. The adsorbent preparation is similar to that explained in "Case study 1", by Verde-lodo clay calcination.

Silver and copper binary adsorption

Kinetic assays have been performed using three different solutions with the same total initial concentration (2 mmol L^{-1}) and different molar fractions, as shown in Table 7.

For the assays, the copper and silver solution (1.0 L) was mixed with the adsorbent (10 g) for 16 h. In pre-determined times, aliquots were collected to

Table 7. Copper and silver concentrations for binary adsorption.

Solution	Silver concentration (mmol L^{-1})	Copper concentration (mmol L^{-1})
1	1.0	1.0
2	1.5	0.5
3	0.5	1.5

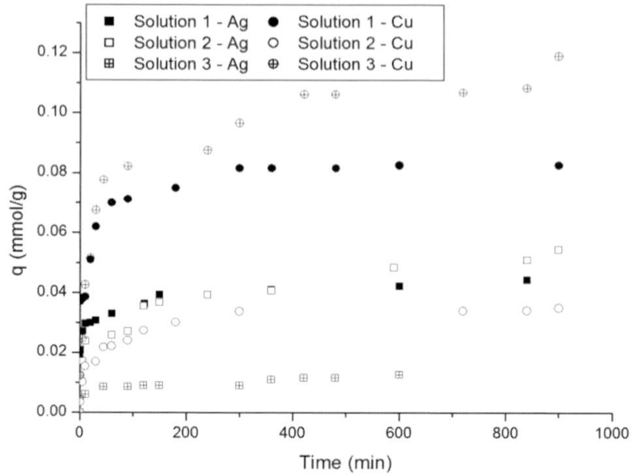

Fig. 5. Kinetic curves obtained from kinetic assays — Solutions 1–3.

verify the metal ion concentration over the course of time. The concentration was determined by absorption spectrophotometer AAnalyst 100 (Perkin Elmer, USA). The adsorbed amount was measured using Eq. (6).

$$q(t) = \frac{V}{m}(C_0 - C(t)) \tag{4}$$

in which $q(t)$ is the adsorbed amount (mmol g^{-1}), V is the solution volume (L), m is the adsorbent mass (g), C_0 is the metal initial concentration (mmol L^{-1}), and $C(t)$ is the metal concentration at time t (mmol L^{-1}).

Adsorption results

Figure 5 presents the kinetic curves obtained for Solutions 1–3.

It can be observed that a 300-min average time is needed for reaching adsorption equilibrium. The adsorbed amount was directly proportional to

Table 8. Parameters obtained from kinetics models to experimental data.

		Solution					
		1		2		3	
Model	Parameter	Ag	Cu	Ag	Cu	Ag	Cu
Experimental	q_e (mmol g^{-1})	0.040	0.082	0.045	0.034	0.010	0.110
Pseudo-first-order	q_e (mmol g^{-1})	0.035	0.077	0.038	0.032	0.010	0.099
	R^2	0.795	0.682	0.754	0.898	0.774	0.954
Pseudo-second-order	q_e (mmol g^{-1})	0.037	0.078	0.041	0.034	0.011	0.106
	R^2	0.859	0.769	0.821	0.954	0.834	0.982

the initial concentration for all solutions. The adsorption of copper was usually higher than that of silver, except for Solution 2. In this case, the initial concentration of silver was three times that of copper, and thus the adsorbed amount of silver was higher than copper. Besides, copper seems to inhibit silver adsorption as its adsorption increases in the binary system.

Pseudo-first- and pseudo-second-order kinetic models have been adjusted to the experimental data. The parameters and determination coefficient (R^2) are shown in Table 8.

It can be observed that R^2 values show reasonable adjustments, due to the binary composition of the solution, which is more complex than single-ion adsorption presented in Case Study 1. The best adjustment was observed for pseudo-second-order model, as the pseudo-first-order model does not describe the number of available active sites properly.

3.8 Considerations

The removal and recovery of toxic metals from water and industrial wastewater are an important issue due to environmental concerns. The toxicity of silver is quite relevant when discharged even in trace levels, since it may cause many negative health effects to humans and general biota. Due to its high commercial value, the recovery of this metal is important not only in environmental but also in economic terms. The adsorption of precious metals has received increasing prominence as an efficient alternative when conventional treatments are not feasible, or because of the selectivity that some adsorbents can present for silver or another specific metal. Besides conventional adsorbents, alternative raw materials have been applied in adsorption processes. The sources of these alternative adsorbents are from bio-mass or industrial/agricultural wastes and, as a result, these are low-cost, earth-abundant materials, which open other avenues for the studies on the selective removal of silver and other noble and toxic metals.

Acknowledgments

The authors thank the BRATAC Company for providing the silk-worm cocoons and CNPq (Grants 470615/2013-3, 473808/2012-9 and 300986/2013-0), CAPES, and São Paulo Research Foundation (FAPESP) (Grants 2015/13505-9, 2014/26355-2, and 2011/51824-8) for financial support. The authors thank Espaço da Escrita, Coordenadoria Geral da Universidade, UNICAMP, for the language services provided.

References

[1] Wang J, Huang CP, Pirestani D (2003) Interactions of silver with wastewater constituents. *Water Research*, **37**: 4444–4452.

[2] Burd, A (2011) Nano silver: Environmental health effects. In *Encyclopedia of Environmental Health*, JO Nriagu (ed.) (Elsevier, Burlington), pp. 22–23.

[3] Zhao Y, Wang D, Xie H, Won SW, Cui L (2015) Adsorption of Ag (I) from aqueous solution by waste yeast: Kinetic, equilibrium and mechanism studies. *Bioprocess and Biosystems Engineering*, **38**(1): 69–77.

[4] Duffus JH (2002) Heavy metals — A meaningless term? (IUPAC Technical Report). *Pure and Applied Chemistry*, **74**(5): 793–807.

[5] Smith DR, Nordberg M (2015) General chemistry, sampling, analytical methods, and speciation. In *Handbook of the Toxicology of Metals*, GF Nordberg, BA Fowler, M Nordberg (eds.) (Academic Press, UK), pp. 15–44.

[6] Wuana, RA, Okieimen, FE (2011) Heavy metals in contaminated soils: A review of sources, chemistry, risks and best available strategies for remediation. *ISRN*, **2011**: 1–20.

[7] Alkorta I, Hernández-Allica J, Becerril JM, Amezaga I, Albizu I, Garbisu C (2004) Recent findings on the phytoremediation of soils contaminated with environmentally toxic heavy metals and metalloids such as zinc, cadmium,

lead, and arsenic. *Reviews in Environmental Science and Biotechnology*, **3**: 71–90.

[8] Nordberg GF, Fowler BA, Nordberg M (2015) Toxicology of metals: Overview, definitions, concepts, and trends. In *Handbook of the Toxicology of Metals*, GF Nordberg, BA Fowler, M Nordberg (eds.) (Academic Press, UK), pp. 3–14.

[9] Govind P, Madhuri S (2014) Heavy metals causing toxicity in animals and fishes. *Research Journal of Animal, Veterinary and Fishery Sciences*, **2**(2): 17–23.

[10] Igwe JC, Abia AA (2006) A bioseparation process for removing heavy metals from waste water using biosorbents. *African Journal of Biotechnology*, **5**(12): 1167–1179.

[11] Arunakumara KKIU, Xuecheng Z (2008) Heavy metal bioaccumulation and toxicity with special reference to microalgae. *Journal of the Ocean University of China*, **7**(1): 60–64.

[12] Elwakeel KZ, El-Sayed GO, Darweesh RS (2013) Fast and selective removal of silver(I) from aqueous media by modified chitosan resins. *International Journal of Mineral Processing*, **120**: 26–34.

[13] Li J, Song Q (2016) Metal sustainability from global E-waste management. In *Metal Sustainability: Global Challenges, Consequences and Prospects*, RM Izatt (ed.) (John Wiley & Sons, UK), pp. 109–123.

[14] Balcerzak M (2016) Quantification of noble metals in biological and environmental samples. In *Handbook of Trace Analysis — Fundamental and Applications*, I Baranowska (ed.) (Springer, USA), pp. 371–394.

[15] Clemente R (2013) Gold. In *Heavy Metal in Soils: Trace Metals and Metalloids in Soils and their Bioavailability*, BJ Alloway (ed.) (Springer Science, UK), pp. 515–525.

[16] Potgieter JH, Potgieter SS, Mbaya RKK, Teodorovic A (2004) Small-scale recovery of noble metals from jewelry wastes. *Journal of South African Institute of Mining and Metallurgy*, **104**(10): 563–571.

[17] Loferski PJ (2014) Minerals Yearbook — Platinum-group Metals. U.S. Geological Survey (Open-File Report), July 2016.

[18] Kozlowski H, Pettit LD (2005) Amino acid and peptide complexes of the platinum group metals. In *Chemistry of the Platinum Group Metals*, FR Hartley (ed.) (Elsevier, New York), pp. 530–545.

[19] Hoppstock K, Sures B (2004) Platinum-group metals. In *Elements and their Compounds in the Environment*, E Merian, M Anke, M Ihnat, M Stoeppler (eds.) (Wiley-VCH, Weinheim), pp. 1047–1086.

[20] Komendová-Vlašánková R (2001) Determination of trace amounts of platinum group metals by inductively coupled plasma atomic emission spectrometry, after separation and preconcentration, in environmental samples. *Chemicke Listy*, **95**: 805–806.

[21] Butterman, WC, Amey EB (2005) Mineral Commodity Profiles — Gold. U.S. Geological Survey (Open-File Report 02–303).

[22] Nam S-H, Lee W-M, Shin Y-J, Yoon S-J, Kim SW, Kwak J II, An Y-J (2014) Derivation of guideline values for gold (III) ion toxicity limits to protect aquatic ecosystems. *Water Resources*, **48**: 126–136.

[23] Butterman WC, Hilliard HE (2005) Mineral Commodity Profiles — Silver. U.S. Geological Survey (Open-File Report 2004–1251).

[24] Ratte HT (1999) Bioaccumulation and toxicity of silver compounds: A review, *Environmental Toxicology and Chemistry*, **18**(1): 89–108.

[25] Drake PL, Hazelwood KJ (2005) Exposure-related health effects of silver and silver compounds: A review. *Annals of Occupational Hygiene*, **49**(7): 575–585.

[26] Clemente R (2013) Silver. In *Heavy Metal in Soils: Trace Metals and Metalloids in Soils and their Bioavailability*, BJ Alloway (ed.) (Springer Science, UK), pp. 535–542.

[27] Marambio-Jones C, Hoek EMV (2010) A review of the antibacterial effects of silver nanomaterials and potential implications for human health and the environment. *Journal of Nanoparticle Research*, **12**: 1531–1551.

[28] Watt S (ed.) (2003) *Silver* (Benchmark Books, New York), 32 p.

[29] Johnson J, Jirikowic J, Bertram M, van Beers D, Gordon RB, Henderson K, Klee RJ, Lanzano T, Lifset R, Oetjen L, Graedel TE (2005) Contemporary anthropogenic silver cycle: A multilevel analysis. *Environmental Science and Technology*, **39**: 4655–4665.

[30] RSC — Royal Society of Chemistry (2016) Available at: http://www.rsc.org/periodic-table/element/47/silver. Last accessed date: 16 October 2017.

[31] Lansdown, ABG (ed.) (2010) *Silver in Healthcare: Its Antimicrobial Efficacy and Safety in Use* (RSC Publishing, London).

[32] Kinraide TB, Yermiyahu U (2007) A scale of metal ion binding strengths correlating with ionic charge, Pauling electronegativity, toxicity, and other physiological effects. *Journal of Inorganic Biochemistry*, **101**: 1201–1213.

[33] World Health Organization (2002) *Silver and Silver Compounds: Environmental Aspects* (WHO, Geneva).

[34] Rosenman KD, Seixas N, Jacobs I (1987) Potential nephrotoxic effects of exposure to silver. *British Journal of Industrial Medicine*, **44**: 267–272.

[35] Leavitt SW, Dueser RD, Goodell HG (1979) Plant regulation of essential and non-essential heavy metals. *Journal of Applied Ecology*, **16**: 203–212.

[36] Holler JS, Fowler BA, Nordberg GF (2015) Silver. In *Handbook of the Toxicology of Metals*, GF Nordberg, BA Fowler, M Nordberg (eds.) (Academic Press, UK), pp. 1209–1216.

[37] Venugopal B, Luckey TD (eds.) (1978) Metal toxicity in mammals. In *Chemical Toxicology of Metals and Metalloids* (Academic Press, New York), pp. 32–36.

[38] World Health Organization (2003) Silver in drinking water. Background document for development of WHO Guidelines for Drinking-water Quality. (WHO/SDE/WSH/03.04/14), Geneva.

[39] EPA — US Environmental Protection Agency. Secondary Drinking Water Standards: Guidance for Nuisance Chemicals. Available at: https://www.epa.gov/dwstandardsregulations/secondary-drinking-water-standards-guidance-nuisance-chemicals. Accessed 2 September 2016.

[40] Chen X, Schluesener HJ (2008) Nanosilver: A nanoproduct in medical application. *Toxicology Letters*, **176**: 1–12.

[41] Benn T, Cavanagh B, Hristovski K, Posner JD, Westerhoff P (2010) The release of nanosilver from consumer products used in the home. *Journal of Environmental Quality*, **39**(6): 1875–1882.

[42] Prado PFA, Ruotolo LAM (2016) Silver recovery from simulated photographic baths by electrochemical deposition avoiding Ag_2S formation. *Journal of Environmental Chemical Engineering*, **4**: 3283–3292.

[43] Pollet B, Lorimer JP, Phull SS, Hihn JY (2000) Sonoelectrochemical recovery of silver from photographic processing solutions. *Ultrasonics Sonochemistry*, **7**: 69–76.

[44] Bas AD, Yazici EY, Deveci H (2012) Recovery of silver from X-ray film processing effluents by hydrogen peroxide treatment. *Hydrometallurgy*, **121–124**: 22–27.

[45] Rivera I, Roca A, Cruells M, Patiño F, Salinas E (2007) Study of silver precipitation in thiosulfate solutions using sodium dithionite. Application to an industrial effluent. *Hydrometallurgy*, **89**: 89–98.

[46] Güvenç A, Karabacakolu B (2005) Use of electrodialysis to remove silver ions from model solutions and wastewater. *Desalination*, **172**: 7–17.

[47] Virolainen S, Tyster M, Haapalainen M, Sainio T (2015) Ion exchange recovery of silver from concentrated base metal-chloride solutions. *Hydrometallurgy*, **152**: 100–106.

[48] Manos MJ, Malliakas CD, Kanatzidis MG (2007) Heavy-metal-ion capture, ion-exchange, and exceptional acid stability of the open-framework chalcogenide $(NH_4)_4In_{12}Se_{20}$. *Chemistry — A European Journal*, **13**: 51–58.

[49] Waitz WH (1982) Ion exchange for recovery of precious metals. *Plating and Surface Finishing*, **69**: 56–59.

[50] Ladhe AR, Frailie P, Hua D, Darsillo M, Bhattacharyya D (2009) Thiol-functionalized silica–mixed matrix membranes for silver capture from aqueous solutions: Experimental results and modeling. *Journal of Membrane Science*, **326**: 460–471.

[51] Cantuaria ML, Nascimento ES, Almeida Neto AF, dos Santos OAA, Vieira MGA (2015) Removal and recovery of silver by dynamic adsorption on bentonite clay using a fixed-bed column system. *Adsorption Science Technology*, **33**(2): 91–103.

[52] Sari A, Tüzen M (2013) Adsorption of silver from aqueous solution onto raw vermiculite and manganese oxide-modified vermiculite. *Microporous and Mesoporous Materials*, **170**: 155–163.

[53] Barakat MA (2011) New trends in removing heavy metals from industrial wastewater. *Arab Journal of Chemistry*, **4**: 361–377.

[54] Amin MT, Alazba AA, Manzoor U (2014) A review of removal of pollutants from water/wastewater using different types of nanomaterials. *Advanced Material Science and Engineering*, **2014**: 1–24.

[55] Zhang C, Hu Z, Li P, Gajaraj S (2016) Governing factors affecting the impacts of silver nanoparticles on wastewater treatment. *Science of the Total Environment*, **572**: 22.

[56] Firlak M, Kahraman MV, Yetimog EK, Zeytuncu B (2013) Adsorption of Ag (I) ions from aqueous solutions using photocured thiol-ene hydrogel. *Separation Science and Technology*, **48**: 2860–2870.

[57] Coelho PCS, Teixeira JPF, Gonçalves ONBSM (2011) Mining activities: Health impacts. In *Encyclopedia of Environmental Health*, JO Nriagu (ed.) (Elsevier), Burlington, pp. 788–802.

[58] Kramer JR, Adams NWH, Manolopoulos H, Collins PV (1996) Fate of silver in sediments and plant material in an old mining camp, Cobalt, Ontario. In *Proceedings of the 4th Argentum International Conference on the Transport, Fate and Effects of Silver in the Environment*, Madison, WI, USA, August 25–28, pp. 79–88.

[59] Su Y, Li Q, Wang Y, Wang H, Huang J, Yan X (2009) Electrochemical reclamation of silver from silver-plating wastewater using static cylinder electrodes and a pulsed electric field. *Journal of Hazardous Materials*, **170**: 1164–1172.

[60] Petrova TM, Tzaneva B, Fachikov L, Hristov J (2013) Silver recovery from spent photographic solutions by a magnetically assisted particle bed. *Chemical Engineering and Processing*, **71**: 83–96.

[61] Sun Z, Xiao Y, Sietsma J, Agterhuis H, Yang Y (2015) A cleaner process for selective recovery of valuable metals from electronic waste of complex mixtures of end-of-life electronic products. *Environmental Science and Technology*, **49**: 7981–7988.

[62] Varga D, Díaz MA, Ruiz I, Soto M (2013) Heavy metal removal in an UASB-CW system treating municipal wastewater. *Chemosphere*, **93**: 1317–1323.

[63] Smith GJ, Flegal AR (1993) Silver in San Francisco Bay Estuarine waters. *Estuaries*, **16**(3A): 547–558.

[64] Olutoye MA, Alhamdu JA (2014) Electrochemical separation of metal silver from industrial wastewater. *ACES*, **4**: 396–400.

[65] Patil YB, Paknikar KM (2000) Biodetoxification of silver–cyanide from electroplating industry wastewater. *Journal of Applied Microbiology*, **30**: 33–37.

[66] Pavlostathis SG, Maeng SK (1998) Aerobic biodegradation of a silver-bearing photoprocessing wastewater. *Environmental and Toxicology Chemistry*, **17**(4): 617–624.

[67] Bortoletto EC, Mafra LI, Sorbo ACAC, Galliani NA, Barros MASD, Tavares CRG (2007) Remoção da prata em efluentes radiográficos, *Acta Scientiarum Technology*, **29**(1): 37–41.

[68] Farkas J, Peter H, Christian P, Urrea JAG, Hassellöv M, Tuoriniemi J, Gustafsson S, Olsson E, Hylland K, Thomas KV (2011) Characterization of the effluent from a nanosilver producing washing machine. *Environment International*, **37**: 1057–1062.

[69] Brooks RR (ed.) (1992) *Noble Metals and Biological Systems: Their Role in Medicine, Mineral Exploration, and the Environment* (CRC Press, USA), 416 p.

[70] Kim JS, Kuk E, Yu KN, Kim J-H, Park SJ, Lee HJ, Kim SH, Park YK, Park YH, Hwang C-Y, Kim Y-K, Lee Y-S, Jeong DH, Cho M-H (2007) Antimicrobial effects of silver nanoparticles. *Nanomedicine*, **3**: 95–101.

[71] Chiu HSS, Tsang KL, Lee RML (1987) Treatment of electroplating wastes. *Water Pollution Control*, **86**: 12–19.

[72] Reynel-Avila HE, Mendoza-Castillo DI, Hernández-Montoya V, Bonilla-Petriciolet A (2011) Multicomponent removal of heavy metals from aqueous solution using low-cost sorbents. In *Water Production and Wastewater Treatment*, B Antizar-Ladislao, R Sheikholeslami (eds.) (Nova Science Publishers, New York), pp. 69–99.

[73] Naeemullah KTG, Afridi HI, Shah F, Arain SS, Brahman KD, Ali J, Arain MS (2016) Simultaneous determination of silver and other heavy metals in aquatic environment receiving wastewater from industrial area, applying an enrichment method. *Arab Journal of Chemistry*, **9**: 105–113.

[74] Chen C, Wang J (2008) Removal of Pb^{2+}, Ag^{+}, Cs^{+} and Sr^{2+} from aqueous solution by brewery's waste biomass. *Journal of Hazardous Materials*, **151**(1): 65–70.

[75] El-Shafey ESI, Al-Kindy SMZ (2013) Removal of Cu^{2+} and Ag^{+} from aqueous solution on a chemically-carbonized sorbent from date palm leaflets. *Environmental Technology*, **34**(3): 395–406.

[76] Badmus MAO, Audu TOK, Anyata BU (2007) Removal of heavy metal from industrial wastewater using hydrogen peroxide. *African Journal of Biotechnology*, **6**(3): 238–242.

[77] Fu F, Wang Q (2011) Removal of heavy metal ions from wastewaters: A review. *Journal of Environmental Management*, **92**: 407–418.

[78] Kurniawan TA, Chan GYS, Lo W-H, Babel S (2006) Physico–chemical treatment techniques for wastewater laden with heavy metals. *Chemical Engineering Journal*, **118**: 83–98.

[79] Özverdi A, Erdem M (2006) Cu^{2+}, Cd^{2+} and Pb^{2+} adsorption from aqueous solutions by pyrite and synthetic iron sulphide. *Journal of Hazardous Materials*, **B137**: 626–632.

[80] Koelmel J, Prasad MNV, Velvizhi G, Butti SK, Mohan SV (2016) Metalliferous waste in India and knowledge explosion in metal recovery techniques and processes for the prevention of pollution. In *Environmental Materials and Waste. Resource Recovery and Pollution Prevention*, MNV Prasad, K Shih (eds.) (Elsevier, London), pp. 339–390.

[81] Gunatilak SK (2015) Methods of removing heavy metals from industrial wastewater. *Journal of Multidisciplinary Engineering Science and Technology*, **1**: 12–18.

[82] Gode F (2007) Removal of chromium ions from aqueous solutions by adsorption method. In *Hazardous Materials and Wastewater: Treatment, Removal and Analysis*, AA Lewinsky (ed.) (Nova Science Publishers, Inc., New York), pp. 275–308.

[83] Kadirvelu K, Goel J (2007) Eco-friendly technologies for removal of hazardous heavy metal from water and industrial wastewater. In *Hazardous Materials and Wastewater: Treatment, Removal and Analysis*, AA Lewinsky (ed.) (Nova Science Publishers, Inc., New York), pp. 127–148.

[84] Černá M (1995) Use of solvent extraction for the removal of heavy metals from liquid wastes. *Environmental Monitoring and Assessment*, **34**: 151–162.

[85] Kit NK, Ismail N, Padmesh TVN (2013) Removal of Zn^{2+} from wastewater using membrane electrolysis. In *Proceedings of the Engineering Undergraduate Research Catalyst Conference — EURECA*, Jalan Taylors, Malasia, pp. 111–112.
[86] Brauckmann BM (1990) Industrial solutions amenable to biosorption. In *Biosorption of Heavy Metals*, B Volesky (ed.) (CRC Press, USA), pp. 51–64.
[87] Bar N, Das SK (2016) Applicability of ANN in adsorptive removal of Cd(II) from aqueous solution. In *Handbook of Research on Natural Computing for Optimization Problems*, JK Mandal, S Mukhopadhyay, T Pal (eds.) (Information Science Reference, USA), pp. 523–560.
[88] Nassef E, El-Taweel YA (2015) Removal of copper from wastewater by cementation from simulated leach liquors. *Chemical Engineering Process Technology*, **6**(1): 1–6.
[89] Noyes R (ed.) (1994) *Unit Operations in Environmental Engineering* (Noyes Publications, New Jersey, USA), pp. 239–263.
[90] Masebinu SO, Muzenda E (2014) Review of silver recovery techniques from radiographic effluent and X-ray film waste. *Proceedings of the World Congress on Engineering and Computer Science*, San Francisco, USA, pp. 613–617.
[91] Ratanamskul C, Lertviriyaprapa K, Saiyasitpanich P (2001) Appropriate technology for silver removal and recovery from spent bleach-fix photographic solution. In *Ninth International Conference of Greening of Industry Network*, Bangkok, Thailand, pp. 1–7.
[92] Gao Y, Zhou Y, Wang H, Lin W, Wang Y, Sun D, Hong J, Li Q (2013) Simultaneous silver recovery and cyanide removal from electroplating wastewater by pulse current electrolysis using static cylinder electrodes. *Industrial Engineering and Chemical Research*, **52**: 5871–5879.
[93] Goshima T, Hori K, Yamamoto A (1994) Recovery of silver from radiographic fixer. *Oral Surgery, Oral Medicine, Oral Pathology*, **77**(6): 684–688.
[94] Figueroa G, Valenzuela JL, Parga JR, Vazquez V, Valenzuela A (2015) Recovery of gold and silver and removal of copper, zinc and lead ions in pregnant and barren cyanide solutions. *Materials Science and Applications*, **6**: 171–182.
[95] Shimojo K, Goto M (2004) Solvent extraction and stripping of silver ions in room-temperature ionic liquids containing calixarenes. *Analytical Chemistry*, **76**: 5039–5044.
[96] Lee Y-III, Park S-H, Kim J-S, Kim D-W, Choi K-Y (1999) Selective extraction of silver(I) ion by a new acyclic diazapolyether compound bearing

dicarboxylate functional end-groups. *Bulletin of Korean Chemical Society*, **20**(4): 487–490.

[97] Tchobanoglous G, Burton FL, Stensel HD (eds.) (2003) *Wastewater Engineering: Treatment and Reuse* (Metcalf & Eddy, Inc., New York, USA), pp. 1138–1162.

[98] Bajpai AK, Rajpoot M (1999) Adsorption techniques — A review. *Journal of Scientific Industrial Research*, **58**: 844–860.

[99] Atkins P, de Paula J (eds.) (2006) *Physical Chemistry*, 8th Edn. (W.H. Freeman, New York, USA).

[100] Ruthven DM (ed.) (1984) *Principles of Adsorption and Adsorption Processes* (John Wiley & Sons, New York, USA).

[101] Kwon J-S, Yun S-T, Lee J-H, Kim S-O, Jo HY (2010) Removal of divalent heavy metals (Cd, Cu, Pb, and Zn) and arsenic(III) from aqueous solutions using scoria: Kinetics and equilibria of sorption. *Journal of Hazardous Materials*, **174**: 307–313.

[102] Abas SNA, Ismail MHS, Kamal ML, Izhar S (2013) Adsorption process of heavy metals by low-cost adsorbent: A review. *World Applied Sciences Journal*, **28**(11): 1518–1530.

[103] Song X, Gunawan P, Jiang R, Leong SSJ, Wang K, Xu R (2011) Surface activated carbon nanospheres for fast adsorption of silver ions from aqueous solutions. *Journal of Hazardous Materials*, **194**: 162–168.

[104] Xu Q, Yin P, Zhao G, Sun Y, Qu R (2010) Adsorption selectivity and dynamic adsorption behaviors of Cu(II), Ag(I), and Au(III) on silica gel encapsulated by amino functionalized polystyrene. *Journal of Applied Polymer Science*, **117**: 3645–3650.

[105] Wang Y, Ma X, Li Y, Li X, Yang L, Ji L, He Y (2012) Preparation of a novel chelating resin containing amidoxime–guanidin group and its recovery properties for silver ions in aqueous solution. *Chemical Engineering Journal*, **209**: 394–400.

[106] Begum, S (2003) Silver removal from aqueous solution by adsorption on concrete particles. *Turkish Journal of Chemistry*, **27**: 609–617.

[107] Silva TL, Meinerz VH, Vidart JM, Gimenes ML, Vieira MGA, Silva MGC (2016) Metallic affinity of toxic and noble metals by particles produced from sericin, alginate and poly(ethylene glycol). *Chemical Engineering Transactions*, **51**: 1–6.

[108] El-Ghaffar MAA, Mohamed MH, Elwakeel KZ (2009) Adsorption of silver(I) on synthetic chelating polymer derived from 3-amino-1,2,4-triazole-5-thiol and glutaradehyde. *Chemical Engineering Journal*, **151**: 30–38.

[109] Dias JM, Alvim-Ferraz MCM, Almeida MF, Rivera-Utrilla J, Sánchez-Polo, M (2007) Waste materials for activated carbon preparation and its use in aqueous-phase treatment: A review. *Journal of Environmental Management*, **85**: 833–846.

[110] Cho EH, Pitt CH (1979) The adsorption of silver cyanide on activated charcoal, *Metallurgical and Materials Transactions B*, **10B**: 159–164.

[111] Jia YF, Steele CJ, Havward IP, Thomas KM (1998) Mechanism of adsorption of gold and silver species on activated carbons. *Carbon*, **36**(9): 1299–1308.

[112] Tripathi, A, Ranjan, MR (2015) Heavy metal removal from wastewater using low cost adsorbents. *Journal of Bioremediation and Biodegradation*, **6(6)**: 1–5.

[113] Ackley MW, Rege SU, Saxena H (2003) Application of natural zeolites in the purification and separation of gases. *Microporous and Mesoporous Materials*, **61**: 25–42.

[114] Vakili M, Rafatullah M, Ibrahim MH, Abdullah AZ, Salamatinia B, Gholami Z (2014) Oil palm biomass as an adsorbent for heavy metals. In *Reviews of Environmental Contamination and Toxicology*, DM Whitacre (ed.) (Springer, Switzerland), pp. 61–88.

[115] Bhatnagar A, Minocha AK (2006) Conventional and non-conventional adsorbents for removal of pollutants from water — A review. *Indian Journal of Chemical Technology*, **13**: 203–217.

[116] O'Connell DW, Birkinshaw C, O'Dwyer TF (2008) Heavy metal adsorbents prepared from the modification of cellulose: A review. *Bioresource Technology*, **99**: 6709–6724.

[117] Uddin, MK (2016) A review on the adsorption of heavy metals by clay minerals, with special focus on the past decade. *Chemical Engineering Journal*, **308**: 438–462.

[118] Cantuaria ML, Almeida Neto AF, Vieira MGA (2014) Biosorption of silver by macrophyte *Salvinia Cucullata*. *Chemical Engineering Transactions*, **38**: 109–114.

[119] Wang J, Chen C (2009) Biosorbents for heavy metals removal and their future. *Biotechnology Advances*, **27**: 195–226.

[120] Sud D, Mahajan G, Kaur MP (2008) Agricultural waste material as potential adsorbent for sequestering heavy metal ions from aqueous solutions — A review. *Bioresource Technology*, **99**: 6017–6027.

[121] Kumar, U (2006) Agricultural products and by-products as a low cost adsorbent for heavy metal removal from water and wastewater: A review. *Scientific Research & Essays*, **1**(2): 33–37.

[122] Silva TL, Silva Jr, AC, Vieira, MGA, Gimenes, ML, Silva, MGC (2014) Production and physicochemical characterization of microspheres made from sericin and alginate blend. *Chemical Engineering Transactions*, **39**: 643–648.

[123] Carvalho J, Araujo J, Castro F (2011) Alternative low-cost adsorbent for water and wastewater decontamination derived from eggshell waste: An overview. *Waste Biomass Valorization*, **2**: 157–167.

[124] Jeon, C (2015) Adsorption behavior of silver ions from industrial wastewater onto immobilized crab shell beads. *Journal of Industrial Engineering Chemistry*, **32**: 195–200.

[125] Luo X, Lin X, Luo X, Luo A, Liang X (2010) Adsorption of Ag^{+} from aqueous solutions by rice husk. *Materials Science Forum*, **658**: 45–48.

[126] Akgül M, Karabakan A, Acar O, Yürüm Y (2006) Removal of silver (I) from aqueous solutions with clinoptilolite. *Microporous Mesoporous Mater*, **94**: 99–104.

[127] Quang DV, Lee JE, Kim, J-K, Kim, YN, Shao, GN, Kim, HT (2013) A gentle method to graft thiol-functional groups onto silica gel for adsorption of silver ions and immobilization of silver nanoparticles. *Powder Technology*, **235**: 221–227.

[128] Yin P, Xu Q, Qu R, Zhao G (2009) Removal of transition metal ions from aqueous solutions by adsorption onto a novel silica gel matrix composite adsorbent. *Journal of Hazardous Materials*, **169**: 228–232.

[129] Liu P, Sehaqui H, Tingaut P, Wichser A, Oksman K, Mathew AP (2014) Cellulose and chitin nanomaterials for capturing silver ions (Ag^{+}) from water via surface adsorption. *Cellulose*, **21**: 449–461.

[130] Lasko CL, Hurst MP (1999) An investigation into the use of chitosan for the removal of soluble silver from industrial wastewater. *Environmental and Scientific Technology*, **33**: 3622–3626.

[131] Praus P, Turicová M, Valášková M (2008) Study of silver adsorption on montmorillonite. *Journal of Brazilian Chemical Society*, **19**(3): 549–556.

[132] Liang X, Luo X, Lin X, Mei Q (2012) The behaviors of removing silver ions by low-cost expanded rice husk, nature diatomite and nature bentonite as sorbents. *Materials Science Forum*, **724**: 472–475.

[133] Akthar N, Sastry S, Mohan M (1995) Biosorption of silver ions by processed *Aspergillus niger* biomass. *Biotechnology Letters*, **17**(5): 551–556.

[134] Simmons P, Singleton I (1996) A method to increase silver biosorption by an industrial strain of *Saccharomyces cerevisiae*. *Applied Microbiology and Biotechnology*, **45**: 278–285.

[135] Elmachliy S, Chefetz B, Tel-Or E, Vidal L, Canals A, Gedanken A (2011) Removal of silver and lead ions from water wastes using azolla filiculoides, an aquatic plant, which adsorbs and reduces the ions into the corresponding metallic nanoparticles under microwave radiation in 5 min. *Water, Air, & Soil Pollution*, **218**: 365–370.

[136] Saman N, Johari K, Song ST, Mat H (2015) Silver adsorption enhancement from aqueous and photographic waste solutions by mercerized coconut fiber. *Separation Science and Technology*, **50**: 937–946.

[137] Zafar, S, Khalid, N, Mirza, ML (2012) Potential of rice husk for the decontamination of silver ions from aqueous media. *Separation Science and Technology*, **47**: 1793–1801.

[138] Zhao G, Wu X, Tan X, Wang X (2011) Sorption of heavy metal ions from aqueous solutions: A review. *Open Colloid Science Journal*, **4**: 19–31.

[139] Çoruh S, Elevli S, Şenel G, Ergun ON (2011) Adsorption of silver from aqueous solution onto fly ash and phosphogypsum using full factorial design. *Environmental Progress and Sustainable Energy*, **30**(4): 609–619.

[140] Calace N, Nardi E, Petronio BM, Pietroletti M, Tosti G (2003) Metal ion removal from water by sorption on paper mill sludge. *Chemosphere*, **51**: 797–803.

[141] Kumar D, Pandey LK, Gaur JP (2016) Metal sorption by algal biomass: From batch to continuous system. *Algal Research*, **18**: 95–109.

[142] Lagergren S (1898) Zur theorie der sogenannten adsorption geloster stoffe. *K Svenska Vetenskapsakad Handl*, **24**(4): 1–39.

[143] Ho YS, McKay G (1999) Pseudo-second-order model for sorption processes. *Process Biochemistry*, **34**(5): 451–465.

[144] Rudzinski W, Plazinski W (2006) Kinetics of solute adsorption at solid/solution interfaces: A theoretical development of the empirical pseudo-first and pseudo-second order kinetic rate equations, based on applying the statistical rate theory of interfacial transport. *Journal of Physical Chemistry B*, **110**: 16514–16525.

[145] Limousin G, Gaudet J-P, Charlet L, Szenknect S, Barthès V, Krimissa M (2007) Sorption isotherms: A review on physical bases, modeling and measurement. *Applied Geochemistry*, **22**: 249–275.

[146] Silva-Medeiros FV, Consolin-Filho N, Lima MX, Bazzo FP, Barros MASD, Bergamasco R, Tavares CRG (2016) Kinetics and thermodynamics studies of silver ions adsorption onto coconut shell activated carbon. *Environmental Technology*, **37**(24): 3087–3093.

[147] Al-Khawaja, ES, Al-Fatlawi, AH (2015) Removal of cadmium Cd(II) and silver Ag (II) from aqueous solutions by nano activated alumina.

Part II: Continuous flow esperiments. *International Journal of Computer Engineering and Technology*, **6**: 48–57.

[148] Vakula NI, Kuramshina GM, Gorb LG, Hill F, Leszczynski J (2013) Adsorption and diffusion of a silver atom and its cation on α-SiO_2 (001): Comparison of a pure surface with a surface containing an AI defect. *Chemical Physics Letters*, **567**: 27–33.

[149] Kulkkarni SJ, Goswami AK (2013) Adsorption studies of organic matter removal from wastewater by using bagasse flyash in batch and column operations. *International Journal of Scientific Research*, **2**: 180–183.

[150] Beigzadeh P, Moeinpour F (2016) Fast and efficient removal of silver (I) from aqueous solutions using aloe vera shell ash supported $Ni_{0.5}Zn_{0.5}Fe_2O_4$ magnetic nanoparticles. *Transactions of Nonferrous Metals Society of China*, **26**: 2238–2246.

[151] Yurtsever M, Şengil A (2012) Adsorption and desorption behavior of silver ions onto valonia tannin resin. *Transaction of Nonferrous Metals Society of China*, **22**: 2846–2854.

[152] Cantuaria ML, Almeida Neto AF, Nascimento ES, Vieira MGA (2016) Adsorption of silver from aqueous solution onto pre-treated bentonite clay: Complete batch system evaluation. *Journal of Clean Production*, **112**: 1112–1121.

[153] Silva TL, Silva Jr AC, Vieira MGA, Gimenes ML, Silva MGC (2016) Biosorption study of copper and zinc by particles produced from silk sericin e alginate blend: Evaluation of blend proportion and thermal cross-linking process in particles production. *Journal of Clean Production*, **137**: 1470–1478.

[154] Do Carmo ACR, De Freitas ED, De Almeida Neto AF, Vieira MGA Efeito da fração molar na cinética de adsorção bicomposto de íons prata e cobre em argila bentonítica, *Anais do XI Congresso Brasileiro de Engenharia Química em Iniciação Científica* [*Blucher Chemical Engineering Proceedings*, 1(3)] Blucher, São Paulo pp. 481–486 (in Portuguese).

Chapter 4
The Biogenic Synthesis of Silver Nanoparticles as a Method for Recovering Silver from Secondary Sources Using Extracts from Indigenous Australian Plants

Derek Fawcett, Sridevi Brundavanam and Gérrard Eddy Jai Poinern*

Department of Physics and Nanotechnology
Murdoch University, Australia
**g.poinern@murdoch.edu.au*

4.1 Silver: An Introductory Perspective

The natural metallic luster and physical beauty of silver (Ag) has made it a highly desirable precious metal for many civilizations past and present. This naturally occurring metal is ductile, malleable, and slightly harder than gold. In its pure form, Ag has the highest electrical and thermal conductivity of all metals, which makes it suitable for a wide range of industrial applications [1]. Archaeological evidence suggests the metal's ductile and malleable nature was adopted by civilizations as early as 3000 BC, where it was used to fashion jewelry and manufacture elaborate ornaments for decorations and religious ceremonies [2]. These early civilizations also used Ag for fabricating silverware, coinage, and building assets through physical bar investment, thus ensuring the demand for this desirable metal was always high. The demand for this precious metal during the early part of the nineteenth century dramatically increased with rapid industrialization. In particular, the advent of photography during this period saw a dramatic increase in the demand for Ag and silver-based compounds. In recent years, the global demand for Ag has continued to increase in spite of the decline in the usage of camera-based photographic film emulsions. Meanwhile, the use of radiographic film for various healthcare-based radiographic imaging techniques has continued to grow. Current estimates indicate that around 2 billion radiographic images are taken globally each year, with radiographic film containing 5–15 g Ag/kg of film [3]. Furthermore, because metallic silver is less expensive than either gold or platinum, there is an increasing trend to use silver in a variety of electronic products and industrial applications [4, 5].

Like other precious metals such as gold (Au) and platinum (Pt), Ag is found in limited quantities in the earth's crust. Generally, Ag can be obtained from primary silver mining or is generated as a by-product of other metallurgical metal processing industries. For example, Ag can be generated as a by-product from the production of Au, copper (Cu), lead (Pb), and zinc (Zn) [6, 7]. In 2015, the global Ag production reached a record output value of 886.7 million ounces (Moz) and reflected a 2% increase on the 2014 output value. Production from primary Ag mining accounted for 264.7 Moz, while Ag produced as a by-product from other metal processing industries accounted for 662 Moz. The silver by-product output by source metal was Au: 118.8 Moz (13%); Pb and Zn: 305.9 Moz (34%); Cu: 192.5 Moz (22%); and other metals: 4.8 Moz (1%). Other sources of Ag included recycling (146.1 Moz) and net hedging supply (7.8 Moz), while the sales of Ag by governments were essentially non-existent. Thus, the estimated total global supply for 2015 was tallied around 1,040.6 Moz [8]. The largest user of Ag in 2015 was the industrial sector, which consumed around 588.7 Moz. Notable users in this sector were electrical and electronics (246.7 Moz), brazing alloys and solders (61.1 Moz), photography (46.7 Moz), photovoltaics (77.6 Moz), and ethylene oxide catalyst demand (10.2 Moz). Other sectors that consumed Ag included jewelry fabrication (226.5 Moz), silverware manufacturing (62.9 Moz), and coinage and bar production (292.3 Moz) [8].

In recent years, there has also seen a significant increase in the use of nanometer-scale Ag or Ag particles in a number of fields, such as biomedical, pharmaceuticals, and microelectronics. Current estimates indicate that around 320 tons of nanometer-scale Ag are produced and used each year worldwide [9]. In 2014, the value of the global nanosilver market was estimated to be around US$ 700.9 million, while an annual growth rate of 14.8% was forecast between 2015 and 2023, resulting in an anticipated global value of US$ 2,415.5 million by 2023 [10]. Key factors contributing to the predicted increase in nanosilver usage during this period are the unique physiochemical properties that make it suitable for a wide range of applications in new emerging technologies and processes. The physiochemical properties of bulk Ag are largely understood, but recent studies have revealed materials in the nanometer scale range (1–100 nm, where 1 nm equals 1×10^{-9} m) possess large surface area to volume ratios that result in changes to surface chemistry and optoelectronic and chemical reactivity [11, 12]. Interestingly, many ancient cultures considered metals to have therapeutic properties [13], with Ag and Ag compounds being used as natural anti-microbial agents in numerous traditional medicinal preparations [14]. Recent studies have also shown that nanosilver possesses enhanced anti-bacterial, anti-fungal, and anti-inflammatory properties that can promote faster wound healing. For example, nanosilver has been incorporated into clothing textiles to inhibit bacterial growth and reduce body odor [15]. In addition, nanosilver-based anti-bacterial coatings have been applied to hemodialysis catheters to reduce infection during treatment and on surgical mesh products [16, 17]. Furthermore, nanosilver has been extensively incorporated into wound dressings and pharmaceutical preparations [18–20]. The effectiveness of nanosilver as an anti-microbial agent results from membrane damage and cellular uptake that causes toxicological damage to cellular DNA [21, 22]. Because of the toxicological damage caused by nanosilver, it is currently being investigated for potential use as an agent against various cancer cells and plasmodial pathogens [23–27]. Another potential use for nanosilver is in crop protection and the management of agricultural plant diseases [28, 29]. This is because recent studies have shown nanosilver to have anti-fungal properties, and it can act as a fungicide against a number of plant pathogens [30, 31]. Furthermore, nanosilver has also been used as a disinfectant agent in food/beverages containers and in water purification systems [32, 33].

Together, the increasing demand for Ag and the development of more Ag-based technologies will mean that new sources will be needed to meet future demands. Meanwhile, the cost of sourcing Ag from primary mining or through by-product generation will continue to increase. Future operating costs for both Ag sources will significantly increase as natural resources progressively diminish. In recent years, concern over the increasing demand and limited natural resources currently available have prompted researchers to investigate various new, cost-effective, and eco-friendly methods for recovering Ag from spent secondary sources [34, 35]. This is of particular importance since, unlike other high-value precious metals like Au and Pt that are heavily recycled, the much lower-valued Ag is typically recycled at much lower levels, particularly when the market values are low [36]. This behavior can easily be seen by inspecting Fig. 1(a), which shows the steady growth in mine production since 2000. However, from 2000 until 2009, Ag scrap usage was basically static (mean value of 181.6 Moz). Then in 2010, there was a short period of increasing Ag scrap usage that coincided with increasing Ag prices that peaked in 2011 (~35.12 US$/oz.). Since reaching this high value, Ag prices have declined, and by 2015, it had reached a

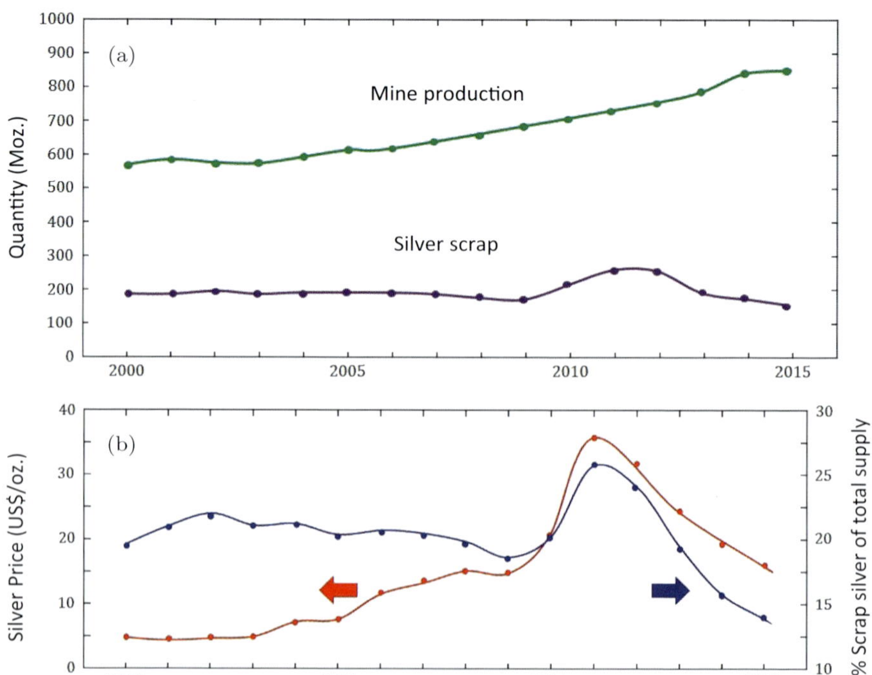

Fig. 1. (a) Total global mine production and silver scrap and (b) silver price and silver scrap as percentage (%) of global total supply between 2000 and 2015 [8, 37–40].

market value of around US$ 15.68/oz. [8]. The impact of the declining Ag prices on scrap usage can be seen in Fig. 1(b). Between 2000 and 2009, scrap accounted for around 20% of the total global Ag supply. However, between 2010 and 2011, with increasing Ag prices, the percentage contribution of scrap reached a maximum value of around 25.7%. Unfortunately, since the high prices experienced in 2011, there has been a significant decline in scrap usage. And by 2015, it had dropped to around 14% of the total supply, which makes recovery and recycling using current strategies less attractive.

One aspect of Ag usage is the level of waste Ag finding its way into the environment. Recent regulations targeting environmental pollution caused by metallic wastes have highlighted the toxic effects of monovalent Ag^+ ions in both terrestrial and marine environments. Ag^+ ions from industrial effluents and spent sources can enter various food chains and cause numerous diseases and disorders in both marine and terrestrial organisms [41–44]. Accordingly, in recent years, there has been considerable interest in reducing environment pollution levels. Thus, using eco-friendly and cost-effective recovery methods has the potential to reduce toxicity levels in the environment and provide an important secondary source of Ag. The need to economically recover Ag from the environment offers new opportunities for using alternative eco-friendly routes. The most attractive feature of these alternative routes is the ability to recover Ag from low-level sources (soil, mine tailings, or other secondary sources such electronic wastes) that are not economically viable using conventional extraction and recycling strategies but may be viable using alternative eco-friendly strategies. To begin, Sec. 4.2 reinforces the need for recovering Ag^+ ions by highlighting and discussing the toxic effect of Ag^+ ions on the environment. This is followed by a brief discussion of phytoextraction, a process where plants can favorably absorb metallic and mineral species from groundwater, soil, and sediments [45]. Using plants to extract and recover Ag from mine tailings, contaminated soils, and wastes from other sources, (i.e., radiographic film and electronics) is different from conventional mining and extraction processes. Bio-mining using plants is an attractive alternative that uses green-chemistry–based techniques and processes that are eco-friendly. Section 4.3 discusses the biogenic synthesis of Ag nanoparticles by plants, the possible formation mechanisms, and the processing parameters that govern nanoparticle formation and growth. Section 4.4, by way of an example, discusses the use of three indigenous Australian plant species for synthesizing Ag nanoparticles.

4.2 Toxicology of Silver: Environmental and Human Risk Assessment

4.2.1 *Introduction*

From a historical perspective, it was estimated that during the 1970s, the United States discharged around 2.5 million kg of Ag back into the environment [46], with the majority of Ag entering surface waters and soils due to the effects of natural leaching mechanisms produced by waste disposal, mining, and industry [47]. One of the biggest contributors to this environmental contamination was the photographic industry. However, the advent and popularity of digital photography in recent years has seen a significant reduction in Ag emissions [46]. In spite of the decline in Ag usage by the photographic industry, there has been an increase in its utilization in other industries. For example, the strong growth in the Chinese photovoltaic solar panel installations has resulted in Ag usage in this sector increasing by 23% in 2015, which increased the end-of-year Ag consumption to 77.6 Moz. Ethylene oxide catalyst demand saw an impressive 103% increase that equated to an end-of-year total consumption of 10.2 Moz for this industrial application [8]. The increasing growth of Ag-based industrial technologies has also seen growing concerns by community groups and government bodies regarding environment pollution and possible health effects [48, 49]. Thus, in recent years there has been a strong interest in reducing environmental pollution and developing alternative precious metal recovery strategies [50, 51]. Sections 4.2.2 and 4.2.3 summarize what is currently known about the effects of Ag in the environment and its possible health effects.

4.2.2 *Environmental risk assessment*

Ag from industrial effluents and spent sources can enter both marine and terrestrial environments. Silver can exist in four different oxidation states: namely Ag^0, Ag^{1+}, Ag^{2+}, and Ag^{3+}. The predominant oxidation states are Ag^0 and Ag^{1+}, while both Ag^{2+} and Ag^{3+} are unstable in aquatic environments [52]. Metallic Ag is insoluble in water, but Ag salts have varying solubilities such as silver chloride (AgCl, largely insoluble: $K_{sp} = 1.77 \times 10^{-10}$) and silver nitrate ($AgNO_3$), which is highly water soluble (256 g/100 mL at 25°C) [53]. Thus, dissolved forms of Ag have a greater potential to enter various aquatic eco-systems and interact with freshwater and seawater organisms. Studies indicate that Ag^+ ions are the most toxic form of Ag to many organisms [54, 55]. The interaction between Ag^+ ions and different organisms is complex, since the surrounding environmental chemistry can greatly influence binding, bio-availability, and toxicity [46]. Thus, environmental factors as well as the presence of dissolved Ag^+ ions and other soluble Ag compounds present in the environment can also influence the propensity for the types of diseases and disorders induced in organisms [48, 49, 52, 56]. For example, the uptake of free Ag^+ ions by fish is much greater in the presence of Ag–chloro complexes commonly associated with marine environments. This is because the concentrations of Ag–chloro complexes are higher and much more bio-available. Whereas, high-Ag-affinity sulfides and organic matter present in freshwater environments can significantly reduce the concentrations of Ag in the water column and in turn reduces its bio-availability, thus rendering freshwater organisms less likely to accumulate Ag than marine organisms under similar contamination levels [46, 57]. Furthermore, Ag^+ ions are persistent, toxic, and can bio-accumulate in many organisms [58]. The cells of all living organisms are surrounded by a cell membrane that contains the cell organelles. This cellular membrane controls the passage of many materials into and out of the cell, via pores, protein carriers, and ion transporters. The nature of Ag^+ ions makes them compatible for uptake by cell ion transporters that normally regulate the passage of Na^+ and Cu^+ ions into the cell [57]. Importantly, there are a number of factors that can influence the ability of Ag^+ ions or Ag compounds to induce toxic effects. These factors include: (1) concentration and solubility of the Ag ions or Ag compound; (2) the ability of Ag or Ag compounds to bind and interact with biological sites; (3) the degree to which Ag and Ag compounds tend to form complexes and their interaction with the cell, and (4) bio-accumulation or metabolism and excretion of the complexes.

Ag toxicity toward microbes has been known for decades and, as a result, metallic Ag was used to manufacture surgical prosthesis and splints, while soluble Ag compounds were extensively used in bactericides, fungicides, and in the treatment of infectious diseases [49, 59]. Recent developments have seen the inclusion of new nanometer-scale Ag formulations in an ever-increasing number of consumer and medical products. These products include food packaging, printing, electronics, household appliances, textiles, pharmaceuticals, and medical devices [60–65]. However, a number of these new nanosilver-based products have created uneasiness within government regulatory departments and wider public anxiety regarding increasing levels of Ag entering the environment. For example, a study by Benn *et al.* [66] revealed that some brands of socks treated with nanosilver to kill bacteria and prevent foot odor tended to lose almost all of their Ag treatment within four washings. In this case, the major concern was nanosilver entering wastewater treatment facilities and killing or disturbing the activity of helpful bacteria, while similar concerns were expressed about nanosilver entering lakes and streams to threaten freshwater organisms. Similar concerns have been expressed by farming groups who fear the anti-microbial activity of nanosilver will disrupt the activities of beneficial bacteria in soils used for crop production [67]. Several studies have suggested that the widespread use of new consumer products, pharmaceuticals, and water treatments incorporating nanosilver formulations will ultimately end up with Ag being deposited into the environment. The anti-microbial properties of nanosilver, as mentioned earlier, result from interactions between the enzymes and proteins effecting cellular respiration and ion transport across the cell membrane. Membrane damage and cellular uptake of Ag result in major toxicological damage, ultimately leading to cell death [48, 57, 68, 69]. Furthermore, there is increasing evidence that indicates higher levels of Ag are being released into the environment and could produce a detrimental environmental impact [70–72]. For instance, some studies have reported fish with accumulated levels of nanosilver particles in their gills and liver tissues tended to experience oxidative stress when exposed to waters with low oxygen levels [73–75]. Other studies have revealed the presence of nanosilver (10–80 nm) can influence the early development of zebra fish embryos by inducing chromosomal anomalies, DNA damage, and spinal cord deformities [41, 76–78]. However, further studies are needed to fully quantify the influence of Ag and nanosilver in the environment. In particular, information is needed on the various forms of Ag being used, their current usage, and release rates from spent consumer products, waste levels from manufacturers, and discharges from industrial processes. This data would make it easier to appraise and better understand the exposure, bioavailability, toxicity, and risks associated with the biological interactions of spent Ag in terrestrial, freshwater, and marine environments. This is very important, since the impact of nanosilver in the environment is still largely unknown.

4.2.3 *Human risk assessment*

For centuries, dilute solutions of $AgNO_3$ and other Ag compounds were used by medical practitioners as antiseptics, disinfectants, and anti-bacterial agents. And prior to the introduction of antibiotics, these Ag compounds were commonly used in the treatment of open wounds and burns. For instance, Ag compounds combined with sulfadiazine form an effective topical anti-bacterial agent for the treatment of burns [79, 80]. In recent years, nanometer-scale forms of Ag have been widely used in wound dressings and incorporated into medical devices that have been implanted into patients [81–84]. Importantly, most humans are exposed to nanosilver that is contained in consumer products such as personal care, cosmetics, clothing, electronics, household products, and shoes [85]. As a result of the wide range and number of Ag- and nanosilver-based products currently available in the marketplace, there are numerous opportunities for exposure. Besides direct exposure, there are also other avenues such as working in or living near industrial facilities (or disposal sites) where poor handling management occurs. Generally, the main types of human exposure are inhalation via the respiratory system, ingestion via the gastrointestinal system, and through dermal contact via the skin. These three interfaces provide the main routes between the environment and the inner body tissues and organs [86]. Exposure to high levels of Ag for extended periods of time can result in an irreversible pigmentation of the skin (argyria) and/or the eyes (argyrosis). The condition produces a bluish-gray or ash-gray discoloration of the skin and other tissues and is most pronounced when regions of the body are exposed to sunlight [87]. Both argyria and argyrosis can be produced when Ag and Ag-based materials come into contact with skin or other body tissues [88]. However, much less Ag will enter the body through the skin than through the respiratory system or the gastrointestinal tract. Metallic Ag and many inorganic Ag compounds ionize to release biologically active Ag^+ ions in aqueous solutions, body secretions and fluids, and secretions. Thus, when Ag^+ ion absorption takes place over an extended period of time and exceeds the capacity of either the liver or kidney to eliminate them, there is a build-up and ultimate overload of Ag^+ ions in the connective tissues surrounding the vascular tissue and glands of the papillary layer of the dermis [89, 90]. Similarly, argyrosis is the accumulation of Ag^+ ions, resulting in the pigmentation of the cornea and conjunctiva of the eyes [91, 92]. Argyrosis is generally considered a more sensitive outward sign of Ag exposure than argyria [90, 93]. Importantly, both argyria and argyrosis are considered to be a cosmetic problem, since there does not appear to be any signs of pathological damage in body tissues [94]. Moreover, numerous studies have failed to clearly identify and link Ag to any adverse health problems due to the influence of other agents [93]. But exposure to high levels of Ag in the air can result in respiratory problems and lung and throat irritation [95–97], while Ag in close contact with the skin can cause allergic reactions in the form of rashes, swelling, and inflammation in some people [98–100]. Furthermore, cytotoxicity testing of metallic Ag, Ag sulfadiazine, and other Ag compounds with cultured fibroblasts, keratinocytes, and other human cell lines have shown Ag^+ ions can interact with sulfhydryl groups and other proteins and enzymes associated with cell membranes. The interaction results in cellular denaturation and damage to cellular organelles in a similar manner seen bacterial and fungal cells [101–103]. The studies also revealed that fibroblasts were more sensitive to Ag^+ ions than keratinocytes, but factors such as the age of the donor cells and composition of the culture medium could also influence Ag^+ ions interactions [104]. At present, there is no conclusive evidence to demonstrate the level of risk associated with Ag. But respiratory issues and allergic responses display by pre-disposed persons indicates that Ag is a potential toxic hazard [49, 98]. Current data indicates that argyria and argyrosis are the main issues associated with long-term exposure to Ag^+ ions, and that Ag is not carcinogenic [105, 106].

4.3 Biological Mining via Phytoextraction and Phytosynthesis

A key factor for the continuing growth of Ag usage in the industrial sector is largely dependent on developing strategies for insuring a sustainable source of supply. Sources of natural high-quality resources such as Ag will continue to decline in the future, and the sources that will be available will incur increasing recovery and refining costs. For future sustainable growth in the industrial sector to take place, there needs to be effective strategies for the recovery and recycling of Ag from mining wastes, manufacturing wastes, processing operations, and spent Ag-based products that end up in disposal sites. For example, it is estimated that around 60–70 million metric tons of electronic and electrical equipment waste end up in landfill or are incinerated. Approximately 63% of this waste consists of precious metals that could have been recovered and recycled if effective recovery strategies were in place [107, 108]. Therefore, new, economically efficient, and eco-friendly clean technologies are need to develop sustainable recovery and recycling processes for maximizing Ag utilization, reducing energy consumption, minimizing operating costs, and significantly decreasing waste levels going into the environment [50, 109]. An alternative solution to conventional mining and hydrometallurgical processes is to use biological entities to mine for precious metals. The use of micro-organisms to biologically mineralize metal-containing ores and concentrates to obtain a variety of metals has been utilized by the mining sector [110, 111]. Biological mining, or bio-mining, can also be achieved using plants to extract and recover Ag from ores and spent wastes such as electronics [112, 113]. Bio-mining with plants (phytomining) is an *in situ*, passive, and eco-friendly approach that only uses naturally occurring solar energy to drive the process. This makes it more economical in terms of operating costs compared to high-energy-consuming mining and refining processes [114]. Silver phytomining can be considered as the process of using plants and their associated micro-organisms to absorb Ag and Ag mineral species from soil and groundwater environments. Furthermore, plant scientists have reported that some plant species can hyper-accumulate certain metals to concentrations several hundreds of times greater than non-accumulating plants [115]. Hyper-accumulation in some plant species is generally believed to provide some degree of protection to the plant from herbivores and insects [116].

Phytomining has been widely investigated because the process uses particular plant species to concentrate low concentrations of metals from soil or mine tailings where conventional mining methods are not economically feasible [117, 118]. For phytomining, plants should have the following characteristics: (1) tolerate high metal concentrations; (2) undergo rapid growth; (3) possess an extensive root system; (4) precipitate biomass production; (5) accumulate high concentrations of metals in their aerial tissues, and (6) be easy to cultivate and harvest [119–121]. However, for phytomining to be effective, precious metals must accumulate in parts of the plant (leaves, shoots, etc.) that can be readily harvested using conventional farming practices. The harvested biomass can then be processed to extract the precious metal before leaf-fall or plant death. If the biomass is not harvested at the appropriate time, fallen leaves and dead plants can decompose and allow the accumulated metals to return to the soil and groundwater [122]. Furthermore, a number of studies have also suggested introducing chemical agents into the soil environment to improve metal bio-availability and promote better metal uptake by the plants root system [123–125]. In addition, a number of studies have also shown that plants used in the remediation of contaminated soils can accumulate metals and store them as metallic nanoparticles [126–128]. In particular, phytomining and phytosynthesis of precious metal nanoparticles such as Au and Ag have attracted considerable interest in recent years due to their industrial importance in a number of areas. Phytosynthesis uses the plants natural physiochemical processes to act as biological factories to manufacture metallic nanoparticles from a variety of metallic sources [129, 130]. Thus, phytosynthesis is a process that offers the opportunity to recover Ag from a wide range of sources containing Ag and Ag compounds and produce high-value Ag nanoparticles that can be used in a number of industrial sectors, pharmaceutical products, and biomedical applications [129]. Section 4.4 discusses the important aspects of phytosynthesis, including synthesis procedure, factors influencing nanoparticle formation, and the potential use of native Australian plants to synthesize Ag nanoparticles due to the unique nature and evolution of the Australian continent's flora.

4.4 Phytosynthesis/Biogenic Synthesis of Ag Nanoparticles

4.4.1 *Introduction*

The increasing demand for Ag by various industrial sectors in recent years has driven the need to find new sources of this precious metal. However, high-quality natural sources of Ag are declining, and as a result, production costs are increasing. Therefore, to develop an effective and sustainable Ag market, recovering Ag from secondary sources must be fully exploited. Recovery and recycling of spent Ag using alternative technologies and proper waste management strategies offers an effective route for channeling Ag back into the supply chain. Furthermore, scrutiny in recent years of the damage caused by the toxic effects of monovalent Ag^{+} ions from industrial effluents and spent sources in the environment has prompted the development of effective waste management strategies. To address these challenges, phytoextraction and phytosynthesis offer alternative approaches for mitigating environmental risks and providing a unique method for recovering Ag from secondary sources.

4.4.2 *Nanoparticle formation during biogenic synthesis*

Researchers have known for a long time that many plants have properties such as phytoextraction and hyper-accumulation of metals as a method to discourage herbivores and insects [116, 131]. These properties allow metal recovery from the soil and sequestering of metallic materials in various parts of the plant [132]. The ability of many plant species to carry out these two processes has attracted considerable interest, since it offers an eco-friendly and viable green-chemistry-based route for synthesizing metallic Ag nanoparticles and removing Ag from the environment [133, 134]. In recent years, a number of studies have shown that extracts taken from a variety of plant species have the ability to biologically synthesize noble metal nanoparticles such as Au, Ag, Pt, and palladium (Pd) [129, 135, 136]. Studies have shown that a wide variety of both marine and terrestrial plants contain bio-active compounds such as alkaloids, amino acids, enzymes, phenolics, proteins, polysaccharides, polyphenols, saponins, sugars, tannins, terpinoids, and vitamins that can support the formation of nanoparticles [137]. Nanoparticle formation is a bottom-up approach, during which bio-active compounds (reducing agents) first promote the assembly of metallic atoms to form precursor building blocks. These molecular building blocks then self-assemble with the aid of other bio-active compounds that act as modeling and stabilizing agents [138]. Differences in the types of bio-active compounds, their concentration, and composition between plant species are considered to be the major factors that influences the size and shape of the nanoparticles formed [139]. Table 1 presents a selection of plant species that have been investigated and reveals a wide range of particle sizes and morphologies have been generated. Reducing nanoparticles from metal salts using extracts taken from plants is a facile room temperature procedure as schematically presented in Fig. 2. The procedure starts with a metal salt solution being mixed with an aqueous solution containing the plant extract. Metal reduction begins immediately, with the metal ions being converted from their mono- or divalent oxidation

Table 1. A selection of plant species studied for the biosynthesis of Ag nanoparticles.

Plant	Size (nm)	Morphology	References
Allium sativum (Garlic clove)	4–22	Spherical	[148]
Aloe vera	50–350	Spherical, triangular	[144]
Ananas comosus (Pineapple)	5–35	Spherical	[149]
Camelia sinensis	30–40	Spherical, triangular, irregular	[150]
Carica papaya (Pawpaw)	15	Cubic	[151]
Carica papaya (Pawpaw)	60–80	Spherical	[152]
Citrus sinensis (Orange) peel	3–12	Spherical	[153]
Citrus sinensis (Orange) peel	35 ± 2 10 ± 1	Spherical @ 25°C Spherical @ 60°C	[154]
Citrus unshiu (Mandarin) peel	5–20	Spherical	[155]
Codium capitatum	3–44	Spherical and cubic	[156]
Citrullus colocynthis	31	Spherical	[157]
Daucus carrota L (Black Carrot)	4.32–17.65	Spherical	[158]
Eucalyptus macrocarpa	10–100	Spherical and cubic	[159]
Mangifera indica	20	Spherical, triangular, hexagonal	[160]
Padina tetrastromatica	5–35	Spherical	[161]
Psidium guajava (Guava)	0.1–0.5 μm	Granular	[162]
Rhododendron dauricum	25–40	Spherical	[163]
Sorghum spp. (Bran powder)	10	Quasi-spherical	[164]
Spyrogira insignis	30	Spherical	[165]

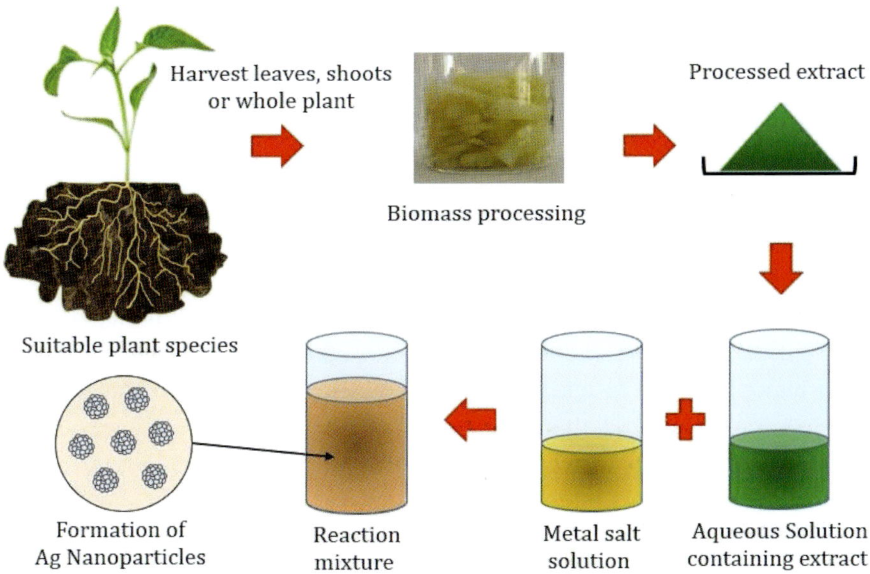

Fig. 2. Schematic representation of the biosynthesis of Ag nanoparticle by plant extracts.

states to their zero-valent states. During this period, nucleation takes place and a color change can be seen taking place in the reaction mixture. Following initial nucleation, smaller neighboring particles assemble to form larger thermodynamically stable nanoparticles. As biological reduction of metal ions continues, the nanoparticle size increases along with a variety of morphologies such as cubes, spheres, triangles, hexagons, pentagons, rods, and wires [137]. During the last stage of the process, stabilizing agents present in the plant extract are able to assist and finalize the most energetically favorable and stable morphology of the nanoparticles [140]. Studies have revealed that nanoparticle formation is governed by several factors that include plant extract concentration, metal salt concentration, reaction solution pH, reaction time, and reaction temperature [141, 142]. These factors and their influence are discussed in Sec. 4.4.3.

4.4.3 *Factors governing nanoparticle formation*

During the formation of metal nanoparticles, several factors influence nucleation, particle growth, and ultimate stable particle morphology. The concentration of reactants in the reactant mixture can have a major role in controlling the nanoparticle morphology. For example, when precursor chloroauric acid was subjected to increasing concentrations of *Cinnamomum camphora* (camphor) leaf extract, the resulting Au nanoparticle morphology changed from triangular to spherical [143]. Varying the concentration of the *Aloe vera* leaf extract in a reaction mixture (containing chloroauric acid) could regulate particle sizes between 50 and 350 nm. Carbonyl compounds present in the extract were found to act as modeling agents, and varying concentrations also influenced the number of Au triangular plates that were formed along with Au spherical nanoparticles. [144]. The pH of the reaction mixture has also been shown to influence both nanoparticle size and morphology [145]. Some studies have found that lower pH (acidic) values tend to produce larger particles. For example, extracts from *Avena sativa* (common oat) biomass and chloroauric acid synthesized with a pH of 2 tended to produce rod-shaped Au nanoparticles ranging in size from 25 to 85 nm. But when synthesis was carried out with reaction mixtures with pH between 3 and 4, there was a tendency to produced smaller-sized particles ranging from 5 to 20 nm [146], while other studies have shown the opposite effect. For instance, when *Cinnamon zeylanicum* bark extract was used to synthesize Pd nanoparticles, the nanoparticles produced when the pH was less than 5 tended to range in size between 15 and 20 nm. But when the nanoparticles were synthesized at pH values greater than 5, the particles ranged in size from 20 to 25 nm. However, further studies are needed to explain this anomaly, but it is believed to result from the types of bio-active compounds present in the respective plant extracts used and the propensity of the compound to take part in the reaction [147].

The influence of reaction time has also been studied by several researchers. A study investigating increasing reaction time using a reaction mixture consisting of *Azadirachta indica* leaf extract and $AgNO_3$ revealed that longer times tended to increase particle size. When the reaction period was adjusted between 30 min and 4 h, the size of the nanoparticles produced ranged from 10 to 35 nm [139]. Similar studies have shown that beyond particular timeframes, few nanoparticles are formed [149, 166]. These studies suggest reaction time is dependent on the reaction rate of the reactants; slower reaction rates tend to produce longer synthesis periods, while reactants such as *Ananas comosus* (pineapple) extract and aqueous

$AgNO_3$ solutions react rapidly to reduce nanoparticles within 2 min, with very little reduction occurring after 5 min [139]. Reaction temperature has also been shown to influence the size, shape, and yield of nanoparticles produced using plant extracts [145, 167]. A recent study by Kaviya *et al.* [154] revealed that synthesizing Ag nanoparticles from a mixture of *Citrus sinensis* (sweet orange) peel extract and $AgNO_3$ at 25°C produced a mean particle size of 35 nm. But when the reaction temperature was raised to 60°C, the mean particle size produced was reduced to 10 nm. Further studies have also shown variations in reaction temperature can result in variations in reaction rates, particle formation rates, and changes in both nanoparticle size and morphology [146]. Both reaction and particle formation rates tend to increase with increasing reaction temperatures, while the average particle size tends to decrease for most nanoparticles [134]. However, more detailed studies are needed to examine the exact nanoparticle formations mechanisms (currently not fully understood) and the influence of the above-mentioned formation factors.

4.4.4 *Ag nanoparticle synthesized by plants*

The biogenic synthesis of Ag nanoparticles using plant sources is a relatively new and widely reported field of research due to the importance of Ag nanoparticles in various industrial, pharmaceutical, and bio-medical sectors of the global economy. The following is a brief overview of the various types of plant species that have been investigated for their ability to synthesize Ag nanoparticles and is presented here to give the reader an appreciation of the current state of research in the field. For further information of the current state of research in this particular field, the reader is referred to several literature reviews [129, 130, 135–139]. The wide variety of plant species has attracted the interest of researchers for centuries. Plants provide an importance food resource and an importance source of medicinal remedies. The medicinal properties of various plant species have been extensively studied and used by humanity since the beginning of recorded time. Herbal remedies for a wide range of human ailments have been reported, but in recent years, the phytosynthesis properties of plants have also been studied. The ability of plants to act as biological factories for the synthesis of precious metals from secondary sources offers an alternative approach to more conventional metal extraction and recovery processes. The most attractive feature of the approach is that concentrations of Ag contained in soil, mine tailings, or secondary sources are at such low levels that conventional recovery methods may not be economical, but plant-based techniques may provide a viable alternative. From a historical point of view, the use of Ag and Ag compounds in medicinal preparations to combat infections has firmly established Ag as an effective anti-microbial agent. Recovering Ag in the form of high-value nanoparticles for pharmaceutical formulations alone is a major economic incentive. The toxicological damage inflicted by Ag nanoparticles toward microbial species means there is a significant and developing market for new nanosilver-based pharmaceuticals and bio-medical products [16–19, 168, 169]. Thus, recovering high-value Ag nanoparticles via plants from secondary sources is an attractive option.

4.5 Potential Bio-factory Capability of Indigenous Australian Plant Species

4.5.1 *Introduction*

The earth's biosphere contains hundreds of thousands of plant species. As mentioned earlier and presented in Table 1, a number of these plant species have been examined for potential bio-factory capabilities toward the biological synthesis of Ag nanoparticles. The results of these studies have shown the compositional make-up of the respective plant extract (bark, flower, leaf, etc.) can directly influence the formation and physiochemical properties of the nanoparticle. Each plant species can contain different types and concentrations of bio-active compounds that can act as metal-reducing and particle-stabilizing agents [170]. At present, the literature reports very few studies of indigenous Australian plant species being used as bio-factories for the synthesis of Ag nanoparticles. The authors have recently undertaken studies to evaluate the bio-factory potential of three native Australian plant species, and some of the results are presented in this section. The three plant species investigated were *Eucalyptus macrocarpa* (rose of the west), *Xanthorrhoea glauca* (grass tree), and *Anigozanthos manglesii* (red-and-green kangaroo paw). Photographic images of the three plant species are presented in Fig. 3 and reveal the dissimilar appearance and nature of the respective species.

Fig. 3. (a) Location of three plant species investigated. Photographic images of the individual plants: (b) *Eucalyptus macrocarpa*, (rose of the west); (c) *Anigozanthos manglesii*, (red-and-green kangaroo paw), and (d) *Xanthorrhoea glauca* (grass tree) all located at the Murdoch University campus (S 32°4′ 6.99″ E115°49′ 58.1″).

4.5.2 *Biosynthesis of Ag nanoparticles using indigenous Australian plants*

The three plant species investigated are all found in the South West corner of the Australian continent (see Fig. 3(a)), which is also recognized as a global bio-diversity hotspot [171]. *Eucalyptus macrocarpa* (commonly known as the rose of the west) is a large plant with distinctive ovate-shaped silvery leaves that can grow to lengths of around 12 cm. The silvery-gray foliage provides an ideal backdrop for the plant's very prominent red flowers, as seen in Fig. 3(b). The profound silvery-gray appearance of the leaves is produced by nanometer-scale surface pillar structures formed by epicuticular waxes. An interesting feature of this surface feature is that it imparts remarkable surface wetting and self-cleaning properties that assist the plant's survival in an arid climate [172]. The South

West corner of Australia is also the home of the exquisite *Anigozanthos manglesii*, which is more commonly known as the red-and-green kangaroo paw. During spring and summer, the plant grows from an underground rhizome to produce a red stalk that reaches a height of around 1 m, and crowning the stalk is an array of red and green tubular flowers as seen in Fig. 3(c). The shapes of the flowers resemble a claw structure that looks very much like the paw of a kangaroo, hence its name. Another floral inhabitant of the region is *Xanthorrhoea glauca*, (commonly known as the grass tree) and is easily recognizable by its vast array of long thin narrow leaves protruding outward from the head of its trunk, as seen in Fig. 3(d).

The biosynthesis process begins by harvesting healthy leaves randomly selected from several plants of each species located at various locations around the Murdoch University campus. The leaves were sorted, inspected for quality, and placed into three processing streams, with each stream containing a particular plant species. The leaves of each stream were initially washed in Milli-Q® water (high-purity water) to remove surface contaminants before being processed in a standard domestic blender. After blending, Milli-Q® water was slowly added to the blend under the influence of slow stirring to form a more fluidic mixture. The blend was then homogenized at 5000 rpm for 10 min at room temperature (24°C). After homogenization, a two-step filtration procedure was followed. The first filtration procedure was carried out using a Hirsch funnel to remove leaf debris. The second filtration consisted of two further purifications using two separate 0.22-μm Millex® (33 mm dia.) syringe filters [148, 173, 174]. The processed leaf extracts were then stored in glass vials ready for the next stage of processing. In the next stage, Ag nanoparticles were produced using a specific plant extract and a secondary source of Ag. In this case, the secondary source of Ag used was $AgNO_3$. The influence of varying reactant concentrations was achieved by varying the amount of leaf extract used in the reaction mixture. In the next stage of the procedure, a fixed amount of processed leaf extract was mixed via stirring with a 1.0 mL solution of 0.1 M $AgNO_3$ for 1 min. After mixing, the reaction mixture was allowed to stand while the metal salt was reduced to form Ag ions that begin to nucleate. Ag reduction continues, and the initial nucleated seeds begin to accumulate and produce thermodynamically stable nanoparticles.

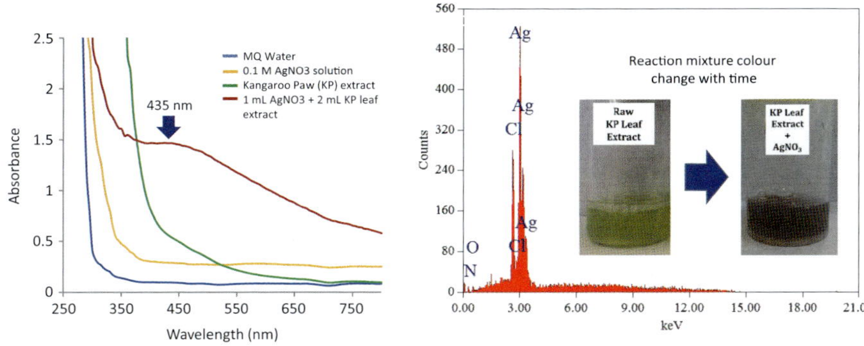

(a) (b)

Fig. 4. (a) UV-visible spectroscopy analysis of Ag nanoparticle synthesis using *Anigozanthos manglesii*, (red-and-green kangaroo paw); (b) Energy dispersive spectroscopy (EDS) was used to identify the presence of metallic Ag in the sample, and the insert shows the color change of the reaction mixture with time [174].

Nanoparticle formation was indicated by a color change in the reaction mixture, and UV-visible spectroscopy was used to monitor nanoparticle formation and growth with time. In the case of *Anigozanthos manglesii*, the pale-green-colored mixture containing the leaf extract and $AgNO_3$ progressively turned dark brown over a 20 min period as seen in a representative sample in Fig. 4(b) inset. A representative UV-visible spectrum is presented in Fig. 4(a) and shows the presence of an optical surface plasmon resonance (SPR) peak at 435 nm. Similar studies in the literature confirm the color change and report SPR peaks between 400 and 440 nm. The presence of this peak indicates the formation of metallic Ag nanoparticles [175, 176].

Nanoparticle growth continues by aggregation, and the final size and morphology is determined by competing growth rates experienced by the various crystallographic facets. The crystalline structure and crystallite sizes of the synthesized nanoparticles was determined using X-ray diffraction (XRD) spectroscopy, and a representative XRD pattern for Ag nanoparticles produced by *Anigozanthos manglesii* is presented in Fig. 5(a). Inspecting the pattern reveals a dominant (111) Bragg reflection peak on the left-hand-side of the pattern and two smaller peaks, namely (200) and (311), on the right-hand-side of the pattern. The occurrence of these peaks in the XRD pattern indicates a face-centred cubic structure was present

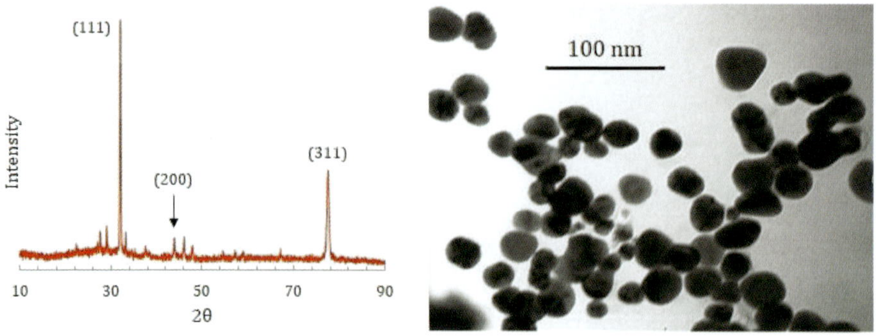

(a) (b)

Fig. 5. *Anigozanthos manglesii* (red-and-green kangaroo paw) synthesized Ag nanoparticles: (a) XRD pattern identifying the presence of Ag nanoparticles and (b) representative TEM image of formed Ag nanoparticles after a 20-min synthesis period [174].

in the produced Ag nanoparticles. Also present in the pattern is peak broadening, which is indicative of nanoparticle formation. This result is consistent with similar studies for biologically synthesized Ag nanoparticles using plant extracts [177, 178]. Another technique that was used to identify elements present in the samples was energy dispersive spectroscopy (EDS). In the present case, EDS was used to identify elements and confirm the synthesis of Ag nanoparticles by the *Anigozanthos manglesii*/$AgNO_3$ reaction mixture. An EDS analysis of a representative sample is presented in Fig. 4(b) and clearly shows a strong Ag peak, thus confirming the presence of metallic Ag in the samples. Furthermore, surface-modifying agents (capping agents) present in the plant extract and the influence of formation factors such as reactant concentrations, reaction mixture pH, reaction time, and temperature all contribute at varying degrees to the growth of individual facets. Thus, the combination of surface-modifying agents and formation factors result in the creation of different morphologies such as cubes, spheres, triangles, hexagons, pentagons, rods, and wires. Images produced by techniques based on scanning electron microscopy (SEM) and transmission electron microscopy (TEM) were used to determine particle size and morphology. In the present case, TEM images were taken after a 20-min synthesis period and reveal particles with varying sizes and morphologies. A representative TEM image is presented in Fig. 5(b) and reveals nanoparticle sizes vary between 25 and 75 nm. The image also shows

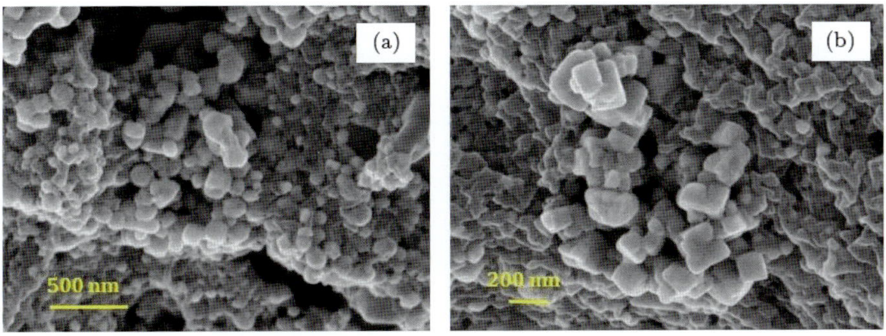

Fig. 6. Representative FESEM images of Ag nanoparticles synthesized using a leaf extract taken from *Eucalyptus macrocarpa* (rose of the west).

particle morphologies that include cubes and triangular and hexagonal plates.

Similarly, a recent study of *Eucalyptus macrocarpa* leaf extracts by the authors revealed that this plant could also bio-synthesize Ag nanoparticles [179]. Representative SEM images taken of the samples reveal the presence of Ag nanoparticles. Figure 6(a) presents a typical SEM image showing the widespread aggregation of Ag nanoparticles across the sample, while Fig. 6(b) shows a representative cluster of nanoparticles. The SEM images reveal the predominant morphology present in the samples was cubic, with side dimensions ranging between 70 and 300 nm. Also present were numerous rounded non-spherical particles and a small number of triangle- and hexagon-like structures.

4.5.3 *Formation and movement of Ag nanoparticles in Xanthorrhoea glauca*

Understanding the formation, penetration, and transportation of metallic ions within plant tissues is important in determining the feasibility of using plants for synthesizing and accumulating Ag nanoparticles for potential harvesting of nanoparticles. The leaves of *Xanthorrhoea glauca* were used to investigate nanoparticle formation sites and movement within the leaf structure. The leaf was selected due to its long, thin, and narrow structure and its highly structured nature, which includes well-defined arrays of vascular bundles as seen in the stained optical microscopy image presented in Fig. 7(a). The vascular bundles are used by the plant to transport water

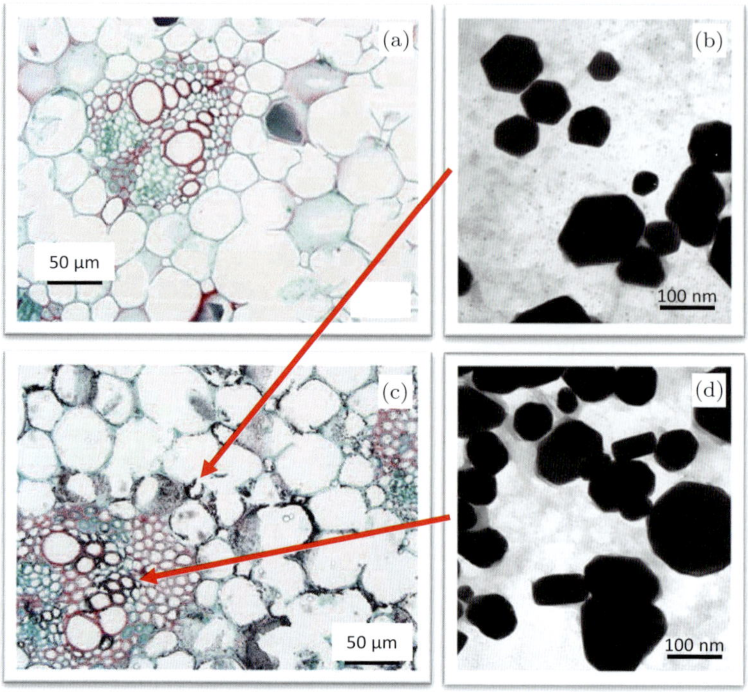

Fig. 7. (a) Optical microscopy image of a typical *Xanthorrhoea glauca* vascular bundle; (b) TEM image of synthesized Ag nanoparticles at cell walls; (c) Ag nanoparticle formation at cell walls at various locations within leaf, and (d) TEM image of Ag nanoparticles formed in and around a vascular bundle.

and minerals (xylem) and nutrients (phloem), and also provide a route for the movement of Ag nanoparticles. After harvesting, the healthy leaves were washed before the base of the leaf was finely cut. The finely cut end was then dipped into a 0.1 M solution of $AgNO_3$. The progress of Ag^+ ions permeation throughout the leaf was monitored using optical-microscopy–based techniques. The treated leaf samples revealed that Ag nanoparticles rapidly formed *in situ* and a distinct color change (dark areas located at the cell walls) could be seen throughout various regions within the leaf tissues, as seen in Fig. 7(c).

TEM images taken of these darkened regions revealed the presence of nanoparticle aggregates consisting of individual nanoparticles with various morphologies that included cubic, hexagonal, and triangular structures. The well-formed Ag nanoparticles seen in Figs. 7(b) and 7(d) ranged in size from 50–200 nm and were predominantly in the darkened regions that border cell walls. The results of this study have shown both Ag^+ ions and Ag nanoparticles were capable of traveling through the vascular bundles and also had the ability to diffuse out from the bundles into the extracellular spaces. Not only were nanoparticles found in the extracellular spaces, but they were also found within some cells, thus indicating that bio-active molecules present in the cell membranes were capable of promoting the reduction of $AgNO_3$ to form Ag nanoparticles. However, further work is needed to identify the bio-active molecules involved in reduction and the efficiency of the reduction process. Also, additional work is needed to explore and determine the exact mechanism(s) of intra-cellular and extra-cellular nanoparticle reduction and plant tolerance levels to increasing concentrations of $AgNO_3$. In particular, further studies are needed to determine the effective methods for using this plant-based metal recovery method for large-scale extraction applications.

4.5.4 *Applications for high-value Ag nanoparticles*

The demand for high-value Ag nanoparticles makes recovering Ag, at the nanometer scale, from secondary sources an attractive option. Using plants as biological factories has the potential to deliver an important eco-friendly source of nanometer-scale Ag via green-chemistry-based techniques. This alternative approach could become an important method for supplying the three major user groups of Ag nanoparticles, namely industrial, consumer products, and pharmaceuticals and biomedicine. In recent years, the use of Ag nanoparticles in consumer products has significantly grown. The main reasons for this is the strong anti-bacterial (and anti-fungal) properties of Ag nanoparticles toward a diverse range of micro-organisms and the emerging resistance of common pathogens to antibiotics [180]. In biomedicine, nanosilver has been used in a variety of applications that include diagnosis, treatments, drug delivery, and coating of medical devices. In particular, it has widespread use in the manufacture of wound dressings and anti-microbial socks and fabrics to inhibit microbial growth and reduce healing periods [25, 180]. Similarly, various forms of nanosilver have been incorporated into tissue scaffolds, bone cements, dental cements, and coating for medical implants [62, 63, 181, 182].

4.5.4.1 *Bacterial toxicity of Ag nanoparticles*

The medicinal properties of silver have been known for centuries, and the ionized form of silver as Ag^+ ions is well known to have anti-bacterial effects. Silver colloids were used in the early twentieth century against a range of bacterial strains, but with the introduction of antibiotics, the use of silver in combating bacterial infections decreased drastically. Since then the widespread and rampant use of a range of antibiotics to kill and prevent bacterial pathogenic human infections and in raising livestock have allowed resistant bacterial strains to develop over time. The ability of bacteria to mutate and share genes has allowed this resistance to antibiotics to be passed onto the next generations.

So the race to develop new anti-microbials does not stop, and once resistance is achieved, novel drugs must be constantly developed at huge R&D costs. Researchers worldwide have refocused their attention to silver, and together with the advent of nanotechnology and nanoscience, better and more robust synthetic methods have been discovered and optimized. As the size and shape play a very important role in the properties of nanoscale matter, it was not surprising to discover that the toxicity of Ag nanoparticles were also related to the shape, size, concentration, as well as the coating of the Ag nanoparticles.

Silver nanoparticles are considered to have broad-spectrum anti-microbial properties with proven anti-fungal, anti-viral, and anti-bacterial activities [183–185]. The exact mechanisms by which silver nanoparticles destroy a particular strain of bacteria are not fully known, although several studies have shown several aspects by which silver NPs or silver ions have halted the growth of certain strains of these pathogens. Bacteria are prokaryotes with a cell membrane and a non-cellulose cell wall with cilia and flagella present in some species. They do not have mitochondria, vacuoles, nuclear membrane, endoplasmic reticulum, and a Golgi body.

Silver nanoparticles can be synthesized in various shapes and sizes as well with particular coatings or capping agents which can affect bacteria in several ways (see Fig. 8). The Ag nanoparticles can directly attach to the bacteria cell wall as well as penetrate the latter directly if it is small enough. Song *et al.* showed bacterial cell death occurs when <10 nm Ag nanoparticles were synthesized and targeted toward Gram-positive and Gram-negative bacteria such as *Salmonella typhi,*

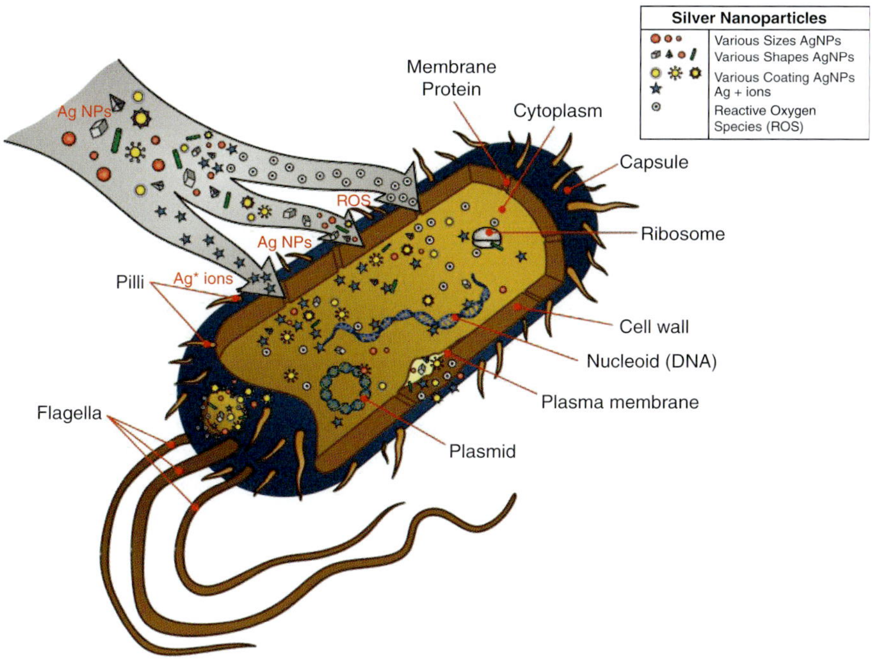

Fig. 8. Representative image of a bacterium and the various pathways that Ag nanoparticles and Ag^+ ions can interact with this organism, leading to impaired vital cell life systems and cell death.

Escherichia coli, Pseudomonas aeruginosa, Staphylococcus aureus, and Mycobacterium tuberculosis [185]. The pathogens were affected in several ways, including plasmolysis, inhibition of cell wall synthesis, as well as metabolic imbalances.

Other studies propose that Ag nanoparticles have the ability to attach directly to the cell membrane and thus disrupt the permeability, leading to respiratory dysfunction. These structural changes to the outer wall can also cause pits. Smaller Ag nanoparticles can cross the cell wall and enter directly inside the bacteria. Inside, the nanoparticles can interact and dissolve further to create Ag^+ ions and bind with thiol groups of many enzymes vital to the bacteria's life processes [186].

Silver is also considered a soft acid and organism with many sulfur and phosphorus containing groups are at particular risk as the Ag–S bond is very strong (216.7 kJ/mol). Furthermore, DNA, the molecule that encodes the blueprint of life and is responsible for the survival of the bacterial strain, is made up of several bases that contain P and S atoms. These bases will bind strongly to Ag NPs and Ag^+ ions and then the DNA cannot function properly and this ultimately leads to cell death.

Normal cellular activities produce reactive oxygen species (ROS) and the organism without any trouble handles these reactive molecules. In aerobic conditions however, Ag NPs have the ability to release Ag^+ ions, and this process is enhanced at lower pH. With silver ions present, these ROS are overproduced to such levels that it interferes with normal cellular activities. DNA contains the biological blueprint of the bacteria and this macromolecule contains several bases with sulfur and phosphorus atoms that can easily bind to Ag and Ag^+ ions. Thus, interfering with the DNA's purpose leads to cell death. For further information, the reader is directed to the review of Rizello and Pompa that summarises most of the bacticeridal effects of nanosilver [186].

Research by the authors has shown the Ag nanoparticles synthesized by the above-mentioned selected indigenous Australian plants were effective anti-microbial agents against a number of bacterial pathogens [173, 174]. For example, Ag nanoparticles produced by *Xanthorrhoea glauca* were found to be effective anti-microbial agents against the bacteria *Deinococcus*. Figure 9 presents a photographic image of the anti-microbial response of Ag nanoparticles produced by the conventional wet chemistry technique. As seen in the photograph, there is a circular region (18 mm dia.) around the test pad containing the Ag nanoparticles. This region is known as the zone of inhibition and within this area no *Deinococcus* bacteria could survive, as seen in Fig. 9(b). Figure 9(c) presents the zone of inhibition for a pad treated with Ag nanoparticles (17 mm dia.) synthesized by *Xanthorrhoea glauca*. The results are graphically presented in Fig. 9(d) and reveal the plant-synthesized Ag nanoparticles have comparable anti-microbial properties to those produced by conventional methods. The outcome of the study clearly demonstrates that *Deinococcus* is susceptible to the anti-microbial action of Ag nanoparticles.

The positive anti-microbial properties displayed by Ag nanoparticles synthesized by the above-mentioned native Australian plants is a clear indication that they have the potential to be incorporated into new pharmaceutical products aimed at combating antibiotic-resistant bacterial pathogens. It also demonstrates the potential of using plants for recovering

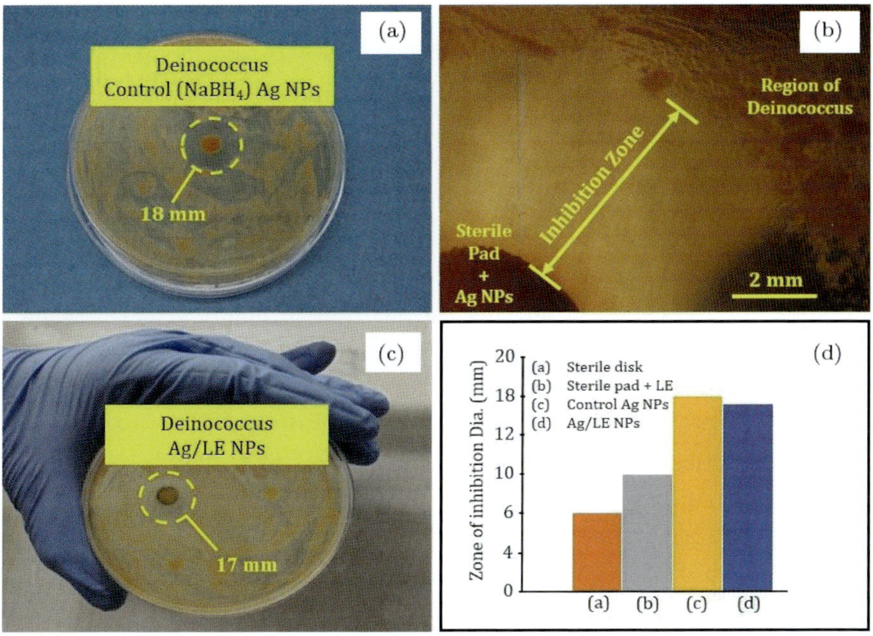

Fig. 9. Anti-microbial activity of Ag nanoparticles synthesized by *Xanthorrhoea glauca* leaf extract: (a) control (conventional wet chemical synthesis); (b) zone of inhibition; (c) anti-microbial action of the leaf-extract-produced Ag nanoparticles, and (d) zone of inhibitions of control and leaf-extract-produced Ag nanoparticles.

Ag from a secondary source. Moreover, Ag recovered as nanoparticles have a much higher value than bulk Ag, thus highlighting the value of using appropriate plant species as bio-factories for the recovery and processing of Ag from suitable secondary sources. However, this Ag recovery method needs further investigation before it can become an economically viable alternative to more traditional recovery methods currently in use.

4.6 Concluding Remarks

Plants can function as bio-factories for the recovery and manufacture of large quantities of Ag nanoparticles. The physiochemical properties of Ag nanoparticles make them highly valued in a number fields such as industrial, consumer products, pharmaceuticals, and biomedicine. Importantly, Ag nanoparticles have a much higher economic value than bulk Ag, which makes plant-based bio-factories an attractive approach. Moreover, studies have shown that high quantities of Ag in the environment can inhibit microbial-based soil-forming processes and influence the early development of fish. Therefore, there is an imperative need to recover Ag wastes from the environment using economically viable and eco-friendly approaches. This is of particular importance, since the Ag used in many consumer products and industrial wastes generally end up in the environment. However, it should be pointed out that there is no conclusive evidence to suggest there are any serious human health risks associated with Ag. But some pre-disposed people have shown respiratory issues and allergic responses to high concentrations of Ag, thus rendering Ag a potential hazard to human health. In terms of recovery, leaf extracts from three indigenous Australian plants species that include *Eucalyptus macrocarpa*, *Anigozanthos manglesii*, and *Xanthorrhoea glauca* have all displayed the ability to reduce Ag^{+} ions from a secondary source. In particular, the uptake of $AgNO_3$ and subsequent reduction of Ag^{+} ions to form Ag nanoparticles within *Xanthorrhoea glauca* leaves indicates the ability to manufacture and accumulate nanoparticles within leaf cellular structures. These studies have shown the possibility of using leaf extracts from the three plant species to recover Ag from a secondary source. At present, only biosynthesis by leaf extracts has been examined, and further experiments are needed in order to examine the overall phytoextraction efficiency and metal reduction efficiency of the plants. In addition, further experiments are needed in order to determine the optimum experimental parameters needed for controlled nanoparticle formation and to elucidate the formation mechanism(s) occurring during synthesis. Furthermore, additional work is needed to develop optimal harvesting and separation processes that are required for the sustainable and commercially successful recovery of Ag from secondary sources.

Acknowledgments

Derek Fawcett would like to thank Horticulture Innovation Australia for their research support through Project AI14003. The authors would also like to thank Dr Mona Shah for assisting in the anti-microbial studies associated with this research work.

References

[1] Nordberg G, Gerhardsson LS (1988) Silver. In *Handbook on Toxicity of Inorganic Compounds*, Seiler HG, Sigel H, Sigel A (eds.) (Marcel Dekker, New York) pp. 619–624.

[2] Syed S (2005) A critical review on the recovery of silver from photographic processing waste. *Journal of Saudi Chemical Society*, **9**: 243–252.

[3] Canda L, Heput T, Ardelean E (2016) Methods for recovering precious metals from industrial waste. *IOP Conference Series on Materials Science and Engineering*, **s**(012020): 1–9.

[4] Vats MC, Singh SK (2015) Assessment of gold and silver in assorted mobile phone printed circuit boards (PCBs). *Waste Management*, **45**: 280–288.

[5] Cui J, Forssberg E (2003) Mechanical recycling of waste electric and electronic equipment: A review. *Journal of Hazard Materials*, **99**: 243–263.

[6] Poinern GEJ, Sharma S, Fawcett D (2016) Adsorption of gold on granular activated carbons and new sources of renewable and eco-friendly activated carbons. In *The Recovery of Gold from Secondary Sources*, S Sabir (ed.) (Imperial College Press, UK), pp. 95–142.

[7] Lanzano T, Bertram M, Palo MD, Wagner C, Zyla K, Graedel TE (2006) The contemporary European silver cycle. *Resources Conservation and Recycling*, **46**: 27–43.

[8] World Silver Survey (2015) GFMS Limited. The Silver Institute. Available at: http://www.silverinstitute.org. Accessed 10 February 2015.

[9] Nowak B, Krug H, Height M (2011) 120 Years of nanosilver history: Implications for policy makers. *Environmental Science and Technology*, **45**(4): 1177–1183.

[10] Nanosilver Market — Global Industry Analysis, Size, Share, Growth, Trends, and Forecast 2015–2023. Available at:http://www.prnewswire.com/news-releases/ Accessed 12 February 2015.

[11] Jefferson DA (2000) The surface activity of ultrafine particles. *Philosophical Transactions of the Royal Society of London, Series A: Mathematical, Physical and Engineering Sciences*, **358**: 2683–2692.

[12] Mansoori GA, George TF, Zhang G, Assoufid L (2007) *Molecular Building Blocks for Nanotechnology* (Springer, New York).

[13] Galib MB, Mashru M, Jagtap C, Patgiri BJ, Prajapati PK (2011) Therapeutic potentials of metals in ancient India: A review through Charaka Samhita. *Journal of Ayurveda & Integrative Medicine*, **2**(2): 55–63.

[14] Moyer CA (1965) A treatment of burns. *Transactions & Studies of the College of Physicians of Philadelphia*, **33**: 53–103.

[15] Clement JL, Jarrett PS (1994) Antibacterial silver. *Metal Based Drugs*, **1**(5–6): 467–482.

[16] Pollini M, Paladini F, Catalno M, Taurino A, Licciulli A, Maffezzoli A, Sannio A (2011) Antibacterial coatings on haemodialysis catheters by photochemical deposition of silver nanoparticles. *Journal of Materials Science: Materials in Medicine*, **22**(9): 2005–2012.

[17] Cohen MS, Stern JM, Vanni AJ, Kelley RS, Baumgart E, Field D, Libertino JA, Summerhayes IC (2007) In vitro analysis of a nanocrystalline silver-coated surgical mesh. *Surgical Infections*, **8**(3): 397–403.

[18] Huang Y, Li X, Liao Z, Zhang G, Liu Q, Tang J, Peng Y, Liu X, Luo Q (2007) A randomized comparative trial between Acticoat and SD-Ag in the treatment of residual burn wounds including safety analysis. *Burns*, **33**(2): 161–166.

[19] Li WR, Xie XB, Shi QS, Duan SS, Ouyang YS, Chen YB (2011) Antibacterial effect of silver nanoparticles on *Staphylococcus aureus*. *Biometals*, **24**(1): 135–141.

[20] Cox SG, Cullingworth L, Rode H (2011) Treatment of paediatric burns with a nanocrystalline silver dressing compared with standard wound care in a burns unit: A cost analysis. *South African Medical Journal*, **101**(10): 728–731.

[21] Hussain SM, Hess KL, Gearhart JM, Geiss KT, Schlager JJ (2005) In vitro toxicity of nanoparticles in BRL 3A rat liver cells, *Toxicology In Vitro*, **19**: 975–983.

[22] Vasir JK, Labhasetwar V (2008) Quantification of the force of nanoparticle–cell membrane interactions and its influence on intracellular trafficking of nanoparticles. *Biomaterials*, **29**: 4244–4252.

[23] Asha Rani PV, Mun GLK, Hande MP, Valiyaveettil S (2009) Cytotoxicity and genotoxicity of silver nanoparticles in human cells. *ACS Nano*, **3**(2): 279–290.

[24] Sukirtha R, Priyanka KM, Antony JJ *et al.* (2011) Cytotoxic effect of green synthesized silver nanoparticles using Meliaazedarach against *in vitro* HeLa cell lines and lymphoma mice model. *Process Biochemistry*, **47**: 273–279.

[25] Seil JT, Webster TJ (2012) Antimicrobial applications of nanotechnology: Methods and literature. *International Journal of Nanomedicine*, **7**: 2767–2781.

[26] Ponarulselvam S, Panneerselvam C, Murugan K, Aarthi N *et al.* (2012) Synthesis of silver nanoparticles using leaves of *Catharanthus roseus* Linn. G. Don and their antiplasmodial activities. *Asian Pacific Journal of Tropical Biomedicine*, **2**: 574–580.

[27] Subramanian V (2012) Green synthesis of silver nanoparticles using Coleus amboinicus lour, antioxidant activity and in vitro cytotoxicity against Ehrlich's Ascite carcinoma. *Journal of Pharmaceutical Research*, **5**: 1268–1272.

[28] Nair R, Varghese SH, Nair BG, Maekawa T, Yoshida Y, Kumar DS (2010) Nanoparticulate material delivery to plants. *Plant Science*, **179**: 154–163.

[29] Khot LR, Sankaran S, Maja JM, Ehsani R, Schuster EW (2012) Applications of nanomaterials in agricultural production and crop protection: A review. *Crop Protection*, **35**: 64–70.

[30] Vivek M, Kumar PS, Steffi S, Sudha S (2011) Biogenic silver nanoparticles by Gelidiella acerosa extract and their antifungal effects. *Avicenna Journal of Medical Biotechnology*, **3**: 143–148.

[31] Park HJ, Kim SH, Kim HJ, Choi SH (2006) A new composition of nanosized silica-silver for control of various plant diseases. *Plant Pathology Journal*, **22**: 295–302.

[32] Huang Y, Chen S, Bing X, Gao C, Wang T, Yuan B (2011) Nanosilver migrated into food-simulating solutions from commercially available food fresh containers. *Packaging Technology and Science*, **24**(5): 291–297.

[33] Tran QH, Nguyen VQ, Le AT (2013) Silver nanoparticles: Synthesis, properties, toxicology, applications and perspectives. *Advances in Natural Science: Nanoscience Nanotechnology*, **4**: 1–21.

[34] Gromov OG, Kuzmin AP, Kunshina GB, Lokshin EP, Kalinnikov VT (2004) Electrochemical recovery of silver from secondary raw materials. *Russian Journal of Applied Chemistry*, **77**(1): 62–66.

[35] Bas AD, Yazici EY, Deveci H (2012) Recovery of silver from X-ray film processing effluents by hydrogen peroxide treatment. *Hydrometallurgy*, **121–124**: 22–27.

[36] World Silver Survey (2015) GFMS Limited. Silver scrap: The forgotten fundamental. Available at: https://www.silverinstitute.org/wp-content/uploads/2017/10/2015WorldSilverSurvey.pdf. Accessed 16 November 2017.

[37] World Silver Survey (2014). GFMS Limited. The Silver Institute. Available at: https://www.silverinstitute.org/wp-content/uploads/2017/10/2014WorldSilverSurvey.pdf. Accessed 16 November 2017.

[38] World Silver Survey (2013). GFMS Limited. The Silver Institute. Available at: https://www.silverinstitute.org/wp-content/uploads/2017/10/2013WorldSilverSurvey.pdf. Accessed 16 November 2017.

[39] World Silver Survey (2012) GFMS Limited. The Silver Institute. Available at: https://www.silverinstitute.org/wp-content/uploads/2017/10/2012WorldSilverSurvey.pdf. Accessed 16 November 2017.

[40] World Silver Survey (2011). GFMS Limited. The Silver Institute. Available at: http://www.silverinstitute.org/site/wp-content/uploads/2011/12/world-silver-survey-2011.Pdf. Summary. Accessed 10 February 2017.

[41] Asharani PV, Wu YL, Gong Z, Valiyaveettil S (2008) Toxicity of silver nanoparticles in zebrafish models. *Nanotechnology*, **19**(25): 255102, 1–8.

[42] Asharani PV, Low Kah Mun G, Hande MP, Valiyaveettil S (2008) Cytotoxicity and genotoxicity of silver nanoparticles in human cells. *ACS Nano*, **3**(2): 279–290.

[43] Liu W, Wu Y, Wang C, Li HC, Wang T *et al.* (2010). Impact of silver nanoparticles on human cells: effect of particle size. *Nanotoxicology*, **4**(3): 319–330.

[44] Piao MJ, Kang KA, Lee IK, Kim HS, Kim S *et al.* (2011) Silver nanoparticles induce oxidative cell damage in human liver cells through inhibition of reduced glutathione and induction of mitochondria-involved apoptosis. *Toxicology Letters*, **201**(1): 92–100.

[45] Brooks RR, Robinson BH (1998) The potential use of hyperaccumulators and other plants for phytomining. In *Plants That Hyperaccumulate Heavy Metals*, RR Brooks (ed.) (CAB International, New York) pp. 327–356.

[46] Purcell TW, Peters JJ (1998) Sources of silver in the environment. *Environmental Toxicology and Chemistry*, **17**: 539–546.

[47] Sanudo-Willhelmy S, Flegal R (1992) Anthropogenic silver in the Southern California bight: A new tracer of sewage in coastal waters. *Environmental Science and Technology*, **26**: 2147–2151.

[48] Ratte HT (1999) Bioaccumulation and toxicity of silver compounds: A review. *Environmental Toxicology and Chemistry*, **18**: 89–108.

[49] Drake PL, Hazelwood KJ (2005) Exposure-related health effects of silver and silver compounds: A review. *Annals of Occupational Hygiene*, **49**(7): 575–585.

[50] Park YJ, Fray DJ (2009) Recovery of high purity precious metals from printed circuit boards. *Journal of Hazardous Materials*, **164**: 1152–1158.

[51] Ali B, Mohammad MS, Francesco V (2013) Process development for recovery of copper and precious metals from waste printed circuit boards with emphasize on palladium and gold leaching and precipitation. *Waste Management*, **33**: 2354–2363.

[52] Wijnhoven SWP, Peijnenburg WJGM, Herberts CA, Hagens WI *et al.* (2009) Nano-silver — A review of available data and knowledge gaps in human and environmental risk assessment. *Nanotoxicology*, **3**(2): 109–138.

[53] World Health Organization (WHO) (2002) Silver and silver compounds: Environmental aspects. Concise International Chemical Assessment Document No. 44. Geneva. Available at: http://aps.who.int/iris/10665/42553 Accessed 27 January 2017.

[54] Eisler R (1996) Silver hazards to fish, wildlife, and invertebrates: A synoptic review. Contaminant Hazard Reviews, Report #32, Pautuxent Wildlife Research Center. US Geological Survey, Reston, USA, pp. 1–63.

[55] Kennedy AJ, Hull MS, Bednar AJ, Goss JD, Gunter JC, Bouldin JL *et al.* (2010) Fractionating nanosilver: Importance for determining toxicity to aquatic test organisms. *Environmental Science & Technology*, **44**(24): 9571–9577.

[56] Varner K, Sanford J, El-Badawy A, Feldhake D, Venkatapathy, R (2010) State of the science literature review: Everything nanosilver and more. US Environmental Protection Agency, EPA/600/R-10/084, Washington DC, USA.

[57] Luoma SN, Rainbow PS (2008) *Metal Contamination in Aquatic Environments: Science and Lateral Management* (Cambridge University Press, Cambridge, UK).

[58] Bianchini A, Wood CM (2008) Does sulfide or water hardness protect against chronic silver toxicity in Daphnia magna? A critical assessment of the acute-to-chronic toxicity ratio for silver. *Ecotoxicology and Environmental Safety*, **71**: 32–40.

[59] Fung MC, Bowen DL (1996) Silver products for medical indication: Risk benefit assessment. *Clinical Toxicology*, **34**: 119–126.

[60] Perelaer J, Hendriks CE, de Laat AWM, Schubert US (2009) One-step inkjet printing of conductive silver tracks on polymer substrates. *Nanotechnology*, **20**: 165303, 1–5.

[61] Wu HP, Liu JF, Wu XJ, Ge MY, Wang YW, Zhang GQ *et al.* (2006) High conductivity of isotropic conductive adhesives filled with silver nanowires. *International Journal of Adhesives*, **26**: 617–621.

[62] Sotiriou GA, Pratsinis SE (2011) Engineering nanosilver as an antibacterial, biosensor and bioimaging material. *Current Opinion in Chemical Engineering*, **1**: 3–10.

[63] Chen X, Schluesener HJ (2008) Nanosilver: A nanoproduct in medical applications. *Toxicology Letters*, **176**: 1–12.

[64] Silver S, Phung LT, Silver G (2005) Silver as biocides in burn and wound dressings and bacterial resistance to silver compounds. *Annual Meeting of the Society for Industrial Microbiology*, Chicago, IL, USA.

[65] Jain PK, Huang XH, El-Sayed IH, El-Sayed MA (2008) Noble metals on the nanoscale: Optical and photothermal properties and some applications in imaging, sensing, biology, and medicine. *Accounts of Chemical Research*, **41**:1578–1586.

[66] Benn TM, Westerhoff P (2008) Nanoparticle silver released into water from commercially available sock fabrics. *Environmental Science and Technology*, **42**: 7025–7026.

[67] Murata T, Kanao-Koshikawa M, Takamatsu T (2005) Effects of Pb, Cu, Sb, In and Ag contamination on the proliferation of soil bacterial colonies, soil dehydrogenase activity, and phospholipid fatty acid profiles of soil microbial communities. *Water, Air & Soil Pollution*, **164**: 103–118.

[68] Myllynen P (2009) Nanotoxicology: Damaging DNA from a distance, *Nature Nanotechnology*, **4**: 795–796.

[69] Vasir JK, Labhasetwar V (2008) Quantification of the force of nanoparticle–cell membrane interactions and its influence on intracellular trafficking of nanoparticles. *Biomaterials*, **29**: 4244–4252.

[70] Liu J, Wang Z, Liu FD, Kane AB, Hurt RH (2012) Chemical transformations of nanosilver in biological environments. *ACS Nano*, **6**: 9887–9899.

[71] Lowry GV, Espinasse BP, Badireddy AR, Richardson CJ, Reinsch BC, Bryant LD *et al.* (2012) Long-term transformation and fate of manufactured Ag nanoparticles in a simulated large scale freshwater emergent wetland. *Environmental Science and Technology*, **46**: 7027–7036.

[72] Marambio-Jones C, Hoek EMV (2010) A review of the antibacterial effects of silver nanomaterials and potential implications for human health and the environment. *Journal of Nanoparticle Research*, **12**: 1531–1551.

[73] Janes N, Playle RC (1995) Modeling silver-binding to gills of rainbow trout (*Onchorrynchus mykiss*). *Environmental Toxicology and Chemistry*, **14**: 1847–1858.

[74] Bilberg K, Malte H, Wang T, Baatrup E (2010) Silver nanoparticles and silver nitrate cause respiratory stress in Eurasian perch (*Perca fluviatilis*). *Aquatic Toxicology*, **96**: 159–165.

[75] Scown T, Santos E, Johnston B, Gaiser B, Baalousha M, Mitov S *et al.* (2010) Effects of aqueous exposure to silver nanoparticles of different sizes in rainbow trout. *Toxicology Science*, **115**: 521–534.

[76] Osborne OJ, Johnston BD, Moger J, Balousha M, Lead JR, Kudoh T *et al.* (2013) Effects of particle size and coating on nanoscale Ag and $TiO_{(2)}$ exposure in zebrafish (*Danio rerio*) embryos. *Nanotoxicology*, **7**: 1315–1324.

[77] van Aerle R, Lange A, Moorhouse A, Paszkiewicz K, Ball K, Johnston BD *et al.* (2013) Molecular mechanisms of toxicity of silver nanoparticles in zebrafish embryos. *Environmental Science and Technology*, **47**: 8005–8014.

[78] Lee KJ, Nallathamby PD, Browning LM, Osgood CJ, Xu XH (2007) In vivo imaging of transport and biocompatibility of single silver nanoparticles in early development of zebrafish embryos. *ACS Nano*, **1**: 133–143.

[79] Klasen HJ (2000) Historical review of the use of silver in the treatment of burns: (I) Early uses. *Burns*, **26**: 117–130.

[80] Pruitt BA, McManus AT, Kim SH *et al.* (1998) Burn wound infections: Current status. *World Journal of Surgery*, **22**: 135–145.

[81] Wilkinson LJ, White RJ, Chipman JK (2011) Silver and nanoparticles of silver in wound dressings: A review of efficacy and safety. *Journal of Wound Care*, **20**: 543–549.

[82] Cheng H, Li Y, Huo K, Gao B, Xiong W (2014) Long-lasting in vivo and in vitro antibacterial ability of nanostructured titania coating incorporated with silver nanoparticles. *Journal of Biomedical and Materials Research A*, **102**: 3488–3499.

[83] Liao J, Anchun M, Zhu Z, Quan Y (2010) Antibacterial titanium plate deposited by silver nanoparticles exhibits cell compatibility. *International Journal of Nanomedicine*, **5**: 337–342.

[84] Taheri S, Cavallaro A, Christo SN, Smith LE, Majewski P, Barton M *et al.* (2014) Substrate independent silver nanoparticle based antibacterial coatings. *Biomaterials*, **35**: 4601–4609.

[85] Dekkers S, De Heer C, De Jong WH, Sips AJAM, Van Engelen JGM (2007) Nanomaterials in consumer products. Availability on the European market and adequacy of the regulatory framework. RIVM/SIR Advisory Report 11014.

[86] Chen X, Schluesener H (2008) Nanosilver: A nanoproduct in medical application. *Toxicology Letters*, **176**: 1–12.

[87] Gulbranson SH, Hud JA, Hansen RC (2000) Argyria following the use of dietary supplements containing colloidal silver protein. *Cutis*, **66**: 373–376.

[88] Espinal ML, Ferrando L, Jimenex DF (1996) Asymptomatic blue nevus-like macule. Diagnosis: localized argyria. *Archives of Dermatology*, **132**: 461–464.

[89] Bleehen SS, Gould DJ, Harrington CI, Durrant TE, Slater DN, Underwood JC (1981) Occupational argyria; Light and electron microsopic studies and X-ray microanalysis. *British Journal of Dermatology*, **104**(1): 19–26.

[90] Lansdown ABG (2010) A pharmacological and toxicological profile of silver as an antimicrobial agent in medical devices. *Advances in Pharmacological Sciences*, **2010**: 1–16.

[91] Moss AP, Sugar A, Hargett NA (1979) The ocular manifestations and functional effects of occupational argyrosis. *Archives of Ophthalmology*, **97**(5): 906–908.

[92] Zografos L, Uffer S, Chamot L (2003) Unilateral conjunctival-corneal argyrosis simulating conjunctival melanoma. *Archives of Ophthalmology*, **121**(10): 1483–1487.

[93] Rosenman KD, Seixas N, Jacobs I (1987) Potential nephrotoxic effects of exposure to silver. *British Journal of Industrial Medicine*, **44**(4): 267–272.

[94] U.S. Environmental Protection Agency: Integrated Risk Information System (IRIS) (1992). Environmental Criteria and Assessment Office of Environmental Assessment, Cincinnati, OH, USA.

[95] Phalen RF, Morrow PE (1973) Experimental inhalation of metallic silver. *Health Physics*, **24**: 509–518.

[96] Pifer JW, Friedlander BR, Kintz RT *et al.* (1989) Absence of toxic effects in silver reclamation workers. *Scandinavian Journal of Work, Environment and Health*, **15**: 210–221.

[97] Sung JH, Ji JH, Yoon JU *et al.* (2008) Lung function changes in Sprague-Dawley rats after prolonged inhalation exposure to silver nanoparticles. *Inhalation Toxicology*, **20**(6): 567–574.

[98] Lansdown ABG (1995) Physiological and toxicological changes in the skin resulting from the action and interaction of metal ions. *Critical Reviews in Toxicology*, **25**(5): 397–462.

[99] Hostýnek JJ, Hinz RS, Lorence CR, Price M, Guy RH (1993) Metals and the skin. *Critical Reviews in Toxicology*, **23**(2): 171–235.

[100] Hipler UC, Elsner P (eds.) (2006) *Biofunctional Textiles and the Skin* (Karger, Basel, Switzerland).

[101] Poon VK, Burd A (2004) *In vitro* cytotoxity of silver: Implication for clinical wound care. *Burns*, **30**(2): 140–147.

[102] Prabhu S, Poulose EK (2012) Silver nanoparticles: Mechanism of antimicrobial action, synthesis, medical applications, and toxicity effects. *International Nano Letters*, **2**(1): 1–10.

[103] Percival SL, Bowler PG, Dolman J (2007). Antimicrobial activity of silver-containing dressings on wound microorganisms using an in vitro biofilm model. *International Wound Journal*, **4**(2): 186–191.

[104] Hidalgo E, Bartolomé R, Barroso C, Moreno A, Domínguez C (1998) Silver nitrate: Antimicrobial activity related to cytotoxicity in cultured human fibroblasts. *Skin Pharmacology and Applied Skin Physiology*, **11**(3): 140–151.

[105] International Agency for Research on Cancer (IARC) (1980). Long-term and short-term screening assays for carcinogens: A critical appraisal. In *IARC Monographs on the Evaluation of the Carcinogenic Risk of Chemicals to Humans*, Suppl 2, pp. 1–426.

[106] U.S. Department of Health and Human Resources (2010). 12th Report on Carcinogens, National Toxicology, Research Program, Research Triangle Park, NC, USA.

[107] OECD Global Forum on Environment (2010) Focusing on Sustainable Materials Management, October 25–27, Mechelen, Belgium. Available at: http://www.oecd.org/env/waste/46132634.pdf.

[108] Cui J, Forssberg E (2003) Mechanical recycling of waste electric and electronic equipment: A review, *Journal of Hazardous Materials*, **99**: 243–263.

[109] Kunda W, Etsell TH (1985) Recovery of silver from X-ray film. In *Precious Metals*, ED Zysk, JA Bonucci (eds.) (Met-Chem Research, Boulder, CO, USA), pp. 289–304.

[110] Rawlings DE, Silver S (1995) Mining with microbes. *Bio/Technology*, **13**: 773–778.

[111] Rawlings DE, Dew D, du Plessis C (2003) Biomineralization of metal-containing ores and concentrates. *TRENDS in Biotechnology*, **21**(1): 38–44.

[112] Dickinson NM, Baker AJM, Doronila A, Laidlaw S, Reeves RD (2009) Phytoremediation of inorganics: Realism and synergies. *International Journal of Phytoremediation*, **11**: 97–114.

[113] Schnoor JL, Licht LA, McCutcheon SC, Wolfe NL, Carreira LH (1995) Phytoremediation of organic and nutrient contaminants. *Environmental Science & Technology*, **29**: 318A–323A.

[114] Van Nevel L, Mertens J, Oorts K, Verheyen K (2007) Phytoextraction of metals from soils: How far from practice? *Environmental Pollution*, **150**: 34–40.

[115] McGrath SP, Zhao FJ (2003) Phytoextraction of metals and metalloids from contaminated soils. *Current Opinion in Biotechnology*, **14**(3): 277–282.

[116] Harris AT, Bali R (2008) On the formation and extent of uptake of silver nanoparticles by live plants. *Journal of Nanoparticle Research*, **10**: 691–695.

[117] Brooks RR, Chambers MF, Nicks LJ, Robinson BH (1998) Phytomining. *Trends in Plant Science*, **3**: 359–362.

[118] Nascimento CWA, Xing B (2006) Phytoextraction: A review on enhanced metal availability and plant accumulation. *Scientia Agricola*, **63**(3): 299–311.

[119] Hernández-Allica J, Becerril JM, Garbisu C (2008) Assessment of the phytoextraction potential of high biomass crop plants. *Environmental Pollution*, **152**: 32–40.

[120] Saraswat S, Rai JPN (2009) Phytoextraction potential of six plant species grown in multimetal contaminated soil. *Chemistry and Ecology*, **25**(1): 1–11.

[121] Shabani N, Sayadi MH (2012) Evaluation of heavy metals accumulation by two emergent macrophytes from the polluted soil: An experimental study. *Environmentalist*, **32**: 91–98.

[122] Hernández-Allica J, Becerril JM, Garbisu C (2008) Assessment of the phytoextraction potential of high biomass crop plants. *Environmental Pollution*, **152**: 32–40.

[123] Pierzynski G, Kulakow P, Erickson L, Jackson L (2002) Plant system technologies for environmental management of metals in soils: Educational materials. *Journal of Natural Resources and Life Science Education*, **31**: 31–37.

[124] Blaylock MJ, Salt DE, Dushenkov S, Zakharova O, Gussman C, Kapulnik Y, Ensley BD, Raskin I (1997) Enhanced accumulation of Pb in Indian mustard by soil-applied chelating agents. *Environmental Science and Technology*, **31**: 860–865.

[125] Saifullah ME, Qadir M, Caritat P, Tack FMG, Laing GD, Zia MH (2009) EDTA–assisted Pb phytoextraction. *Chemosphere*, **74**(10): 1279–1291.

[126] Kumar PBAN, Dushenkov V, Motto H, Raskin I (1995) Phytoextraction: The use of plants to remove heavy metals from soils. *Environmental Science and Technology*, **29**(5): 1232–1238.

[127] Koontz HV, Berle KL (1980) Silver uptake, distribution, and effect on calcium, phosphorus, and sulfur uptake. *Plant Physiology*, **65**(2): 336–339.

[128] Sheoran V, Sheoran AS, Poonia P (2009) Phytomining: A review. *Minerals Engineering*, **22**(12): 1007–1019.

[129] Shah M, Fawcett D, Sharma S, Tripathy SK, Poinern GEJ (2015) Green synthesis of metallic nanoparticles via biological entities. *Materials*, **8**: 7278–7308.

[130] Fawcett D, Verduin JJ, Shah M, Sharma S, Poinern GEJ (2017) A review of current research into the biogenic synthesis of metal and metal oxide nanoparticles via marine algae and sea grasses. *Journal of Nanosciences*, **2017**: 1–15.

[131] Van der Ent A, Baker AJM, Reeves RD, Pollard AJ, Schat H (2013) Hyperaccumulators of metal and metalloid trace elements: Facts and fiction. *Plant Soil*, **362**: 319–334.

[132] Schnoor JL (2000) Phytostabilization of metals using hybrid poplar trees. In *Phytoremediation of Toxic Metals Using Plants to Clean Up the Environment*, I Raskin, BD Ensley (eds.) (John Wiley & Sons, Inc., New York), pp. 133–150.

[133] Mendez MO, Maier RM (2008) Phytoremediation of mine tailings in temperate and arid environments. *Reviews in Environmental Science and Biotechnology*, **7**: 47–59.

[134] Gericke M, Pinches A (2006) Biological synthesis of metal nanoparticles. *Hydrometallurgy*, **83**: 132–140.

[135] Iravani S (2015) Green synthesis of metal nanoparticles using plants. *Green Chemistry*, **13**: 2638–2650.

[136] Kulkarni N, Muddapur U (2014) Biosynthesis of metal nanoparticles: A review. *Journal of Nanotechnology*, **2014**: 1–8.

[137] Akhtar MS, Panwar J, Yun YS (2013) Biogenic synthesis of metallic nanoparticles by plant extracts. *ACS Sustainable Chemical Engineering*, **1**: 591–602.

[138] Thakkar KN, Mhatre SS, Parikh RY (2010) Biological synthesis of metallic nanoparticles. *Nanomedicine, Nanotechnology, Biology and Medicine*, **6**: 257–262.

[139] Prathna TC, Chandrasekaran N, Raichur AM, Mukherjee A (2011) Kinetic evolution studies of silver nanoparticles in a bio-based green synthesis process. *Colloids and Surfaces A*, **377**: 212–216.

[140] Malik P, Shankar R, Malik V, Sharma N, Mukherjee TK (2014) Green chemistry based benign routes for nanoparticle synthesis. *Journal of Nanoparticles*, **2014**: 1–14.

[141] Mittal AK, Chisti Y, Banerjee UC (2013) Synthesis of metallic nanoparticles using plants. *Biotechnology Advances*, **31**: 346–356.

[142] Jha AK, Prasad K, Kulkarni AR (2009) Plant system: Nature's nanofactory. *Colloids and Surfaces B*, **73**: 219–223.

[143] Huang J, Li Q, Sun D, Lu Y, Su Y, Yang X, Wang H, Wang Y, Shao W, He N *et al.* (2007) Biosynthesis of silver and gold nanoparticles by novel sundried *Cinnamomum camphora* leaf. *Nanotechnology*, **18**: 1–11.

[144] Chandran SP, Chaudhary M, Pasricha R, Ahmad A, Sastry M (2006) Synthesis of gold nanotriangles and silver nanoparticles using Aloe vera plant extract. *Biotechnology Progress*, **22**: 577–583.

[145] Sathishkumar M, Krishnamurthy S, Yun YS (2010) Immobilization of silver nanoparticles synthesized using the *Curcuma longa* tuber powder extract on cotton cloth for bactericidal activity. *Bioresource Technology*, **101**: 7958–7965.

[146] Armendariz V, Herrera I, Peralta-Videa JR, Jose-Yacaman M, Troiani H, Santiago P, Gardea-Torresdey JL (2004) Size controlled gold nanoparticle formation by *Avena sativa* biomass: Use of plants in nanobiotechnology. *Journal of Nanoparticle Research*, **6**: 377–382.

[147] Sathishkumar M, Sneha K, Kwak IS, Mao J, Tripathy SJ, Yun YS (2009) Phyto-crystallization of palladium through reduction process using *Cinnamon zeylanicum* bark extract. *Journal of Hazardous Materials*, **171**: 400–404.

[148] Ahamed M, Khan M, Siddiqui M *et al.* (2011) Green synthesis, characterization and evaluation of biocompatibility of silver nanoparticles. *Physica E* (*Low-dimensional Systems and Nanostructures*), **43**: 1266–1271.

[149] Naheed A, Sharma S (2012) Green synthesis of silver nanoparticles using extracts of *Ananas comosus*. *Green and Sustainable Chemistry*, **2**: 141–147.

[150] Vilchis-Nestor AR, Sánchez-Mendieta V, Camacho-López MA, Gómez-Espinosa RM, Arenas-Alatorre JA (2008) Solventless synthesis and optical properties of Au and Ag nanoparticles using *Camellia sinensis* extract. *Materials Letters*, **62**: 3103–3105.

[151] Jain D, Aima HK, Kachhwaha S, Kothari S (2009) Synthesis of plant-mediated silver nanoparticles using papaya fruit extract and evaluation of their antimicrobial activities. *Digest Journal of Nanomaterials and Biostructures*, **4**: 557–563.

[152] Mude N, Ingle A, Gade A, Rai M (2008) Synthesis of silver nanoparticles using Callus extract of *Carica papaya* — A First Report. *Journal of Plant Biochemistry and Biotechnology*, **18**(1): 83–86.

[153] Konwarha R, Gogoia B, Philip R *et al.* (2011) Biomimetic preparation of polymer- supported free radical scavenging, cytocompatible and antimicrobial 'green' silver nanoparticles using aqueous extract of *Citrus sinensis* peel. *Colloids and Surfaces B: Biointerfaces*, **84**: 338–345.

[154] Kaviya S, Santhanalakshmi J, Viswanathan B, Muthumary J, Srinivasan K (2011) Biosynthesis of silver nanoparticles using *Citrus sinensis* peel extract and its antibacterial activity. *Spectrochimica Acta* Part A: *Molecular* and Biomolecular *Spectroscopy*, **79**: 594–598.

[155] Basavegowda N, Rok Lee Y (2013) Synthesis of silver nanoparticles using *Satsuma mandarin* (Citrus unshiu) peel extract: A novel approach towards waste utilization. *Materials Letters*, **109**: 31–33.

[156] Kannan RRR, Stirk WA, Van Staden J (2013) Synthesis of silver nanoparticles using the seaweed *Codium capitatum* P.C. Silva (Chlorophyceae). *South African Journal of Botany*, **86**: 1–4.

[157] Satyavani K, Ramanathan T, Gurudeekan S (2011) Green synthesis of silver nanoparticles using stem dried callus extract of bitter apple (*Citrullus colocynthis*). *Digest Journal of Nanomaterials and Biostructures*, **6**: 1019–1024.

[158] Abubakar AS, Salisu IB, Chahal S *et al.* (2014) Biosynthesis and characterization of silver nano particles using black carrot root extract. *International Journal of Current Research and Review*, **6**(17): 5–8.

[159] Poinern GEJ, Shah M, Chapman P, Fawcett D (2013) Green biosynthesis of silver nanocubes using the leaf extracts from *Eucalyptus macrocarpa*. *Nano Bulletin*, **2**: 1–7.

[160] Philip D (2011) *Mangifera indica* leaf-assisted biosynthesis of well-dispersed silver nanoparticles. *Spectrochimica Acta Part A: Molecular and Biomolecular Spectroscopy*, **78**: 327–331.

[161] Princy KF, Gopinath KF (2015) An eco-friendly synthesis and characterization of silver nanoparticles using marine macroalga padina tetrastromatica. *International Journal of Science and Research*, **4**(6): 1050–1054.

[162] Sriram T, Pandidurai V (2014) Synthesis of silver nanoparticles from leaf extract of *Psidium guajava* and its antibacterial activity against pathogens. *International Journal of Current Microbiology and Application Sciences*, **3**(3): 146–152.

[163] Mittal AK, Kaler A, Banerjee UC (2012) Free radical scavenging and antioxidant activity of silver nanoparticles synthesized from flower extract of *Rhododendron dauricum*. *Nano Biomedicine Engineering*, **4**: 118–124.

[164] Njagi EC, Huang H, Stafford L *et al.* (2010) Biosynthesis of iron and silver nanoparticles at room temperature using aqueous Sorghum bran extracts. *Langmuir*, **27**(1): 264–271.

[165] Castro L, Blázquez ML, Muñoz JA, Gonzaález F, Ballester A (2013) Biological synthesis of metallic nanoparticles using algae. *IET Nanobiotechnology*, **7**: 109–116.

[166] Dwivedi AD, Gopal K (2010) Biosynthesis of silver and gold nanoparticles using *Chenopodium album* leaf extract. *Colloids and Surfaces A*, **369**: 27–33.

[167] Song JY, Jang HK, Kim BS (2009) Biological synthesis of gold nanoparticles using *Magnolia kobus* and *Diopyros kaki* leaf extracts. *Process Biochemistry*, **44**: 1133–1138.

[168] Simon-Deckers A, Loo S, Mayne-L'hermite M. N, Herlin-Boime N, Menguy, N, Reynaud C, Gouget B, Carrière M (2009) Size composition and shape dependent toxicological impact of metal oxide nano-particles and carbon nano-tubes toward bacteria. *Environmental Science and Technology*, **43**: 8423–8429.

[169] Brayner R, Ferrari-Iliou R, Brivois N, Djediat S, Benedetti M, Fiévet F (2006) Toxicological impact studies based on *Escherichia coli* bacteria in ultrafine ZnO nanoparticles colloidal medium. *Nano Letters*, **6**: 866–870.

[170] Mukunthan K, Balaji S (2012) Cashew apple juice (*Anacardium occidentale* L.) speeds up the synthesis of silver nanoparticles. *International Journal of Green Nanotechnology*, **4**(2): 71–79.

[171] Cincotta RP, Wisnewski J, Engelman R (2000) Human population in the biodiversity hotspots. *Nature*, **404**: 990–992.

[172] Poinern GEJ, Le XT, Fawcett D (2011) Super-hydrophobic nature of nanostructures on an indigenous Australian eucalyptus plant and its potential application. *Nanotechnology, Science and Applications*, **4**: 113–121.

[173] Shah M, Poinern GEJ, Fawcett D (2016) Biosynthesis of silver nanoparticles using indigenous *Xanthorrhoea glauca* leaf extract and their antibacterial activity against *Escherichia coli* and *Staphylococcus epidermis*. *International Journal of Research in Medical Sciences*, **4**(7): 2886–2892.

[174] Shah M, Poinern GEJ, Fawcett D (2016) Biogenic synthesis of silver nanoparticles via indigenous *Anigozanthos manglesii*, (red and green kangaroo paw) leaf extract and its potential antibacterial activity. *International Journal of Research in Medical Sciences*, **4**(8): 3427–3432.

[175] Chandan-Singh ML, Vinect-Sharma KR *et al.* (2011) A green biogenic approach for synthesis of gold and silver nanoparticles using *Zingiber officinale*. *Digest Journal of Nanomaterials and Biostructures*, **6**(2): 535–542.

[176] Shanmugavadivu M, Kuppusamy S, Ranjithkumar R (2014) Synthesis of pomegranate peel extract mediated silver nanoparticles and its antibacterial activity. *American Journal of Advanced Drug Delivery*, **2**(2): 174 –182.

[177] Zahir AA, Bagavan A, Kamaraj C, Elango G, Rahuman AA (2012) Efficacy of plant-mediated synthesized silver nanoparticles against *Sitophilus oryzae*. *Journal of Biopesticides*, **288**(Suppl. 5): 95–102.

[178] Ahmad N, Sharma S, Alam MK, Singh VN, Shamsi SF, Mehta BR, Fatma A (2010) Rapid synthesis of silver nanoparticles using dried medicinal plant of basil. *Colloids and Surfaces B: Biointerfaces*, **81**(1): 81–86.

[179] Shah M, Poinern GEJ, Sharma S, Fawcett D (2015) Biogenic synthesis of gold and silver nanoparticles using the leaf extract from *Eucalyptus macrocarpa*. *International Journal of Sciences*, **4**(6): 27–33.
[180] Tolaymat TM, El Badawy AM, Genaidy A, Scheckel KG, Luxton TP, Suidan M (2010). An evidence-based environmental perspective of manufactured silver nanoparticle in syntheses and applications: A systematic review and critical appraisal of peer-reviewed scientific papers. *Science of the Total Environment*, **408**(5): 999–1006.
[181] Shastri JP, Rupani MG, Jain RL (2012) Antimicrobial activity of nanosilver-coated socks fabrics against foot pathogens. *Journal of the Textile Institute*, **103**(11): 1234–1243.
[182] Arora M, Chan EK, Gupta S, Diwan AD (2013) Polymethylmethacrylate bone cements and additives: A review of the literature. *World Journal of Orthopedics*, **4**(2): 67–74.
[183] Kim SW, Jung JH, Lamsal K, Kim YS, Min JS, Lee YS (2012) Antifungal effects of silver nanoparticles (AgNPs) against various plant pathogenic fungi, *Mycobiology*, **40**(1): 53–58.
[184] Elechiguerra JL, Burt JL, Morones JR, Camacho-Bragado A, Gao X, Lara HH, Yacaman MJ (2005) Interaction of silver nanoparticles with HIV-1. *Journal of Nanobiotechnology*, **3**: 6.
[185] Song HY, Ko KK, Oh IH, Lee BT (2006). Fabrication of silver nanoparticles and their antimicrobial mechanisms. *European Cells and Materials*, **11**(1): 58.
[186] Rizzello L, Pompa PP (2014). Nanosilver-based antibacterial drugs and devices: Mechanisms, methodological drawbacks and guidelines, *Chemical Society Reviews*, **43**: 1501.

Chapter 5
Electrochemical Recovery of Silver from Waste Solutions

Victor Reyes-Cruz*, María Aurora Veloz Rodríguez*,‡, José Angel Cobos Murcia† and Gustavo Urbano Reyes†

**Universidad Autónoma del Estado de Hidalgo*
Área Académica de Ciencias de la Tierra y Materiales
Carretera Pachuca-Tulancingo Km 4.5 s/n
Mineral de la Reforma, Hidalgo, C.P. 42184, México

†Consejo Nacional de Ciencia y Tecnología
Departamento de Cátedras, Av. Insurgentes Sur 1582
Col. Crédito Constructor, Delegación Benito Juárez
México D.F., C.P. 03940, México

‡mveloz@uaeh.edu.mx

5.1 Introduction

Silver is a gray glossy metal that is at solid state under standard conditions, and due to its physical properties and economic value, is considered a precious metal. With regard to its applications, most of the silver obtained worldwide is used in the photography industry. Due to its unique characteristics, such as its reaction to light and reproduction of images, it is applied in the components of the photographic paper emulsions. Silver is also utilized in medicine as a disinfectant, and in the field of electronics, it serves to fabricate conduction materials. On the other hand, only one-third of the silver extracted is used to make jewelry.

In last few years, countries such as the United States, Japan, and Germany have become involved in the recovery of silver from pre-consumed products, with radiographic wastes being the main source. Silver is found in the dark areas of the radiography sheets; once their diagnosis function is accomplished, they are discarded. Sulfate solutions are also used in radiography. All together, these wastes represent 46% of the silver employed by the photographic industry, which makes silver recovery an attractive effort in order to diminish production costs and environmental contamination due to unappropriated disposition.

Due to the above-mentioned reasons, it has become a priority to enhance the development of optimal technologies for silver recovery from wastes of the radiographic industry; allowing silver recycling, in the same or in other industries, thereby propitiating the expansion of new industrial processes.

Today, electrochemical processes are the most widely adopted among the methods used to recover silver contained in radiographic wastes, by using electrochemical reactors are used due to their low operational costs. Many kinetic parameters have been studied and optimized, and the results were published [1]. Thus, in this chapter, the assessment of the merit figures (current efficiency, energy consumption, and space–time yield) is undertaken. Evaluation was carried out using a system composed of silver waste solutions (sulfate medium) and a Batch reactor or leaching solutions from solid radiographies (HNO_3 medium) and a Filter Press reactor.

5.2 Antecedents

Silver is one of the metals that have been highly prized by humans since the ancient times because it was used, jointly with other metals, to fabricate war weapons. Later it was employed to manufacture utensils and ornaments, and after that, to fabricate the first coins to commercialize several products and services. Nowadays, silver has different uses and is sold and purchased worldwide. The properties of silver such as ductility, malleability, and high electrical and thermic conductivity have allowed its use in diverse fields, including numismatics, radiography, automotive sector, and jewelry [2].

In 2015, the worldwide production of silver through mining was estimated at 686.1 million pounds [3–5], of which México, China, Peru, Australia, Russia, and Bolivia are the main producers, contributing 24%, 17.19%, 17.13%, 8.59%, 6.58%, 5.97% respectively, of the world production, while the other countries produced the remaining 20.54%.

In the last three years, the prices of the Troy oz. have maintained stability (see Table 1) in spite of the world economic crisis, due to the general

Table 1. Values of the average prices per year of the silver Troy oz [6, 7].

Annual Price (US$)				
2013	2014	2015	2016	2017
31.15	24.85	18.63	17.15	18.18

investment interest in the metal and the increase in its demand in industrial applications.

The production of silver worldwide is aimed at satisfying four main sectors: industrial, jewelry, photography/radiography, and coin minting. Among them, the industrial sector has experienced a greater increase in use of silver due to the demand from the electrical and electronics sector, since silver is used to improve the performance of technological products such as mobiles, music players, and electronic components of cars.

The photography and radiography sector is also important, thanks to the extreme precision and low cost of the images when silver is used, which is why photographic films are still employed in many fields.

One of the main drawbacks of the silver-containing products is that once they have outlived their usefulness, they are discarded. However, they are of the economic interest as they contain silver; and it is possible to recover the metal with processes such as foundry. However, in the specific case of radiography, wastes containing silver are found as solid compounds in radiographic films and as development solutions used in fixing the images.

Although today some new technologies address the problem of the treatment of wastes generated by the photography and radiography sector, they satisfy neither the requirements of maximum recovery nor the required environmental levels, which is less than 5 ppm in effluents [8, 9]. Nowadays, several countries have been involved in the recovery of silver contained in wastes generated by different sectors (scrap or secondary sources); among the main ones are the United States, Japan, India, and Germany. By 2015, these four countries recovered slightly more than 3,286 metric tons of silver, with photography and radiography wastes being the main source of production [10, 11].

However, it is important that other countries, such as México, also become involved in the development of their own technologies through the design and use of increasingly efficient and versatile process units (reactors) that allow them to recover the silver from industrial wastes, such as those generated by the radiography sector.

5.3 Typical Processes for Silver Recovery

Currently, there are some processes employed for silver recovery from the solutions used to develop photographs and radiographs; some of them are chemical or electrochemical. Otherwise, those solutions are discarded without treatment.

Among the most common and universally validated methods for silver recovery from discarded liquids are ion exchange (which occurs in metal replacement cartridges) and electrochemical recovery using process units (reactors) (see Table 2) [12–15]. The latter method is convenient and is more viable to use because of its low cost of recovery.

5.4 Electrochemical Reactors

For more than 30 years, electrochemical reactors have been part of the technological development. The variety of possible applications has resulted in very different configurations, sizes, and even materials used in their manufacture. The development of these process units [16] are mainly based on:

(1) The search for simple designs that generate low costs of operation as well as maintenance.
(2) The need for different flow velocities that allow the mass transport and surface area of the electrodes, per unit volume of the reactor.
(3) The need to design an electrochemical cell (or modify an existing one) to manipulate the type of reagent or product, ensuring its purity and conditioning, as well as the desired frequency of entering and exiting the reactor.

5.5 Batch-type Electrochemical Reactor

The Batch type is one of the most commonly used electrochemical reactors. Typically, its design is an undivided cell with a pair of two-dimensional

Table 2. Comparative technologies for silver treatment and recovery [13].

	Initial concentration of silver (mg L^{-1})	Final concentration of silver (mg L^{-1})	Type of solutions	Investing costs (USD)	Operation costs
Terminal electrolysis	2,000–12,000	50–250	Most of the solutions rich in silver	2,000–30,000 (depending of the size and equipment)	Low
Electrolysis in the line of recirculation	500–7,000	250–1,000	Fixing solutions in several processes	1,500–8,000 (depending of the size and equipment)	Low
Metallic replacement	Variable	0.5–15	Most of the solutions rich in silver	50–3,000 (needs a tank and pumping system)	High
Sulfur precipitation	Variable	0.1–1.0	Most of the solutions rich in silver	2,000–10,000	Medium
Precipitation	> 250	0.3–1.5	Most of the solutions rich in silver	3,000–75,000 (depending on the automating)	Medium
Ion interchange (clean solutions)	< 30	0.1–1.0	Clean and diluted solutions	10,000–100,000	High

(plane) electrodes, vertically submerged within a plastic tank. The cathode is usually a sheet of aluminum, stainless steel, or titanium, while the anode is typically an inert oxygen-producing material, such as lead dioxide ($Pb|PbO_2$) or any type of metal oxide. The incorporation of membranes and separators is desirable in the case where they are needed, particularly for pollution control and metal recovery [17, 18]. Generally, this type of reactor has a uniform concentration throughout the reactor volume, provided a suitable recirculation system is available.

Electrorecovery of metals such as Cu, Zn, Co, Ni, Cr, Mg, Cd, Pb, Ag, and Au can be carried out within Batch reactors. The cells can be opened at the top for easy inspection, electrode insertion, and cathode extraction. The material of construction of the tank is usually plastic, while the most recent ones are constructed from polypropylene. Batch-type reactors, with 0.5 m^2 electrodes, have allowed the treatment of solutions with high metal concentrations between 200 and 500 ppm in an economic manner [17, 19].

5.6 Filter Press Electrochemical Reactors

Filter press reactors using parallel plate electrodes are most commonly used in laboratory and industrial applications. Filter press cells have applications in organic and inorganic synthesis, fuel cells, redox batteries, and effluent treatment processes [15]. A large number of reactors are now available for various processes, which have been well characterized in the literature [20], among which are the Electrocell AB reactor, the FM01-LC reactor, and the FM21 reactor. These have been used in studies at the laboratory level, pilot-plant level, and industrial scale (mainly for the production of chlorine soda) [21].

Electrodes can be placed horizontally or vertically. The geometry of parallel plates is more popular and convenient for many of the electrochemical processes, because they have the simplicity of cell structures construction, electrode connections, and membrane seals, as well as wide variety of electrode materials and separators. Distribution of potential and current is reasonably uniform through the electrodes. The mass transport can be adjusted and controlled using a variety of turbulence promoters and/or by mechanical control of the electrolyte flow rate.

Detailed construction of the electrodes and cell structure is highly varied. Electrodes, electrolyte sections, and insulation plates (to separate the cells electrically when necessary) are sealed in a container that can hold up to 100 individual cells. The components of the cell are often mounted on a filter press. A wide range of values for the electrolyte flow can be handled, and the electrolyte feed can be connected in series or parallel. In most of the filter press reactors, the electrodes are of great length and comprise a single piece [22].

5.7 Merit Figures Assessment

In order to evaluate the behavior of filter press electrochemical reactors type ER01-FP, the merit figures are considered; within them, current efficiency (ϕ^e), energy consumption (E_s), and space–time yield (ρ_{ST}).

In practice, it is important to recognize the possibility of side reactions. This can be seen by defining a current efficiency, ϕ^e, (Eq. (1)), as the fraction of the electric charge used for the primary (desired) reaction [12].

$$\phi^e = \frac{Q}{Q_T} \tag{1}$$

where Q_T is the total electric charge and Q is the charge required for the chemical change of interest. The current efficiency will have values in the range $0 < \phi^e < 1$; it is usually expressed in percentage by multiplying the right side of the equation by 100.

An alternative equation to define the current efficiency ϕ^e (Eq. (2)) is:

$$\phi^e = \frac{W_j}{W_T} \tag{2}$$

where W_j is the metal weight, j is the deposit, and W_T is the weight of the deposit j that should be obtained by imposing a current value during certain time lapse.

Calculation of the performance and reliability will use the following equations to determine energy consumption values (E_s) (Eq. (3)) and space–time yield (ρ_{ST}) (Eq. (4)) for the silver recovery.

$$E_s = -\frac{nFE_{\mathrm{cel}}}{\phi^e} \tag{3}$$

$$\rho_{\mathrm{ST}} = \frac{\phi^e iM}{nFV_R} \tag{4}$$

where n is the electron number, F is the Faraday constant, E_{cel} is the cell potential, ϕ^e is the current efficiency, M is the atomic mass, i is the current, and V_R is the reactor volume.

5.8 Research on Silver Electrorecovery

In 1979, Nageswar *et al.* [23] carried out a study of the morphology of silver that had been electrodeposited in sulfate baths, perchlorates, and ammonium nitrates in a sulfadiazine and sulfadimidine medium. In their work, they found that the nucleation processes depended both on the concentration of the solution and on the current density [23]. In 1988, Adzic *et al.* investigated the granular growth of silver electrodeposited on a substrate of vitreous carbon present in a solution of silver nitrate, using triple-pulse potentiostatic technique to obtain a small number of nuclei (between one and ten) [25]. In 1996, Sánchez *et al.* oshowed that the adsorption of silver on a vitreous carbon electrode from a cyanide solution is presented as a quasi-reversible process and that the kinetic parameters are a function of the temperature, in the range of 10°C to 60°C, showing that the reduction of silver is faster at high temperatures [26].

In 1997, Hubin *et al.* [24] studied the process of silver nuclei formation, starting from a solution of thiosulfate prepared in laboratory. They used a rotary disc silver electrode and developed a mathematical model, in which they considered 2D, followed by 3D growth [25, 27]. In 1998, Palomar *et al.* [25] identified the different processes of silver nucleation in an ammoniacal medium. In their work, it was found that the silver can form 2D or 3D nuclei, which will depend on the composition of the solution [26]. In 1999, Ravaine *et al.* [26] carried out a systematic study of silver ion electrodeposition on the surface of a silver nitrate solution covered by dihexadecyl phosphate (DHP). In this study, two processes are present simultaneously during the formation of the silver deposit: a 2D growth along the interface and a 3D growth [27].

In 2000, Pollet *et al.* [27] studied the electrochemical recovery of silver from a sulfate solution, used in photographic processes, on a cylindrical stainless steel electrode, observing the effect of ultrasound on recovery. The studies showed that the optimal potential to remove the silver from the solution is −500 mV versus Standard Calomel Electrode

(SCE) [28]. In 2002, Reyes-Cruz *et al.* [28] carried out a study of silver and gold electrodeposition from cyanide solutions in the presence of high concentrations of copper (around 100 times higher than silver). In the study, the predominance and Pourbaix-type diagrams were drawn for gold, silver, and copper, and it was found that it is thermodynamically possible to make a selective deposit of silver and gold without depositing copper in an electrochemical reactor type FM01-LC to pilot scale [29]. In 2003, Gutiérrez-Badillo *et al.* [29] developed a process for silver recovery from radiological wastes, using different solutions to dissolve the silver of the radiological material. This study indicated that ammonia has the best properties to leach silver, showing that the parallel plate reactor allows the recovery of silver [30].

In 2005, Adani *et al.* [30] investigated the adsorption of silver from a synthetic solution that simulates the conditions of the effluents generated in the medical and photographic X-ray processes on granulated activated carbon in a batch process. They used sodium thiosulfate and ammonium as a leaching agent, showing that when activated carbon is used without any previous treatment, the recovery is low. However, with pre-treatment with sulfuric acid and nitric acid, a higher recovery of silver, 98.5% and 95% for sodium and ammonium thiosulfate, respectively, was achieved [31]. On the other hand, Zarkadas *et al.* [31] in 2005 investigated the influence of citric acid on the electrodeposition of silver from an aqueous solution of $AgNO_3$. By using a solution of $AgNO_3$ with citric acid and imposing a density of galvanostatically controlled current at 0.02 A cm^{-2}, it is possible to obtain homogeneous and uniform deposits on the surface of the high-purity silver electrode, indicating that the morphology and roughness of the deposits is independent of the composition of the solution. While using the solution of $AgNO_3$ free of citric acid, it was not possible to obtain homogeneous and compact deposits [32].

Alonso *et al.* [32] performed a kinetic study, which serves to determine that titanium is the best electrocatalytic substrate among those selected for the separation of silver from synthetic ammoniacal thiosulfate solutions in the presence of high copper concentrations, which simulate the composition of actual leaching solutions [33]. In 2009, Raju *et al.* [33] studied the recovery of silver from a waste solution in a nitric acid medium at different concentrations on a Dimensionally Stable Anode (DSA)-O2-type electrode.

The study revealed that, at a nitric acid concentration of 0.5 M (41% by volume) and a current density of 12 A dm^{-2}, gives rise to a recovery efficiency of 99% [34]. In 2014, Herver Lara performed a study of silver recovery by means of an electrolytic process using graphite and copper anodes and cathodes of stainless steel or copper in a fixing solution that is used in the photographic industry, at different voltages (2, 4, 0.8 V) at reaction times of 2, 4, 8, 16, and 24 h [35].

As can be observed, there are few studies in the literature that evaluate the behavior of electrochemical reactors, in terms of current efficiency, energy consumption, and space–time yield.

This chapter shows the results of the assessment of merit figures to electrochemical reactors used for silver recovery from effluents and solid wastes of the photography industry in two different electrolytic media.

5.9 Silver Electrorecovery from Effluents and Wastes of the Radiography Industry in a Batch-type Reactor

The electrochemical recovery process of silver present in a thiosulfate solution is carried out by imposing a controlled cell potential in the Batch reactor, which guarantees the selectivity of the process and thus the purity of the reservoir. The cell potentials applied in the A304 stainless steel electrode also come from the energetic conditions of current and potential, determined in the electrochemical studies that precede this research [1, 36].

In those studies, the controlled cell potentials on A304 stainless steel were set in the range of −0.6 to −0.9 V, for 210 min at a linear constant flow velocity of 0.18 ms^{-1}. This range of cell potential was selected under the consideration that the reduction of the medium (secondary silver sulfide reaction, Ag_2S) would not interfere with the silver selective deposition process.

Table 3 shows a concentration decrease of the electroactive species $Ag(S_2O_3)_2^{-3}$ with respect to the electrolysis time, for the five values of cell potentials imposed on the stainless steel electrode A304. It also shows that as the cell potential increases, there is a significant decrease in the $Ag(S_2O_3)_2^{-3}$ concentration. This behavior is due to the increase in the cathodic current as the cell potential increases, thereby boosting the transformation rate of the silver electroactive species.

Table 3. Reduction of $Ag(S_2O_3)_2^{-3}$ concentration.[a]

	Concentration (ppm)				
	Imposed potential, E_{imp} (V)				
Time, t (min)	−0.6 V	−0.7 V	−0.8 V	−0.85 V	−0.9 V
0	910	910	910	910	910
30	905	875	850	835	820
60	860	825	815	790	765
90	840	795	745	745	720
120	800	735	675	630	595
150	795	725	640	590	570
180	720	685	610	580	510
210	745	675	605	560	480
Recovered mass (w)/g	0.2283	0.6621	1.1699	1.3923	1.8393

Note: [a]When a constant potential cell is imposed on the A304 stainless steel electrode, between −0.6 and −0.9 V for 210 min, at a linear flow velocity of 0.18 ms^{-1}.

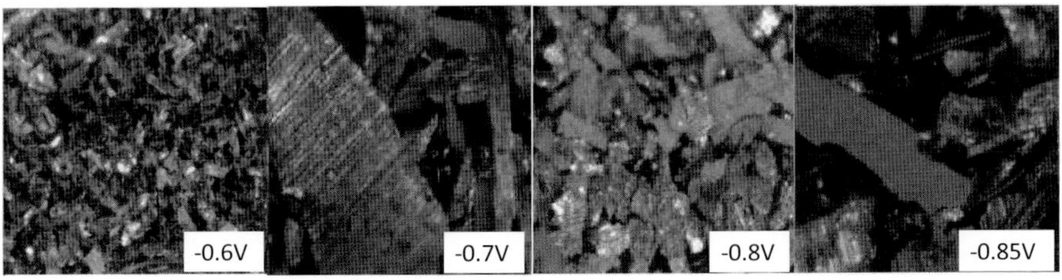

Fig. 1. Images of the mechanically withdrawn deposits from the electrode A304 , after the electrolysis studies on the cell potential range of −0.6 to −0.85 V, holding a linear flow rate of 0.18 ms^{-1} and an electrolysis time of 210 min.

The images in Fig. 1 show the microscopic characterization of the mechanically withdrawn deposits on the electrode A304, after imposing a cell potential in the range of −0.6 to −0.85 V, for 210 min at a constant linear flow velocity of 0.18 ms^{-1}. These images were captured at 200× resolution.

Figure 1 shows that above the potential −0.85 V, a dark deposit is obtained, which is characteristic and indicative that the secondary reaction that produces Ag_2S is taking place on the surface, due to the reduction of the medium.

Table 4. Merit figures assessment of controlled cell potentials for the A304 stainless steel electrode.[a]

Potential, E (V)	Current, i (A)	Theoretical mass, w_{Theor} (g)	Real mass, w_{Real} (g)	$\phi^{\mathbf{e}}$ (%)	E_{S} (kWmol^{-1})	ρ_{ST} (kgs^{-1}m^{-3})
−0.6	0.04	0.5635	0.2283	40.5	0.0397	2.59×10^{-6}
−0.7	0.08	1.1269	0.6621	58.8	0.0319	7.51×10^{-6}
−0.8	0.12	1.6904	1.1699	69.2	0.0310	1.33×10^{-5}
−0.85	0.13	1.8313	1.3923	76.0	0.0300	1.58×10^{-5}
−0.9	0.16	2.2539	1.8393	81.6	0.0296	2.09×10^{-5}

Note: [a]Using a linear flow velocity of 0.18 ms^{-1} and an electrolysis time of 210 min.

Table 4 shows the current efficiency, energy consumption, and space–time yield at constant potentials in the range of −0.6 to −0.9 V on the A304 stainless steel electrode.

In Table 4, it can be observed that as the cell potential increases, the current efficiency ϕ^e increases to 81.6%. This behavior is due to the higher cathodic current (0.16 A) that increases the cathodic potential (−0.9 V), resulting in most of the current being used for silver deposition.

On the other hand, the results also show that the energy consumption, E_{S}, decreases and the space–time yield, ρ_{ST}, increases as the cell potential is more cathodic at the A304 electrode. In addition, a higher recovered mass of silver is obtained as the cell potential increases (w_{real}). Although the values of ϕ^e, E_s, and ρ_{ST} meet the standards for a good operation of the electrochemical reactor (Batch), selective silver recovery is carried out when cell potential values imposed are less than −0.85 V. This is why it is important to know the behavior of the variation of the flow velocity of the thiosulfate solution in the recovery of silver.

Table 5 shows the assessment of the merit figures by imposing a −0.8 V cell potential on the A304 stainless steel electrode during 60 min of electrolysis, as well as the amount of recovered silver mass as the linear flow velocity increases from 0.14 to 0.37 ms^{-1}.

Table 5 shows that as the linear flow velocity increases from 0.14 to 0.18 ms^{-1}, the highest silver recovery (w_{real}) is obtained, which subsequently decreases when the velocity increases above 0.34 ms^{-1}. This behavior was attributed to the fact that at higher linear flow velocities, there

Table 5. Assessment of the merit figures and mass deposited at different linear flow velocities for the A304 stainless steel electrode.

Linear flow (ms^{-1})	Theoretical mass, w_{Theor} (g)	Real mass, w_{real} (g)	ϕ^e (%)	E_s ($kWmol^{-1}$)	ρ_{ST} ($kgs^{-1}m^{-3}$)
0.14	1.17	1.13	96.8	2.21E-02	4.48×10^{-5}
0.16	1.21	1.15	95.2	2.25E-02	4.56×10^{-5}
0.18	1.25	1.46	117.0	1.83E-02	5.79×10^{-5}
0.34	0.27	0.23	84.7	2.53E-02	3.65×10^{-5}
0.37	0.24	0.26	107.7	1.99E-02	4.13×10^{-5}

Note: [a]At −0.8 V, with area of 145 cm^2 and a volume of the solution of 7 L.

is a shorter residence time of the electroactive species ($Ag(S_2O_3)_2^{-3}$) on the surface of the A304 stainless steel electrode, resulting in a lower quantity of silver deposited.

On the other hand, Table 5 also shows that the current efficiency ϕ^e and the space–time performance ρ_{ST} increased as the linear flow velocity increased to 0.18 ms^{-1}, while the energy consumption decreased as the linear flow velocity increased in the same range (from 0.14 to 0.18 ms^{-1}). At linear flow velocity of 0.34 ms^{-1}, the energy consumption increased, while the current efficiency and the space–time yield decreased. Meanwhile, at a flow rate of 0.37 ms^{-1}, power consumption decreased, while current efficiency and space–time yield again increased. This behavior indicates that at higher linear flow velocities (above 0.18 ms^{-1}), the secondary reaction of Ag_2S is carried out.

The results also showed that by imposing a −0.8 V cell potential on the A304 stainless steel electrode for reaction times greater than 60 min, dark deposits occur due to the secondary reaction of Ag_2S (see Fig. 2).

Accordingly, it was decided to use a 0.6 V cell potential at a linear flow velocity of 0.31 ms^{-1} to obtain the silver selective deposit using a 20 L volume of the thiosulfate solution in the Batch reactor.

Table 6 shows the mass variation and current efficiency ϕ^e when a cell potential of −0.6 V is imposed on the A304 stainless steel electrode. It shows that as the electrolysis time increases, the actual mass recovered from silver is increased. It also demonstrates that the current efficiency is

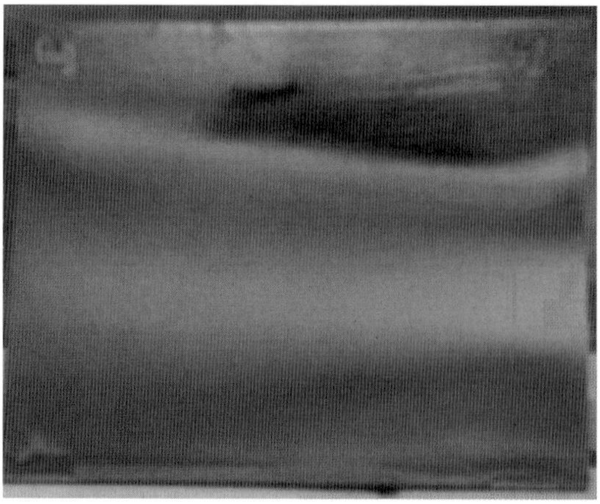

Fig. 2. Deposit obtained on the A304 electrode after the electrolysis studies using the cell potential of −0.8 V, maintaining a linear flow velocity of 0.31 ms^{-1} and 210 min of electrolysis time.

Table 6. Variation of mass and current efficiency ϕ^e respect to the electrolysis time on the stainless steel A304 electrode.[a]

Time, t (h)	Potential, E (V)	Current, i (A)	Theoretical mass, w_{Theor} (g)	Real mass, w_{Real} (g)	ϕ^e (%)
0	−0.6	0.28	0	0	0
8	−0.6	0.282	9.0799	9.0846	100
16	−0.6	0.33	21.2508	21.5709	102
24	−0.6	0.33	31.8762	31.9688	100
32	−0.6	0.32	41.2137	41.4721	101
40	−0.6	0.3	48.2973	49.9316	103
48	−0.6	0.29	56.0249	56.1813	100
56	−0.6	0.27	60.8546	60.9829	100

Note: [a]Area of 145 cm^2, at −0.6 V and a linear flow velocity of 0.31 ms^{-1}.

100% throughout the silver recovery process, indicating that all the current that passes through the Batch reactor is being used for silver deposition. It is important to mention that current efficiency is being calculated from the mass values obtained theoretical and experimentally. That is, the current applied should transform the theoretical mass, however, the actual mass obtained is more than expected; which is attributed to an electroless process.

Finally, after 56 h, the selective recovery of all the silver present in the thiosulfate solution is achieved. Figure 3 shows the images of the selective deposit.

Figure 4 presents the micrographs obtained by scanning electron microscopy (SEM) and the Energy-Dispersive X-ray Spectroscopy (EDS) spectra of the deposits removed from the surface of the A304 stainless steel

Fig. 3. Deposits on A304 stainless steel plates and the morphology obtained of the selective silver deposit on the surface using scanning electron microscopy.

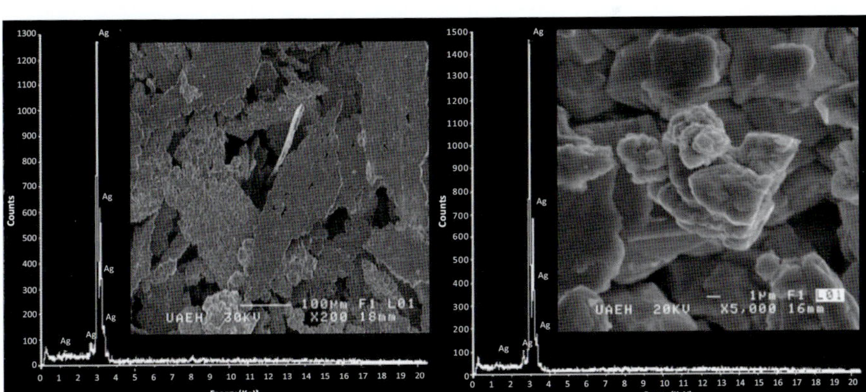

Fig. 4. Micrographs and spectra obtained by SEM and EDS of the deposits removed from the surface of A304 stainless steel electrode after performing the electrolysis study by imposing a −0.6 V cell potential for 120 min and a linear flow velocity of $0.31\ ms^{-1}$. The micrographs were obtained using secondary electrons at 20 and 30 kV.

electrode when a cell potential of −0.6 V with a linear flow velocity of 0.31 ms^{-1} and an electrolysis time of 480 min are used. In the image, it can be observed that the deposits accumulated have a practically laminar-type morphology, which are similar in all the tests performed. On the other hand, the studies of EDS performed on each deposit obtained revealed majority peaks of silver, which demonstrates that the recovery of silver is being carried out properly on the surface of stainless steel A304 electrode.

The results of this section indicate that it is possible to completely recover silver from the thiosulfate solution, by selectively applying a −0.6 V cell potential on a stainless steel A304 electrode and a linear flow velocity of 0.31 ms^{-1}, in a Batch reactor. This also allows complying with the amount of silver (5 ppm) that must be present in wastes of this type, indicated in the environmental regulations. This research also contributes to the development of technologies for silver recovery and has an impact on the economy of silver-producing countries.

5.10 Silver Electrorecovery from Solid Wastes of the Radiography Industry in a Press-type Reactor

In this part of the study, an electrochemical reactor type filter press ER01-FP was used in a vertical position, with a three-electrode system (work, auxiliary, and reference), without an anionic membrane between the cathodic and anodic compartments. A titanium electrode was used as the working electrode, a DSA electrode as the auxiliary electrode, and a saturated calomel electrode (ECS) as the reference electrode.

The working solution used in the electrochemical studies was prepared in the laboratory with 5% v/v nitric acid and deionized water (resistivity of 18 mΩ) to dissolve the radiographic plate, denominated in the text as DSFR, giving a silver concentration of 2,088 ppm.

A chronopotentiometric study was performed by imposing a constant current greater and equal to −150 mA, which was determined in a study prior to this investigation. This is favorable because the industrial level seeks to minimize production costs by imposing currents instead of potentials in the deposit processes. This study allowed the characterization of the reactor ER01-FP in an upright position.

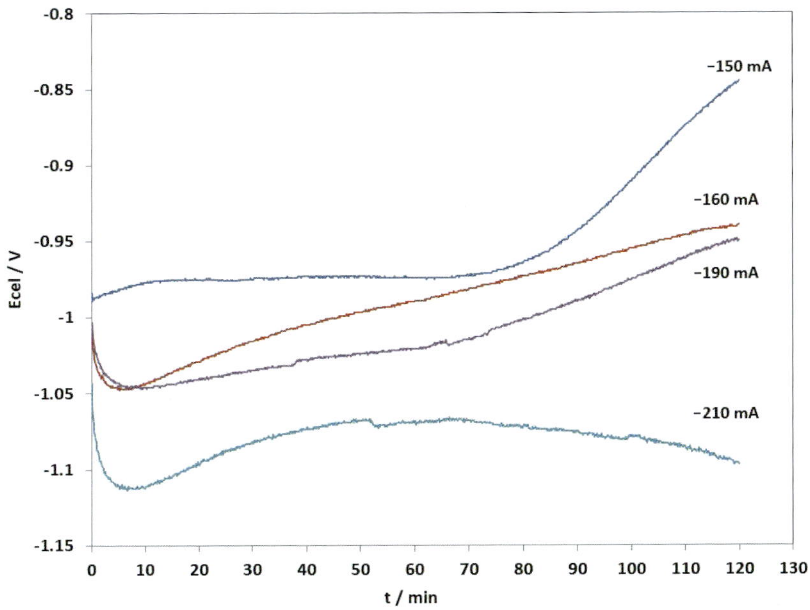

Fig. 5. Chronopotentiograms of the silver reduction from DSFR solution on the titanium electrode (area: 64.3 cm^2, vertical) in the current range of −150 to −210 mA, maintaining a constant flow rate of 15 $L min^{-1}$.

Figure 5 shows the galvanostatic transients obtained by imposing a constant current in the range of −150 to −210 mA on the Ti electrode in a vertical position, for 120 min of electrolysis, at a flow rate of 15 $L min^{-1}$.

Figure 5 shows that the potentials, as expected, tend to be more cathodic as the cathodic current increases. Also, it can be seen that, as the electrolysis time elapses, there is a decrease in the cathodic potential for currents of −150, −160, and −190 mA, while for the −210 mA current, the potential becomes more cathodic when the electrolysis time elapses. This behavior indicates that the silver deposit can be carried out in the range between −150 and −190 mA, while at a current of −210 mA, the hydrogen evolution reaction is also coupled.

Figure 6 shows the response of the variation of the silver concentration with respect to the initial concentration, $C(t)/C(0)$, as a function of the electrolysis time in the current range of −150 to −210 mA, on the titanium electrode in an upright position, maintaining a constant flow rate of 15 $L min^{-1}$ for 120 min.

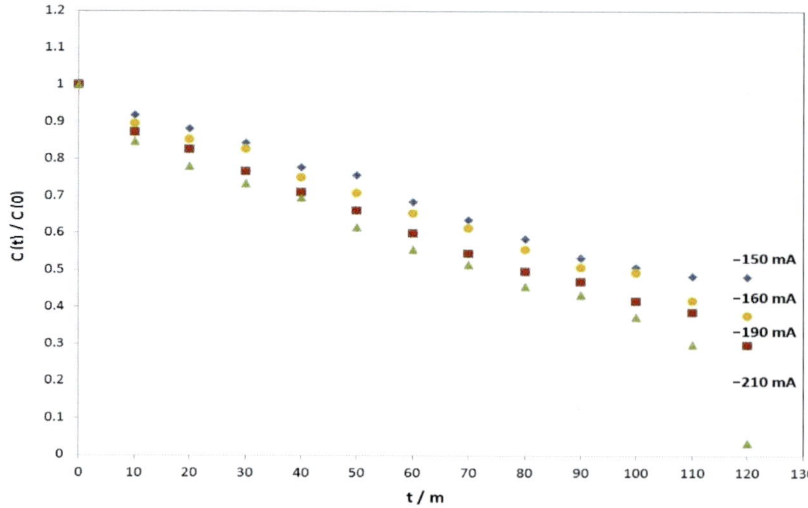

Fig. 6. Variation of the normalized concentration of the silver ion species present in the DSFR solution as a function of the electrolysis time, when a controlled current in the range of -150 to -210 mA is imposed on a titanium electrode (area: 64.3 cm^2, vertical position) at a velocity of 15 $Lmin^{-1}$.

Figure 6 shows that, in the whole range of studied currents, the variation of the normalized concentration of silver decreases with respect to electrolysis time. In addition, by imposing more cathodic currents, there is a greater decrease of the silver ion species in the DSFR solution, indicating that a massive deposit of silver is occurring on the surface of the titanium electrode.

It is also important to mention that the percentage of silver recovery achieved at 120 min of electrolysis, when the ER01-FP reactor operates horizontally was 70% [37], while for the vertical operation it was 50% [38]. The decrease in the recovery percentage of silver was attributed to the fact that the electroless process that occurs when the ER01-FP reactor operates horizontally does not occur in the same way in a vertical position, due to a different nucleation process of the silver deposits in the Ti electrode and the different hydrodynamic conditions.

In order to identify the real nature of the deposits obtained on the surface of the Ti electrode when the ER01-FP reactor operates in an upright position, SEM and SEM/EDS were performed on the deposits. Before carrying out

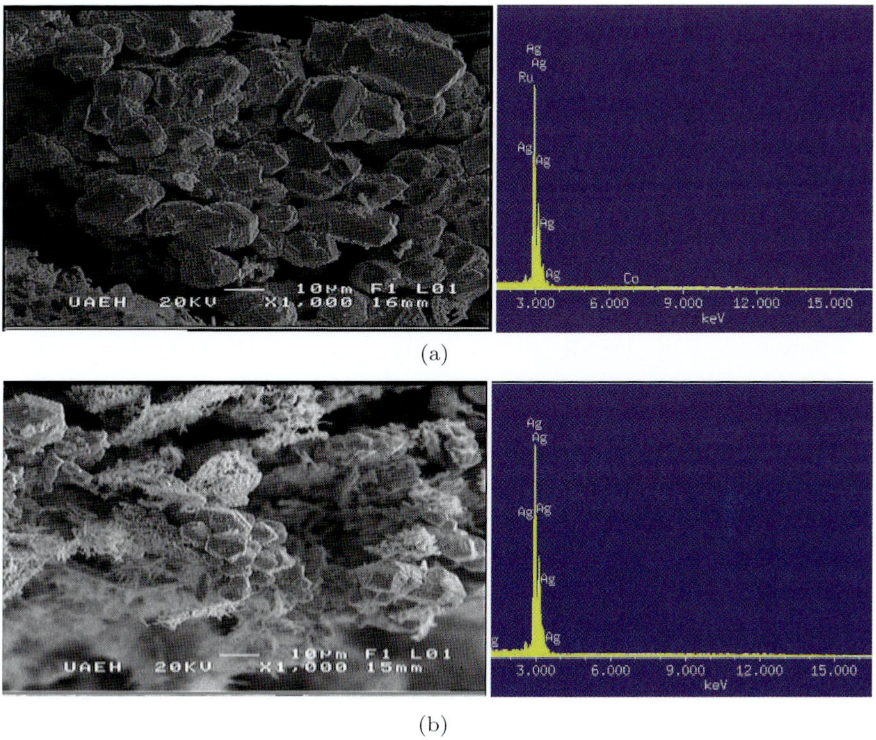

Fig. 7. SEM/EDS images obtained on deposits removed from the surface of the Ti electrode when operated vertically, after performing the electrolysis study on the DSFR solution, by imposing a current of: (a) −150 mA and (b) −210 mA. The micrographs were obtained using secondary electrons at 20 kV.

such studies, the deposits were mechanically removed from the surface of the Ti electrode with the help of a Teflon spatula.

Figure 7 shows two micrographs and the SEM/EDS spectra of the deposits obtained on the titanium electrode when the ER01-FP reactor operates upright, for two imposed currents of −150 and −210 mA, at 120 min of electrolysis. Also, it is observed that the massive deposition on the Ti electrode surface, when the ER01-FP reactor operates vertically, can present two types of morphology: a granular morphology in a lower current (−150 mA; Fig. 7(a)) and a more dendritic growth when the cathodic current is −210 mA (Fig. 7(b)). This behavior is attributed to the fact that the silver nucleation on the Ti electrode is carried out in a 3D form by the coupling of the hydrogen evolution reaction that modifies the hydrodynamic conditions on the electrode surface.

The results obtained by EDS in each of the deposits presented majority peaks of silver, indicating that the deposits obtained on the surface of the Ti electrode are due only to the silver recovered from the DSFR solution.

On the other hand, Fig. 8 shows the galvanostatic transients obtained by imposing a constant current of $-150\,\text{mA}$ on the titanium electrode in a vertical position, by varying the flow rate in the range of $10\text{–}20\,\text{Lmin}^{-1}$. It can be observed that the potential values are more negative by increasing the flow rate of the DSFR solution in the range of $10\text{–}17\,\text{Lmin}^{-1}$, while for $20\,\text{Lmin}^{-1}$ the potential becomes more positive. On the other hand, it is observed that, in each of the transients, the potential takes less negative values as the electrolysis time increases. This behavior indicates that silver nucleation is being performed on the silver already deposited on the surface of the Ti electrode, which leads to a lower energy demand to carry out the silver deposit.

Figure 9 shows the variation of the normalized concentration of silver with respect to the initial concentration, $C(\text{t})/C(0)$, as a function of the

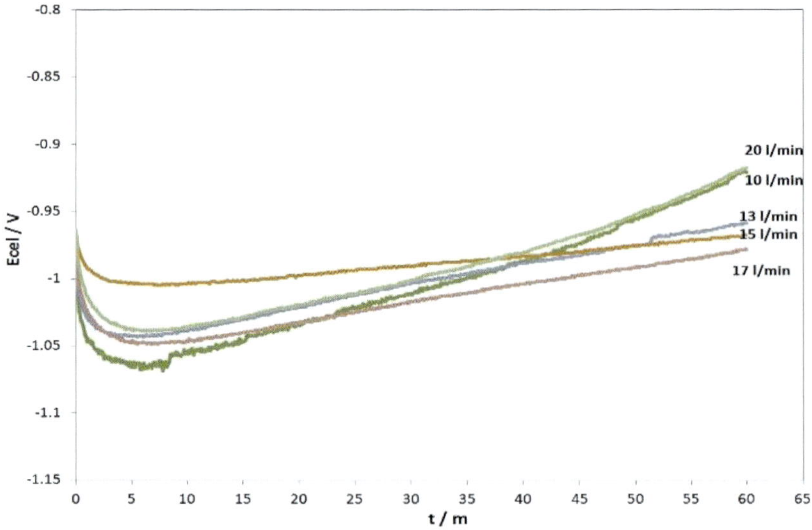

Fig. 8. Chronopotentiograms of the silver reduction from the DSFR solution on the Ti electrode (area: $64.3\,\text{cm}^2$, in vertical position), when a constant current of $-150\,\text{mA}$ is imposed and the flow rate is in the range of $10\text{–}20\,\text{Lmin}^{-1}$.

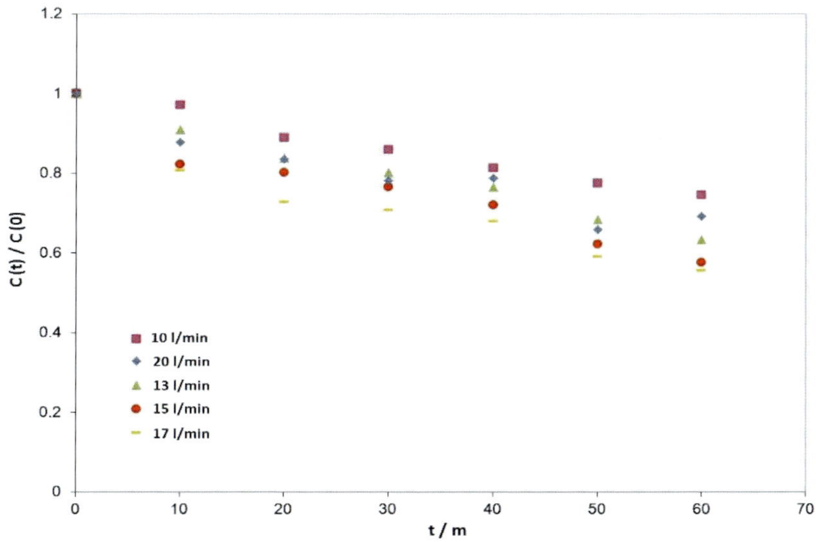

Fig. 9. Variation of the normalized concentration of the silver ion species present in the DSFR solution as a function of the electrolysis time, when a constant current of -150 mA is imposed on the Ti electrode (area: $64.3\,cm^2$, in vertical position) varying the flow rate of the DSFR solution in the range of 10–20 $Lmin^{-1}$.

electrolysis time, when the flow rate of the DSFR solution is varied in the range of 10–20 $Lmin^{-1}$ and a constant current of -150 mA is imposed on the Ti electrode in an upright position. It is observed that, in the range of flow rates from 10 to 17 $Lmin^{-1}$, there is a decrease in the variation of the normalized concentration of silver with respect to the electrolysis time. As the flow velocity increases in this range, there is a greater decrease in the concentration of the silver ion species in the DSFR solution, indicating that there is a greater recovery of silver on the surface of the Ti electrode. On the other hand, at a flow velocity of 20 $Lmin^{-1}$, the variation of the normalized concentration of silver with respect to the time of electrolysis is smaller than that of 17 $Lmin^{-1}$ at all electrolysis times. This behavior is attributed to the fact that the evolution of hydrogen is being carried out in addition to the silver deposit.

Electrochemical studies using chronopotentiometries in the ER01-FP reactor in a vertical position showed that at an electrolysis time of 60 min and a flow rate of 10 $Lmin^{-1}$, the recovery of silver was 25.4%, and that as the flow rate increases, a greater recovery occurs, since a recovery of

44% was obtained at 17 Lmin^{-1}. When comparing the variation of the silver concentration while the ER01-FP reactor operates horizontally at 60 min of electrolysis and flow rate of 15 Lmin^{-1} (35%, research prior to this work [9]) a greater recovery is observed by operating the ER01-FP reactor vertically (42%). This indicates that there is a greater recovery of silver.

The evaluation of the parameters of current efficiency (ϕ^e), space–time yield (ρ_{ST}), and energy consumption (E_s) associated with the reactor ER01-FP in a vertical position and with different cathodic currents imposed on the Ti electrode and flow rates are shown in Tables 7–10.

In Table 7, it can be observed that as the current takes more cathodic values, the amount of real mass and the current efficiency increase. For a current of −210 mA, there is a silver recovery of 1.62 g and a current efficiency of 95%.

The current efficiencies above 100% observed in the work prior to this investigation [39] do not occur when the reactor ER01-FP operates in an

Table 7. Current efficiencies respect to the weight obtained at different current values for an electrolysis time of 120 min.

Current, i (A)	Theoretical mass, w_{Theor} (g)	Real mass, w_{real} (g)	ϕ^{e} (%)
−150	1.20	1.08	90
−160	1.29	1.20	93
−190	1.53	1.44	94
−210	1.70	1.62	95

Table 8. Energy consumption E_s and space–time yield ρ_{ST} for the silver obtained in the ER01-FP reactor in vertical position.[a]

Current, i (A)	Potential, E (V)	E_{s} (kWhkg^{-1})	ρ_{ST} (%)
−150	−0.974	0.2361	6.0×10^{-5}
−160	−0.99	0.2507	6.7×10^{-5}
−190	−1.02	0.2507	8×10^{-5}
−210	−1.07	0.2852	9×10^{-5}

Note: [a]When a current is imposed in the intervals of −150 to −210 mA on Ti electrode for 120 min of electrolysis, in a reactor volume of 2.5 L

Table 9. Current efficiencies with respect to the weight of silver obtained at different flow rates.[a]

Linear flow (ms^{-1})	Theoretical mass, w_{Theor} (g)	Real mass, w_{real} (g)	ϕ^e (%)
10	0.604	0.555	92
15	0.604	0.590	97
17	0.604	0.604	100
20	0.604	0.53	88

Note: [a]For a 60 min electrolysis time and −150 mA current.

Table 10. Energy consumption E_s and space–time yield ρ_{ST}.

Linear flow (ms^{-1})	Potential, E (V)	E_s ($kWhkg^{-1}$)	ρ_{ST} (%)
10	−0.92	0.2486	6.17×10^{-5}
15	−0.959	0.2438	6.55×10^{-5}
17	−0.979	0.2432	6.70×10^{-5}
20	−0.918	0.2598	5.90×10^{-5}

upright position, which confirms that the recovery of silver was through an electrochemical process and not through an electroless process.

Table 8 shows that the energy consumption E_s and the space–time yield ρ_{ST} increase as the current acquires more cathodic values at the Ti electrode. The increase in energy consumption, E_s, is due to the increase in cell potential, while the increase in space–time yield ρ_{ST} is due to a higher reduction current and current efficiency of the silver recovery process.

On the other hand, in Table 9, it is observed that as the flow rate increases in the range from 10 to 17 $Lmin^{-1}$, the actual mass and the current efficiency increase. That is, for a flow rate of 17 $Lmin^{-1}$, there is a maximum silver recovery of 0.604 g with a current efficiency of 100%, while for a flow velocity of 20 $Lmin^{-1}$, there is a lower recovered silver mass (0.53 g) and lower current efficiency (88%). The mass decrease and lower current efficiency is attributed to the fact that the flow is more turbulent and part of the current is consumed in secondary reactions such as hydrogen evolution.

Table 10 shows that the energy consumption is decreased as the flow rate increases in the range from 10 to 17 $Lmin^{-1}$; this decrease is due to the increase in current efficiency. While the flow rate of 20 $Lmin^{-1}$ increases the energy consumption E_s, the increase is due to the decrease in cell potential and current efficiency.

On the other hand, it is observed that the space–time yield ρ_{ST}, also increases in the interval of 10–17 $Lmin^{-1}$, which is due to the increase in current efficiencies, while for the flow velocity of 20 $Lmin^{-1}$, the space–time yield ρ_{ST} decreases due to the decrease in current efficiency at this flow velocity.

The obtained results of energy consumption E_s and of space–time yield ρ_{ST}, for the Ti electrode, indicate that the ER01-FP reactor in the vertical position is the one that presents the best behavior for the recovery of silver contained in solid wastes from the photography and radiography industry. This is due to the fact that, under the same operational conditions (−150 mA, 15 $Lmin^{-1}$, and 120 min), it presented lower energy consumption (0.2361 $kWhkg^{-1}$), higher space–time yield (ρ_{ST} 6×10^{-5} $kgs^{-1}m^{-3}$), and greater recovery of silver (1.08 g) compared to the ER01-FP reactor in the horizontal position, which presented $E_s = 0.4402$ $kWhkg^{-1}$, $\rho_{ST} = 4.7 \times 10^{-5}$ $kgs^{-1}m^{-3}$, and 0.845 g of silver.

As can be observed, the electrochemical recovery of silver from waste solutions is possible using electrochemical reactors. To achieve good results, it is important to take into account the chemical composition of the solutions, purity of the electrodes used, and the disposition of the flow through the reactors. Studies on this type of parameters are of the main interest, since they provide high energy savings, require low consumption of different reactants, and can serve as an environmental-friendly way of silver recovery.

References

[1] El Financiero (2014) México consolida el primer lugar en producción de plata. Available at: http://www.elfinanciero.com.mx/mercados/commodities/mexico-consolida-el-primer-lugar-en-produccion-de-plata.html. Accessed 1 May 2017.

[2] Oro y Finanzas (2016) Los 10 principales países mineros de plata y las fuentes de demanda en 2015. Available at: https://www.oroyfinanzas.com/2016/06/10-principales-paises-mineros-plata-fuentes-demanda-2015/. Accessed 1 May 2017.

[3] Diario de Yucatán en Línea (2016) México rompe record mundial en producción de planta en 2015. Available at: http://yucatan.com.mx/economia/industria/mexico-rompe-record-mundial-en-produccion-de-plata-en-2015. Accessed 1 May 2017.
[4] Sociedad Nacional de Minería Petróleo y Energía (2017). Available at: www.snmpe.org.pe. Accessed 1 May 2017.
[5] Cámara Minera de México (2017) La industria minera de México, retos y oportunidades. Cámara Minera de México. Available at: www.cammimex.org.mx/informe/laindustriamineramex.pdf. Accessed 1 May 2017.
[6] CAMIMEX (2015) Informe anual 2015, del grupo de metales no ferrosos, excluyendo al cobre. Available at: www.camimex.org.mx/directorio/gproduct.html. Accessed 1 May 2017.
[7] U.S. Geological Survey (2017) Silver. Available at: https://www.usgs.gov/. Accessed 1 May 2017.
[8] CAMIMEX (2016) Boletín mensual Marzo. Available at: http://www.camimex.org.mx/boletin/mensuales/marzo16.pdf. Accessed 1 May 2017.
[9] Ramírez-Ortega PA (2010) Recuperación de Ag en un reactor electroquímico tipo prensa (ER01-FP) proveniente de los desechos sólidos de la industria fotográfica y radiográfica. Doctoral thesis, Universidad Autónoma del Estado de Hidalgo, Hidalgo, México.
[10] El Economista (2016) Producción de oro, cobre y plata suben en el 2008. Available at: http://eleconomista.com.mx/industrias/2016/06/30/produccion-cobre-mexico-sube-51-abril-plata-oro-caen. Accessed 1 May 2017.
[11] Oropeza MT, Ponce de León C and González I (Eds.) (1995) *Ingeniería Electroquímica Principios y Aplicaciones* (Sociedad Mexicana de Electroquímica, México, D. F).
[12] Raju T, Joon SI, Shik M (2009) Electrochemical recovery of silver from aqueous Ag (I)|Ag (II) redox mediator solution used in mediated electro oxidation process. *Korean Journal of Chemical Engineering*, **26**(4): 1053–1057.
[13] Environment Information from Kodak (1999) *The Technology of Silver Recovery for Photographic Processing Facilities- Referencia J-212* (Eastman Kodak Company) at http://www.kodak.info/eknec/documents/f7/0900688a80f8f0f7/J212ENG-0311.pdf.
[14] Oropeza MT (1991) Alternativas a la recuperación no contaminante de plata: Disolución y formación de electrodepósitos (electrocristalización), Master Thesis, UAM-I, México, D.F.
[15] Pletcher D, Walsh FC (1990) *Industrial Electrochemistry*, 2nd edn. (Chapman & Hall, London).

[16] Coeuret F (1992) *Introducción a la ingeniería electroquímica*, Edit. (Reverté, Barcelona, España).

[17] Pletcher D (1991) *A First Course in Electrode Processes* (RSC Publishing, UK).

[18] Newman J, Tiedemann W (1978) Flow-through porous electrodes. *Advances in Electrochemistry and Electrochemical Engineering*, **11**: 353.

[19] Gunter K (1981) Silver recycling from photographic solutions: Review of the methods. *Chemical Labor Betr.*, **32**(2): 40–48.

[20] Chikhi M, Rakib M, Viers Ph, Laborie P, Hita A, Durand G (2002) Current distribution in a chlor-alkali membrane cell experimental study and modeling. *Desalination*, **149**: 375–381.

[21] Walsh FC (1993) *A first course in electrochemical engineering*, Electrochemical Consultancy, Hants, England, 381 pp.

[22] Griffiths M, Ponce de León C, Walsh FC (2005) Mass transport in the rectangular channel of a filter-press electrolyzer. *AIChE Journal*, **51**: 682.

[23] Nageswar S (1979) Electrodeposition of silver on silver single crystals. *Materials Chemistry*, **4**(2): 175–188.

[24] Hubin A, Simons W, Gonissen D (1997) Study of the initial stages of silver electrocrystallisation from silver thiosulphate complex. Part II: Analysis of current transients. *J. Electroanal. Chem.*, **435**(1–2): 149–155.

[25] Palomar M (1998) *Influencia de las propiedades fisicoquímicas sobre el proceso de electronucleación de metales sobre sustratos de diferente naturaleza (superficies mono y policristalinas). Desarrollo de modelos teóricos que describen la formación y crecimiento de núcleos*. Doctoral Thesis, Universidad Autónoma Metropolitana-Iztapalapa, México D.F.

[26] Ravaine S, Breton C, Mingotaud C, Argoul F (1999) Electrodeposition of two-dimensional silver under dihexadecyl phosphate monolayers. *Materials Science & Engineering*, **8–9**: 589.

[27] Pollet B, Lorimer JP, Phull SS, Hihn JY (2000) Sonoelectrochemical recovery of silver from photographic processing solution. *Ultrasonic Sonochemistry*, **7**: 69–79.

[28] Reyes-Cruz VE, Ponce de León C, González I, Oropeza MT (2002) Electrochemical deposition of silver and gold from cyanide leaching solution. *Journal of Hydrometallurgy*, **65**(2):187–203.

[29] Gutiérrez-Badillo J (2003) *Recuperación de plata*. Bachelor Thesis, Universidad Autónoma Metropolitana, México, D.F.

[30] Adani KG, Barley RW, Pascoes RD (2005) Silver recovery from synthetic photographic and medical X-Ray process effluents using activated carbon. *Journal of Minerals Engineering*, **18**: 1269–1276.

[31] Zarkadas GM, Stergiou A, Papanastasiou G (2005) Influence of citric acid on the silver electrodeposition from aqueous $AgNO_3$. *Journal of Electrochimica*, **50**(25–26): 5022–5031.

[32] Alonso A (2007) *Electroseparación selectiva de plata a partir de soluciones amoniacales de tiosulfato*. Doctoral Thesis, UAM-I, México, D.F.

[33] Raju T, Joon S, Shik IM (2009) Elecetrochemical recovery of silver from aqueous Ag(I)/Ag(II) redox mediator solution used in mediated electro oxidation process. *Korean Journal of Chemical Engineering*, **26**(4): 1053–1057.

[34] González I, Nava Montes de Oca JL, Benavides R, Reséndiz A, Butrón EJ, Lara-Valenzuela C (2009) Universidad Autónoma Metropolitana and Servicios Industriales Peñoles SA de CV. MX2009/000062 and WO/2010/002235.

[35] Matías-Hernández DJ (2012) *Evaluación de figuras de mérito y estudio económico de la electrorecuperación de Ag en un reactor tipo Batch*. Master Thesis, Universidad Autónoma del Estado de Hidalgo, México.

[36] Palacios-Reséndiz JA (2005) *Estudio electroquímico del depósito de Ag proveniente de los desechos de la industria fotográfica y radiográfica*. Bachelor Thesis, Universidad Autónoma del Estado de Hidalgo, México.

[37] Ramírez-Ortega PA (2006) *Estudio electroquímico preliminar en un reactor tipo prensa para la recuperación de Ag proveniente de los desechos sólidos de la industria fotográfica y radiográfica*. Bachelor Thesis, Universidad Autónoma del Estado de Hidalgo, México.

[38] Rodríguez-González I (2017) *Recuperación de plata proveniente de radiografías en un reactor electroquímico tipo prensa en posición vertical*. Bachelor Thesis, Universidad Autónoma del Estado de Hidalgo, México.

[39] Picket DJ (1979) *Electrochemical Reactor Design*, 2nd edn. (Elsevier, Amsterdam).

[40] Adzic GD, Drazic DM and Despic AR (1988) *A method of obtaining a small number of nucleation sites and growing large cristal grains*. J. Electroanal. Chem., **239**: 107–111.

[41] Sánchez H, Chainet E, Nguyen B, Ozil P and Meas Y (1996) *Electrochemical deposition of silver from a low cyanide concentration bath*; J. Electrochem. Soc. **143**(9): 2799-–2815.

[42] Herver-Lara, C. G.(2014); "Recuperación de plata de líquido fijador cansado por medio de un proceso electrolítico"; Bachelor Thesis; Universidad Veracruzana; http://cdigital.uv.mx/handle/123456789/38338.

Chapter 6
Recovery of Silver from Industrial Wastes: Strategies and Technologies

M. Chakankar*, **U. Jadhav**† **and H. Hocheng***,‡

*Department of Power Mechanical Engineering
National Tsing Hua University, No. 101, Sector 2
Kuang Fu Road, 30013, Hsinchu, Taiwan ROC

†Department of Microbiology
Savitribai Phule Pune University, Pune, India
‡hochengh@gmail.com

6.1 Introduction

The progression of industrialization has resulted in an ever-increasing demand for precious metals; however, natural reserves of precious metals are diminishing. Precious metals can be successfully recovered from different waste materials, which act as secondary sources of raw material [1, 2]. Rigorous environmental protection legislation has increased the need to develop more efficient methods for metal recovery. Precious metals have extensive applications. Various secondary sources of precious metals include jewelry and electronics for gold; jewelry, X-ray film, and photographic emulsions, and industrial materials such as catalysts, batteries, and glass for silver; and catalytic converters, industrial catalysts, e-waste, etc. for platinum-group metals [3].

Silver is one of the most widely used metals in industry. Nearly 40–50% of silver is used in the film processing industry, followed by 20–30%

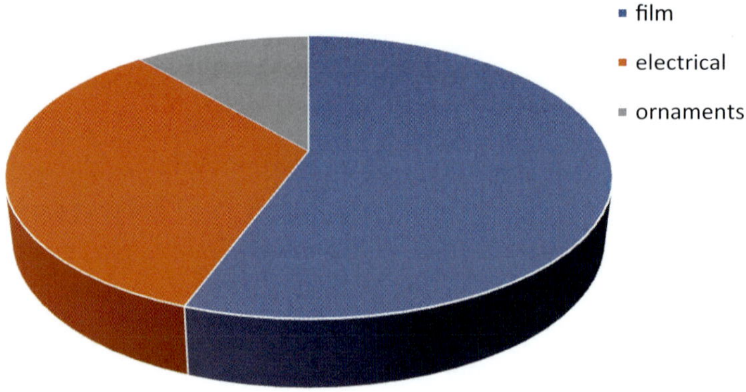

Fig. 1. Consumption of silver by industries.

in the electrical industry, while 10% is used for ornaments and jewelry, as shown in Fig. 1 [4]. Silver is also used in dental materials, medicines, disinfectants, etc. [5, 6]. According to World Silver Survey [7], although the global silver supply for 2015 was 1,040.6 million ounces, this amount could not meet the escalating yearly demand (Table 1). Reports state that the yearly demand for silver is steadily increasing by ~2–2.5% [8, 9]. Silver is generally obtained from natural sources and is found together with other metals such as gold, zinc, copper, and lead [4]. The deployment of natural silver resources, along with the high cost of silver regeneration from waste, has resulted in a rapid rise in the cost of silver production [10]. Because of the increased use and demand for silver, the recovery of silver from various wastes is of major importance for the silver market, in addition to being the most environmentally beneficial and economically efficient solution. To meet the present need for the development of a cost-effective and environmentally friendly process for silver recovery from various wastes, it is necessary to compare the present methods and look forward to developing new ones.

The literature is burgeoned with numerous studies on physical, chemical, and biological methods for the recovery of metals. This chapter partially congregates the technologies for silver recovery from various sources and describes the bio-hydrometallurgical process in detail, evaluating the current mechanisms for silver recovery and their further technological evolution.

Table 1. World silver supply and demand [7].

(million ounces)	2006	2007	2008	2009	2010	2011	2012	2013	2014	2015
Supply										
Mine Production	643.4	667.7	684.7	717.3	753.0	757.6	790.8	823.7	868.3	886.7
Net Government Sales	78.5	42.5	30.5	15.6	44.2	12.0	7.4	7.9	—	—
Scrap	207.1	204.2	202.0	201.2	227.5	261.5	255.5	192.1	163.3	146.1
Net Hedging Supply	−11.6	−24.1	−8.7	−17.4	50.4	12.2	−47.1	−34.8	16.8	7.8
Total Supply	**917.3**	**890.3**	**908.6**	**916.7**	**1,075.2**	**1,043.3**	**1,006.6**	**988.9**	**1,053.3**	**1,040.6**
Demand										
Jewelry	174.5	182.3	177.6	176.9	190.0	187.9	185.4	217.8	224.0	226.5
Coins & Bars	50.7	56.1	192.3	91.6	144.4	210.4	160.5	242.1	236.1	292.3
Silverware	62.1	60.2	58.4	53.2	51.6	47.2	43.7	58.8	60.7	62.9
Industrial Fabrication	648.9	661.0	656.8	542.6	650.3	676.3	615.0	619.1	611.2	588.7
....of which Electrical & Electronics	242.3	262.5	271.7	227.4	301.2	290.8	266.7	266.0	263.4	246.7
...of which Brazing Alloys & Solders	54.7	58.3	61.6	53.6	60.9	62.7	60.6	63.2	66.1	61.1
...of which Photography	142.2	117.0	98.2	76.4	67.5	61.2	54.2	50.5	48.5	46.7
...of which Photovoltaic[a]	—	—	—	—	—	75.8	62.9	62.5	63.2	77.6
...of which Ethylene Oxide	6.6	7.9	7.4	4.8	8.7	6.2	4.7	7.7	5.0	10.2
...of which Other Industrial[a]	203.1	215.3	217.9	180.4	212.1	179.4	165.8	169.3	165.1	146.4
Physical Demand	936.3	959.6	1,085.1	864.2	1,036.4	1,121.8	1,004.6	1,137.9	1,131.9	1,170.5
Physical Surplus/Deficit	**−19.0**	**−69.3**	**−176.6**	**52.5**	**38.8**	**−78.5**	**2.0**	**−149.0**	**−78.6**	**−129.8**
ETP Inventory Build	126.8	54.8	101.3	156.9	129.5	−24.0	55.3	2.5	1.5	−17.7
Exchange Inventory Build	−9.0	21.5	−7.1	−15.3	−7.4	12.2	62.2	8.8	−8.8	0.3
Net Balance	**−136.8**	**−145.5**	**−270.7**	**−89.2**	**−83.3**	**−66.7**	**−115.5**	**−160.2**	**−71.3**	**−112.5**
Silver Price, $ per oz.	11.55	13.38	14.99	14.67	20.19	35.12	31.15	23.79	19.08	15.68

Note: [a]Photovoltaic demand included in "Other Industrial" prior to 2011.
Source: ©GFMS, Thomson Reuters/The Silver Institute.

It is important for the protection of natural deposits and the environment that metals in various industrial wastes are retrieved by a systematic approach utilizing the research conducted to date. Most of the research done so far has been performed at the lab scale using synthetic media, which although feasible, is not cost-effective at the industrial level. Although a large amount of data is available on metal recovery, there are some challenges that require more investigation to develop feasible methods at industrial and commercial levels. More studies need to be conducted using different wastes containing various metals to assess the feasibility of a given technique in real-world conditions. Processes for the selective recovery and sequential separation of mixed metals from aqueous solutions by optimizing different parameters must be developed.

Numerous technologies have been reported for the recovery and recycling of silver from various wastes. Figure 2 shows a process cycle for silver recovery [11].

The recovery of silver from waste solutions of medical and industrial processing of X-ray and photographic films has been done for over 100 years. Silver has widespread uses in modern life, from eye drops to

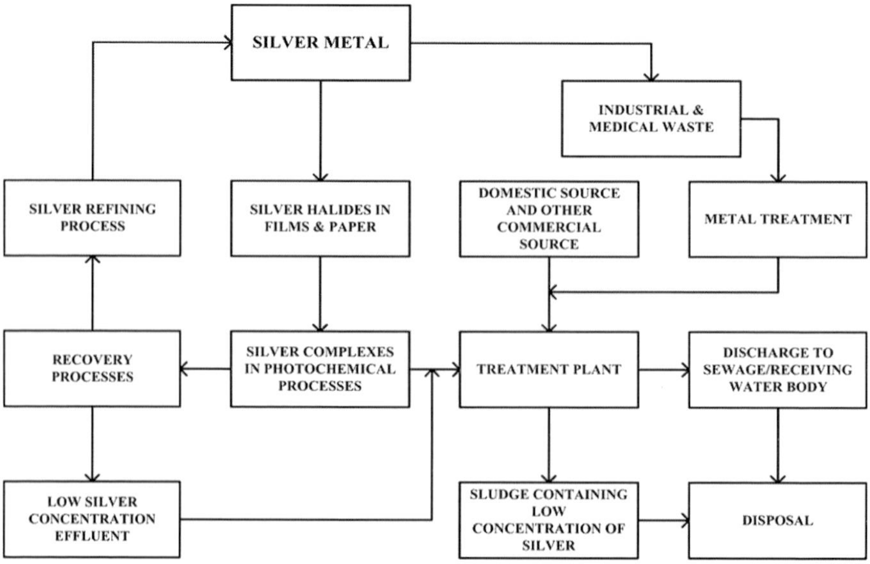

Fig. 2. Process cycle for silver recovery [11].

lubricants for jet engines. Silver is used by hospitals, medical clinics, and dental surgeons, as well as by photographic processing departments, and waste from these sources is extremely important for silver recovery efforts [2]. Silver is used in radiographic/photographic industries due to its high photosensitivity [10]. The radiographic film consists of a polyester sheet coated on both sides with a photosensitive, radioactive material that is light-sensitive [12]. Nearly two billion radiographs, including chest X-rays, mammograms, and CT scans are taken every year [8, 12]. Medical fields take 94–98% of the X-rays images that generate photographic chemicals and scrap film waste [12]. One kilogram of developed X-ray film uses approximately 14–17 g of silver, equating to a significant secondary source of silver [13]. The effluents from X-ray film processing facilities can reach a silver content of 1–12 g/L [8]. Of the global silver demand, 25% is met through recycling efforts, of which 75% is acquired from photographic waste [14].

Although silver is mostly recovered from scraps of X-ray films, photographic films, and jewelry [15], recently, electronic equipment such as printed circuit boards (PCBs), cell phones, and batteries have also been examined [3]. In this technological era, the use of electronic equipment has increased considerably, as has the rate at which electronics are replaced and the amount of e-waste generated. E-waste represents the fastest growing waste streams in the world. As described in the United Nations Environment Programme (UNEP) report, 50 million tonnes of e-waste is generated annually worldwide, of which only 10% is recycled [16]. According to the EPA, recycling 1 million cell phones can recover approximately 24 kg (50 lb) of gold, 250 kg (550 lb) of silver, 9 kg (20 lb) of palladium, and more than 9,000 kg (20,000 lb) of copper. Waste processing is necessary for environmental protection as well as for the economic benefit achieved by the recovery of valuable components [17]. In addition, the amount of valuable material in e-waste makes it attractive in terms of metal recovery. In particular, PCBs of electronic devices have copper as a dominant component. The metal content of PCBs (especially precious metals) varies with their age, origin, and manufacturer [18]. PCB waste has a valuable metal content of approximately 0.3–0.4%, (Ag, Pt, Au) [19]. The presence of these metals determines the value of electronic scrap and its processing cost.

6.2 Silver Recovery Techniques

Various physicochemical treatment techniques have been effectively applied to extract toxic heavy metals from wastes. Silver recovery methods fall into two broad categories, hydrometallurgical or pyrometallurgical techniques [13].

6.2.1 *Pyrometallurgical techniques*

Pyrometallurgical techniques involve the use of high-temperature treatments that physically and chemically alter the waste to recover the valuable metals. Such techniques are traditionally used for silver recovery from various wastes and have several steps, including incineration, smelting, roasting, drossing, sintering, and melting at high temperature [18].

The conventional method of burning film to recover silver is crude and generates air pollution. This technique has been reported to be relatively ineffective, with a recovery rate of $<95\%$ and requires temperatures higher than 950°C to destroy the X-ray film polymers [12, 13]. While this process is fast, it is rather expensive because of the high energy requirement. During the process, the extracted silver is covered with carbon, reducing the purity that can be achieved. Moreover, it produces silver emissions ($\sim$3–25%), which results in the loss of metals during combustion [10, 20–22].

6.2.2 *Hydrometallurgical techniques*

Hydrometallurgical techniques include the use of acids and other agents that are further subjected to various processes such as electrolysis, metallic replacement, chemical precipitation, adsorption, and liquid membrane [5, 6, 8, 12, 23]. Hydrometallurgical methods used for silver recycling are based on incorporating the silver into solution, in an ionic form, and then extracting it through electrolysis [24]. In this technique, metals are recovered in aqueous or organic solutions using chemical reactions [25]. The process is performed in three steps, viz., leaching, solution concentration and purification, and metal recovery. Leaching agents used include nitric acid; mixtures of nitric, hydrochloric, and sulfuric acids; sulfuric acid; nitric acid and hydrogen peroxide; aqua-regia; ferric chloride; thiourea; potassium iso-cyanate; potassium iodide and iodine; iodide–nitrite mixture; thiosulfate; and cyanides [26–29]. During the second step, the solutions are separated by solvent extraction, precipitation, cementation, ion exchange, filtration, and distillation to isolate and concentrate the metals of interest [30, 31]. Finally, the metal is recovered by electrolysis, gaseous reduction, and precipitation [32].

Each method has its advantages and disadvantages that must be taken into account in selecting a suitable recycling process. The hydrometallurgical processes have >99% recovery efficiencies and require lower power consumption depending on the method chosen. Furthermore, the lower cost of hydrometallurgical processes compared to pyrometallurgical techniques makes their application feasible in small plants. However, pre-treatment of waste to separate the metal fraction is required to increase the effectiveness of each phase of the hydrometallurgical processes. The main problem of hydrometallurgical processes is the high consumption of water and/or chemicals that then need further treatment.

The main issues regarding silver recovery are the reduction of pollutant emissions during the recovery processes and the overall economic benefit. However, some health and environmental impacts of silver exposure also exist. In the case of X-ray films, stripping the gelatin–silver layer by chemical methods using ammonium thiosulfate, sodium thiosulfate, nitric acid, or reagents such as sodium cyanide, NaOH, nitric acid, or organic compounds cause environmental hazards and are either time-consuming or very expensive, while the use of NaOH at high temperatures poses a serious industrial safety problem [33]. The use of a hybrid process that combines the chemical and biological methods will be more efficient and more ecologically friendly.

6.2.3 *Bio-hydrometallurgical techniques*

Bio-hydrometallurgy, also referred to as bio-leaching, is a novel biotechnological process used to eliminate or retrieve heavy metals from contaminated sediments and poor-quality ore and mineral compounds by acidification and solubilization of heavy metals using micro-organisms or their products. Bio-hydrometallurgical processes employing micro-organisms are currently a topic of interest and may result in long-term solutions for metal recovery, gradually replacing the traditional ones. They are also a natural method for cleaning the environment [34].

Microbial leaching is the interaction between metals and microbes, where insoluble metal sulfides are converted to metal sulfates with the aid of organic and inorganic acids secreted by the microbes into the aqueous media [35, 36]. Bio-hydrometallurgical processing of solid waste is similar to natural bio-geochemical metal cycles and decreases the demand for resources, such as ores, energy, and landfill space [37]. Therefore, microbial leaching is environmentally friendly, flexible (microbes can easily adapt to varied conditions), and easier to conduct. It is also relatively inexpensive, which has encouraged its use [18, 38] and is considered a green technology (generates less waste) [39, 40]. Bio-leaching has been applied successfully and commercially to bio-hydrometallurgy for the extraction of precious metals [41].

Many scientists are currently interested in bio-hydrometallurgy, a technology that is an attractive alternative to currently used physical and chemical methods to recover valuable metals from waste. These microbiological processes have the capability to be used for waste treatment and recovery of valuable metals and are also a green and efficient technology compared to the alternatives [42]. Another beneficial aspect of using bio-processing for the treatment of wastes is the efficient recovery of valuable and strategic metals with the simultaneous detoxification of the processed wastes. The recovery of valuable metals such as silver from different wastes using bio-hydrometallurgical methods would be beneficial in developing both a recycling culture and a circular economy based on recycling with the aim of reducing waste generation and using waste as a resource.

6.2.3.1 *Bioleaching*

Considering the disadvantages of conventional recovery methods, the ideal methods used to recover silver from various wastes should be cost-effective and eco-friendly, and enzyme-based methods are an attractive alternative. Gelatin is an animal collagen protein that is largely composed of amino acid residues such as glycine, proline, and 4-hydroxyproline. X-ray film has an emulsion layer of silver and gelatin that can be stripped using proteolytic enzymes to release the silver [43, 44]. Thus, both silver and the polyester base can be recovered and recycled by enzymatic hydrolysis [45]. In recent years, enzymatic methods using microbial proteases have been investigated as alternatives to the more conventional methods of silver recovery [33, 43, 46–49]. The enzymatic process is more specific, and the gelatin layer is removed without damaging the polyester base, taking only a few minutes. Gelatin molecules are crosslinked with hardeners, and it is difficult for standard proteases to degrade it in a short time [33]. Most of the proteases used for silver recovery are of bacterial origin, however, Shankar *et al.* [33] reported the potential use of a fungal alkaline protease from *Conidiobolus coronatus* for this purpose.

Nakiboglu *et al.* [44] developed an enzymatic method for the recovery of silver from X-ray photographic film waste. Enzyme extract was obtained from *Bacillus subtilis* ATCC 6633 grown on modified Horikoshi medium. This bacterium produces a mixture of neutral and alkaline proteases that effectively recovers the silver by degrading gelatin layers on the films. Process optimization revealed that the optimum conditions for gelatin degradation were at pH 8.0 and 50°C. Furthermore, the recovered silver had a purity of >99% by smelting the obtained slurry in the presence of borax.

Alkaline protease from *Conidiobolus coronatus* was explored for enzymatic hydrolysis of gelatin from X-ray film waste [33]. The process efficiently removed the gelatin layer from the polyester film, and silver was recovered from the hydrolysate. Various process parameters were optimized for the recovery of silver from the film. The enzymatic hydrolysis of gelatin was measured in terms of increased turbidity in the hydrolysate as a result of the release of protein and hydroxyproline. The gelatin layer was stripped completely within 6 min with 1.35 U/mL of the protease at 40°C, pH 10. The rate of gelatin hydrolysis increased with an increase in protease concentration. The enzyme could be effectively reused for four cycles of gelatin hydrolysis. Silver in the hydrolysate was approximately 3.87% (w/w) based on total weight of the sludge. Using 0.9 U/mL of protease, it could be reused up to four times [33].

Cavello *et al.* [50], utilized keratinolytic serine proteases from *Purpureocillium lilacinum* LPS # 876 for the enzymatic decomposition of gelatin layers on used X-ray film and described their repeated utilization for silver recovery. At the optimum conditions of 6.9 U/mL enzyme, 60°C, and pH 9.0, hydrolysis of the gelatin layers and the resulting release of silver particles was achieved within 6 min. Figure 3 shows the enzyme hydrolysis of X-ray film and gelatin and the effect of temperature on this process. The

(a)

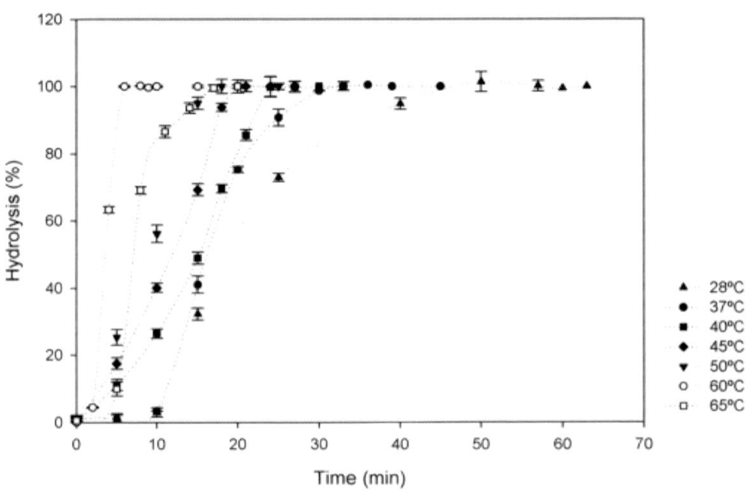

(b)

Fig. 3. Enzymatic hydrolysis of used X-ray film and gelatin. (a) Enzymatic hydrolysis of used X-ray films at 37°C. The control solution remains clear whereas the enzyme solution turns turbid, owing to the release of black metallic silver. Clean PET film was obtained after an enzymatic treatment of X-ray film, whereas the control film remains unchanged; (b) Effect of temperature on time course of gelatin hydrolysis: (▲) 28°C, (•) 37°C, (■) 40°C, (♦) 45°C, (▼) 50°C, (◯) 60°C, and (□) 65°C. The enzyme concentration used was 6.9 U/mL.

Table 2. Reuse of *P. lilacinum* protease for gelatin hydrolysis from used X-ray films.

	Time required for complete gelatin removal (min)							
	Cycle 1	Cycle 2	Cycle 3	Cycle 4	Cycle 5	Cycle 6	Cycle 7	Cycle 8
Protease	6	15	23	Ir	—	—	—	—
Propylene glycol	6	12	25	Ir	—	—	—	—
Glycerol	6	9	15	20	25	28	32	Ir

Note: Ir, incomplete removal after 45 min.

addition of polyols, such as glycerol and propylene glycol, protected the enzyme from thermal denaturation, thus increasing the enzyme stability. When compared to the use of the enzyme without protective agents, the glycerol-protected enzyme was reused effectively for more cycles, which suggests the feasibility of a continuous process of recycling silver from used X-ray film (Table 2). Researchers successfully utilized three different wastes, viz., hair waste, glycerol (a biodiesel waste product), and X-ray film waste, to produce serine proteases, enhance enzyme thermal stability, and recover silver and PET films, respectively. Thus, the wastes were used in an environment-friendly way and not wasted.

Al-abdalall and Al-Khaldi [51] employed an alkaline protease from *Bacillus subtilis* to retrieve silver from X-ray film. This enzyme was able to extract silver in 30 min, however, its activity decreased with an increasing incubation period. The gelatin layer was stripped completely within 30 min using 97 U/mL protease at 50°C and pH 8, leaving the polyester film clean, and the silver was recovered in the hydrolysate. The continuous exposure of the enzyme to high temperature was suggested as a plausible reason for its decreased activity.

Recycling of natural mineral resources, especially silver, remains the most practical option to slow down the exhaustion caused by their overuse. This study showed that the alkaline protease of *B. subtilis* sub sp. *subtilis* has the potential of being reused for extracting silver from used X-ray film in an eco-friendly manner [51].

In the presence of cyanide, many metals and metalloids (such as Ti, V, Cr, Mn, Fe, Co, Ni, Cu, Zn, Ge, Mo, Tc Ru, Rh, Pd, Ag, Cd, W, Re, Os, Ir, Pt, Au, Hg, Tl, Po, and U) form well-defined cyanide complexes that are water soluble and exhibit high chemical stability [52, 53]. Industries use this principle for the extraction of gold from low-grade ores. Biological cyanide is formed as a secondary metabolite by a variety of bacteria e.g., *Chromobacterium violaceum*, *Pseudomonas fluorescens*, or *P. aeruginosa*, many of them belonging to the soil microflora.

Brandl *et al.* [54] evaluated the ability of cyanogenic microorganisms such as *C. violaceum*, *P. fluorescens*, and *P. plecoglossicida* to solubilize metals from metal-containing solids such as gold-containing electronic scrap, silver-containing jewelry waste, or platinum-containing automobile catalytic converters. All the organisms were supplemented with glycine in the growth medium, and metal-containing solids were added at different concentrations. Both *C. violaceum* and *P. plecoglossicida* were found to be effective cyanide producers. The pH increased from 9 to 9.5 during incubation, which allows cyanogenic bacteria to mobilize the metals from solids under alkaline conditions. This is of interest at the industrial level since metal chemistry (e.g., solubility, mobility, sorption, precipitation, the formation of secondary minerals) under acidic conditions might change under alkaline conditions (especially in the presence of cyanide) and the metal might be recovered more easily. The biological mobilization of silver, gold, and platinum as cyano-complexes was identified by high-pressure liquid chromatography.

Gold was efficiently mobilized by *C. violaceum*, which produced high concentrations of dicyanoaurate, whereas platinum proved to be very resistant to mobilization by cyanogenic micro-organisms.

Pseudomonas plecoglossicida mobilized the silver effectively in the presence of powdered jewelry waste (in a suspension of 286 mg/L). Approximately 5% of silver was mobilized as dicyanoargentate, $[Ag(CN)_2]^-$ in one day. After this rapid initial growth phase, the cyanide-complexed silver concentration increased only slightly until the end of the incubation period. The authors further added non-biological cyanide (10 mM KCN solution) for silver leaching, which resulted in the complete mobilization of silver, but they found that dicyanoargentate at concentrations >20 mg/L inhibited the growth of *P. plecoglossicida*, resulting in a low mobilization efficiency. The observed half-maximal effective concentration (EC50) was in the range of 12–15 mg/L.

In spite of the toxicity of dicyanoargentate, the mobilization of silver by cyanogenic micro-organisms proved to be promising. However, the authors suggest the requirement of a process to reduce inhibitory effects, for example, by trapping the dicyanoargentate formed or by applying a sequential two-step process where the biological cyanide formation is separated from the metal mobilization.

These findings represent a novel type of microbial metal mobilization (termed "bio-cyanidation"), which might have an industrial application. Additionally, metal recovery from these solutions, cyanide-complexed metals might be easily separated by chromatographic methods and adsorbed onto activated carbon [54].

The use of button cell batteries has increased recently due to their small size, long operating life, and high capacity per unit mass, and they are abundantly used in electric watches, portable medical monitors, calculators, hearing aids, digital thermometers, toys, etc. [55, 56]. They are generally discarded as waste after use. To avoid environmental issues and for economic reasons, they must be recycled [55, 57]. Various pyro- and hydrometallurgical techniques used for the recovery of silver have disadvantages such as the requirement of high energy and emission of toxic gases [55]. Figure 4 shows a cross-sectional view of a silver oxide–zinc button cell.

Jadhav and Hocheng [58] investigated the use of *Acidithiobacillus ferrooxidans* culture supernatant for the extraction of silver from a spent silver oxide zinc button cell battery. Various process parameters, such as

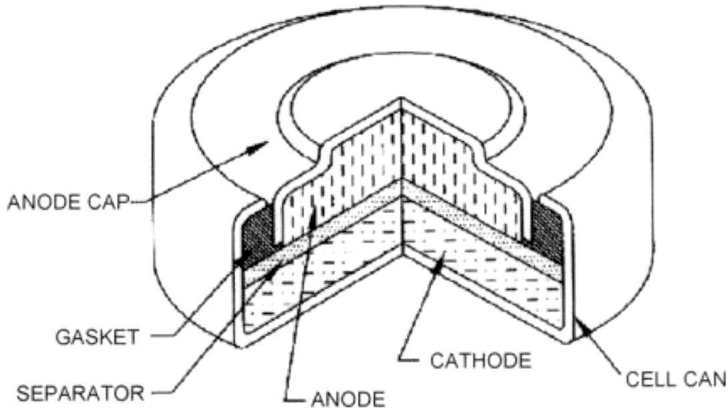

Fig. 4. A cross-sectional view of a silver oxide-zinc button cell [55].

time, the volume of culture supernatant, pH, initial shaking speed, and temperature were optimized for the silver extraction. The authors carried out preliminary silver extraction experiments to assess the practicality of a non-contact leaching process. Comparisons of one-step, two-step, and spent bio-leaching using microbial culture have been studied in detail and revealed the feasibility of the latter process [40, 59, 60]. For industrial applications, bio-leaching by spent culture is desirable to increase leaching efficiency [61].

Acidithiobacillus ferrooxidans was cultivated in basal 9 K medium supplemented with 1% sulfuric acid and $FeSO_4$ as an energy source. Figure 3 shows the spent silver oxide cell and its powder. An indirect bio-leaching process was used, in which the battery powder was added to the bacterial culture supernatant. A total of 45.8% of the silver was dissolved in 1 h, which further increased to 98% at 3 h (Fig. 5), illustrating the importance of reaction time. The dissolution of silver in the culture supernatant suggests the role of *A. ferrooxidans* was most likely to regenerate Fe^{3+} as an oxidant. The authors also confirmed the role of Fe^{3+} in silver dissolution by comparing the ferrous/ferric ion concentrations and found that, after 3 h of bio-leaching, 80% Fe^{3+} was converted into Fe^{2+}

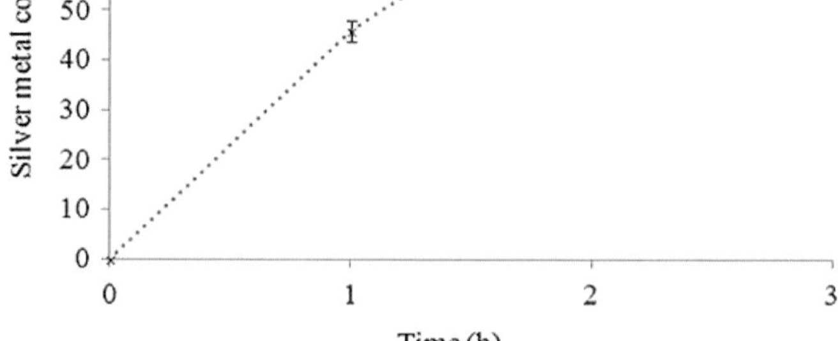

Fig. 5. Time period evaluation of silver dissolution by using 100 mL culture supernatant (with 20 g/L initial $FeSO_4$ concentration), at 30°C and 150 rpm.

Table 3. The status of Fe^{3+} during silver extraction from spent oxide-zinc button cell batteries at various initial ferrous sulfate concentrations.

	Various initial ferrous sulfate concentrations (g/L)							
	Silver extraction (%)				Fe^{3+}(M)			
Time (h)	20	40	60	80	20	40	60	80
0	0.0	0	0	0	0.0719	0.1438	0.2158	0.2877
1	45.8	62.78	98.5	98.5	0.0523	0.0589	0.0845	0.1040
2	78.2	98.5	—	—	0.0385	0.0181	—	—
3	98.5	—	—	—	0.0143	—	—	—

(Table 3). The metal is thus released due to chemical oxidation by ferric ions. The ferrous–ferric couple serves as a redox mediator and catalyzes the reaction.

The authors also carried out the silver dissolution experiment using whole microbial culture and observed that the microbial cells required 30 h for 98.5% silver solubilization, which may have been due to the lower availability of Fe^{3+}. These results demonstrate the leaching efficiency of culture supernatant and can be useful to set a two-stage reactor system. The advantages of separating microbial growth from the leaching process are (1) the reactor can be operated in a continuous mode, and (2) microbial activity is not affected by possible mass transfer limitations or the toxicity of metals [62]. During silver recovery, the biologically produced ferric ions take part in silver dissolution. The ferric ions react with silver, thus dissolving it, and are in turn converted into ferrous ions, which are reused for the growth of bacteria. Thus, a cleaner metal recovery process can be developed.

The effect of the volume of culture supernatant used in silver extraction was studied, and it was noted that the silver extraction increased with an increase in volume. It was observed that the leaching efficiency increased gradually with an increase of initial $FeSO_4$ concentration. The use of 40 g/L and 60 g/L $FeSO_4$ culture supernatants decreased the time of silver dissolution to 2 h and 1 h, respectively. A pH of 2.5 and a shaking speed of 150 rpm were found to be optimum for the extraction of silver. Temperature has an important role in most reactions, but in the present study, no such effect on silver extraction was observed. For all the temperatures studied, similar silver extraction results were obtained, which might be due to the use of the culture supernatant.

Fe^{3+} ions play an important role in metal dissolution, which are supplied by *A. ferrooxidans*, and is proposed to be the oxidizing agent for the dissolution and removal of metals. During silver recovery, these ferric ions produced by the bacteria react with silver, dissolve it and get converted into ferrous ions, which are further reused for bacterial growth.

Furthermore, the authors compared these bio-leaching results with the previously reported chemical leaching results using nitric acid [55, 56]. The comparison revealed that the leaching of silver by using biologically produced ferric ions was as efficient as nitric acid leaching. Approximately 98% of the silver was dissolved in 60 min. Therefore, the use of biologically produced ferric sulfate is advantageous over chemical leaching, since it is an efficient process and has minimal impact on the environment.

For industrial applications, a two-step process is believed to be appropriate to increase leaching efficiency. In the two-step process, the bacterially generated acidic solution is treated as a lixiviant for metal dissolution. In general, the advantages of such processes are that independent lixiviant generation removes the link between the bio-process and the chemical process and thus makes it possible to optimize each process independently to maximize productivity. This strategy can be used when the ore/waste does not contain the necessary mineral components in sufficient quantities to sustain a viable bacterial population. Furthermore, higher waste concentrations can be treated compared to the one-step process, which results in increased metal yields.

The generation of e-waste is increasing daily. Solders are a crucial part of any electronic product, providing electrical connection and mechanical support in modules [63]. Tin–copper (Sn–Cu), tin–copper–silver (Sn–Cu–Ag), and tin–lead (Sn–Pb) solders are used extensively in electronic devices. Hocheng *et al.* [64] utilized the culture supernatant of *Aspergillus niger* and *Acidithiobacillus ferrooxidans* for effective bio-leaching of metals from solder using a two-step process. Compared to the *A. ferrooxidans* culture supernatant, the *A. niger* culture supernatant showed a faster metal dissolution (99%) in 60 h, at a 200 rpm shaking speed at 30°C. Metals were removed faster from Sn–Cu–Ag solder than Sn–Cu and Sn–Pb solders using the *A. niger* culture supernatant. Bioleaching of metals by the *A. niger* culture supernatant is attributed to the organic acids produced by the fungus. The addition of sodium chloride into the bio-leached solution from Sn–Cu–Ag precipitated 80% of silver.

6.2.3.2 *Bio-sorption*

Bio-sorption is another biological process that has been claimed as a promising process for pollutant removal and/or metal recovery, because of its simplicity, efficiency, and availability of biomass and waste bio-products [65]. Most researchers define bio-sorption as a passive and metabolically independent process that can be performed either by dead biomass or fragments of cells and tissues, for example, the passive uptake of metals by microbial biomass [66–68]. However, it can also be performed by live cells through passive uptake or metabolically independent adsorption of a sorbate by surface complexation onto cell walls and/or other outer layers as the first, fast, and reversible adsorption step operating within a much slower and complex overall bio-accumulation mechanism [66, 67]. Bio-sorption has also been reported for silver retrieval.

Merroun *et al.* [69] utilized the biomass of *Myxococcus xanthus* for silver sorption. The dry biomass acted as a good sorbent for silver recovery even at low silver concentrations. At silver concentrations of 2–0.05 M, the percentage of silver accumulation ranged from 8.12% to 75%. Transmission electron microscopy (TEM) of the *Myxococcus* biomass revealed the silver sorption within the extracellular polysaccharide, on the cell wall, and in the cytoplasm. The authors proposed two mechanisms for the silver sorption, as silver was bound to the cell surface and extracellular polysaccharide and was deposited intracellularly. Furthermore, the desorption of silver from *M. xanthus* was executed using sodium carbonate or ammonia, with ammonia being found to be an efficient desorbent, recovering 78% of the silver.

Pethkar and Paknikar [70] attempted the removal and recovery of silver from photofilm processing unit wastewater using a bio-sorption process. *Thiobacillus thioparus* was used to treat this wastewater that was further passed in a continuous up-flow mode through a column of beads of a fungal culture *Cladosporium cladosporioides*. The maximum silver biosorption was 43 mg/g at pH 7–9, a biomass concentration of 2.5% (w/v), and a contact time of 20 min. The silver was eluted from the column using a 0.1% alcoholic sodium cyanide solution and subjected to electrowinning, and a recovery efficiency of 70% was observed.

Sneha *et al.* [71] determined the capacity of active and inactive cells of *Corynebacterium glutamicum* to adsorb and/or reduce silver ions. The silver bio-sorption elevated with an increase in pH, and equilibrium was achieved in 30 min. At an initial silver concentration of 1,000 mg/L, the maximum uptake was found to be 50.1 mg/g and 52.5 mg/g for active and passive cells, respectively. Furthermore, the investigation of the bio-reduction capacity of both live and dormant cells of *C. glutamicum* revealed a good amount of reduction on the inactive cell surfaces. Moreover, inactive *C. glutamicum* cells are easy to handle, and their surplus availability as a fermentation waste make them a good choice for the recovery of silver from industrial effluents. The crystallization of silver ions from aqueous solution by both active and inactive biomass can be used to develop a cost-effective and eco-friendly technique to remove or recover noble metals.

Das *et al.* [72] applied dead biomass of macro-fungus *Pleurotus platypus* for sorption of Ag(I). The highest silver uptake observed was 46.7 mg/g of biomass at pH 6 in the presence of 200 mg/L Ag(I) at 20°C. Kinetic studies using various expressions showed a very good accordance with a pseudo-first-order model. The thermodynamic calculations showed the feasibility, exothermic nature, and spontaneity of the bio-sorption of Ag(I) onto *P. platypus* at 20°C. Thus, the study described the use of the macro-fungus *P. platypus* as a cost-effective and efficient bio-sorbent for Ag(I) removal from silver-containing waste solutions.

Li *et al.* [73] assessed bio-sorption of silver ions on *Bacillus cereus* biomass. The result showed that more than 90% of silver bio-sorption was attained within 90 min. The process of bio-sorption was well described by a pseudo-second-order kinetic model. The scanning electron microscopy and X-ray photoelectron spectroscopy (XPS) studies suggested that precipitation was the plausible mechanism of bio-sorption. Figure 6 shows the XPS analysis of silver-loaded biomass. The researchers suggested that this phenomenon might be the result of the chemical interactions between metal ions and bio-polymers from the cell wall (such as polysaccharides,

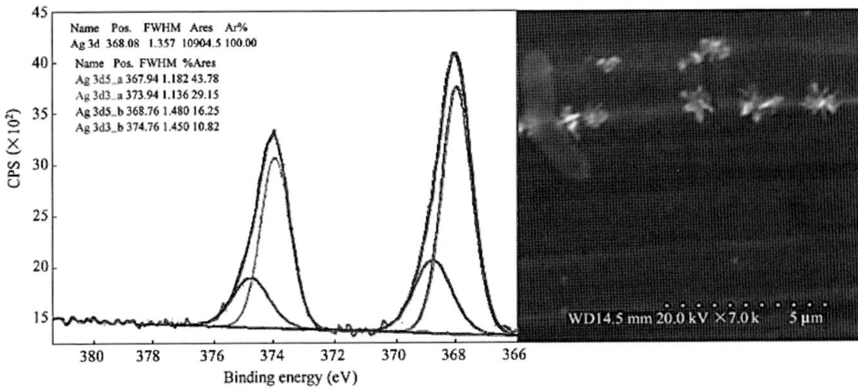

Fig. 6. Photoelectron spectrum of biomass challenged with Ag^+ for 24 h and micrograph of crystal from the silver ions loaded within biomass [73].

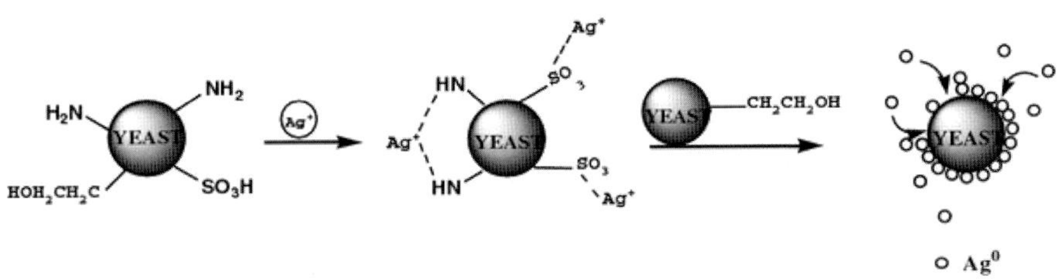

Fig. 7. The process of adsorption of Ag(I) by the yeast waste [76].

proteins, and some hydrolysis) and the entrapment of metals in the form of insoluble micro-deposits [74, 75]. Additionally, the molecular genetics of silver resistance of *B. cereus* biomass was detected and elucidated by a whole cell sensor tool.

Zhao *et al.* [76] investigated the use of brewer fermentation yeast waste as a bio-sorbent to adsorb silver from aqueous solution. The FTIR — Fourier Transform Infra-red Spectroscopy spectrum confirmed the presence of many functional groups on the yeast surface with reduction abilities that could reduce silver. Thus, the possible mechanism for sorption between the silver ions and the binding sites on the surface of yeast biomass were proposed to be ion exchange, chelating, and reduction. Figure 7 shows the proposed mechanism for silver adsorption by waste yeast. With the reduction groups, which might be formed during the

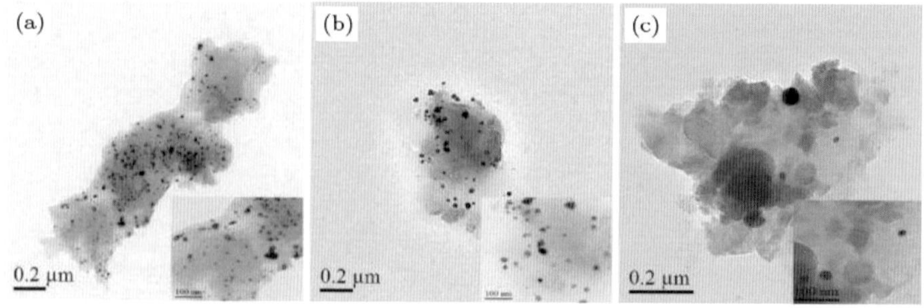

Fig. 8. TEM images of the yeast waste after adsorption of Ag(I) at various temperatures: (a) 288 K, (b) 298 K, and (c) 308 K [76].

process of the brewing process, the waste yeast could easily absorb Ag(I) on the surface and reduce them to silver nanoparticles (Ag^0). The deposition of silver nanoparticles on the surface of yeast was observed through TEM, X-ray diffraction (XRD), and XPS results. Figure 8 shows the TEM images of yeast waste after adsorption of silver at different temperatures. Consequently, the kinetic experiments demonstrated that sorption equilibrium was reached within 60 min, and a 93% Ag^+ removal efficiency was attained when the initial concentration of Ag(I) was below 100 mg/L. The thermodynamic parameters of the adsorption process (DG, DH, and DS) demonstrated that the adsorption was spontaneous and exothermic. Thus, silver can be adsorbed on yeast waste without any modification and can be used for waste solutions with small amounts of silver.

6.3 Conclusions and Future Implications

Recycling of secondary raw materials is a requisite of modern life. The importance of recycling is that, even though industrial wastes are polluting the environment, many useful constituents could be acquired from their processing. Thus recycling protects the environment, conserves natural resources, and reduces waste. The recycling of precious metals, such as silver, results in the recovery of the metal without loss of quality, minimizing pollution and energy consumption and reduces mining and waste. Methods such as pyrometallurgical processes, hydrometallurgical processes, and bio-leaching have been successfully used to recover precious metals from wastes. However, when compared with other processes, bio-leaching is more eco-friendly technique. However, at present, bio-leaching methods are not as effective as other methods and are still used at the lab scale. Finding new precious metal recovery technologies can balance the economic pressure with the need of preserving natural ecosystems and can have a positive global impact. Economic analyses to understand costs and benefits of the process are also desired.

To improve bio-hydrometallurgical techniques for metal recovery to be comparable to their counterparts, new strategies must be implemented. Instead of using traditional screening and adaptation practices for microbes, the use of improved techniques such as mutation and genetic engineering must be employed. There is a need to define and understand the potential interactions among the components of microflora. Increased knowledge of the potential microbial communities in the area and vicinity of different mining and wastes site using DNA and PCR — Polymerase Chain Reaction techniques could be fruitful. This will reduce the required time and improve the economics of the process. In the case of low concentration and mixed metal conditions, new reactor configurations, materials, and system integration processes need to be developed to improve metal recovery yields and rates. For sustainable environmental protection to occur, this emerging process must be adopted and used to replace the currently used methods.

References

[1] Kowalska S, Lukomska A, Los P, Chmielewski T, Wozniak B (2015) Potential-controlled electrolysis as an effective method of selective silver electrowinning from complex matrix leaching solutions of copper concentrate. *International Journal of Electrochemical Science*, **10**: 1186–1198.

[2] Canda L, Heput T, Ardelean E (2016) Methods for recovering precious metals from industrial waste. *International Conference on Applied Sciences 2015 (ICAS2015), IOP Conference Series: Materials Science and Engineering*, **106**: 012020.

[3] Shirahase T, Kida A (2009) Metals content in one waste personal computer by detailed dismantling. *Japan Society of Material Cycles and Waste Management*, **20**(4): 217–230.

[4] Oncel SM, Mahir I, Mahmut B (2005) Leaching of silver from solid waste using ultrasound assisted thiourea method. *Ultrasonics Sonochemistry*, **12**(3): 237–242.

[5] Elwakeel KZ, El-Sayed GO, Darweesh RS (2013) Fast and selective removal of silver(I) from aqueous media by modified chitosan resins. *International Journal of Mineral Processing*, **120**: 26–34.

[6] Tang B, Yu G, Fang J, Shi T (2010) Recovery of high-purity silver directly from dilute effluents by an emulsion liquid membranecrystallization process. *Journal of Hazardous Materials*, **177** (1–3): 377–383.

[7] World Silver Survey (2016). Available at: https://www.silverinstitute.org. Accessed 22 September 2016.

[8] Bas AD, Yazici EY, Deveci H (2012) Recovery of silver from X-ray film processing effluents by hydrogen peroxide treatment. *Hydrometallurgy*, **121–124**: 22–27.

[9] Butterman WC, Hilliard HE (2005) Silver. U.S. Geological Survey (USGS), Reston, Virginia.

[10] Zhouxiang H, Jianying W, Ma Z, Jifan H (2008) A method to recover silver from waste X-ray films with spent fixing bath. *Hydrometallurgy*, **92**(3–4): 148–151.

[11] Kodak (2003). The fate and effects of silver in the environment. Kodak, USA J-216(ENG).

[12] Khunprasert P, Grisdanurak N, Thaveesri J, Danutra V, Puttitavorn W (2008) Radiographic film waste management in Thailand and cleaner technology for silver leaching. *Journal of Cleaner Production*, **16**(1): 28–36.

[13] Abdel-Aal EA, Farghaly FE (2007) Preparation of silver powders in micron size from used photographic films via leaching–cementation technique. *Powder Technology*, **178**(1): 51–55.

[14] Masebinu SO, Muzenda E (2014) Review of Silver Recovery Techniques from Radiographic Effluent and X-ray Film Waste. *IAENG Proceedings of the World Congress on Engineering and Computer Science (WCECS 2014)*, San Francisco, USA.

[15] Ajiwe VIE, Anyadiegwu IE (2000) Recovery of silver from industrial wastes, cassava solution effects. *Separation and Purification Technology*, **18**(2): 89–92.

[16] Bulletin Waste Management World (2011). Undercover investigations into e-waste smuggling. May 16.

[17] Pawłowski L (2011) Effect of mercury and lead on the total environment. *Environmental Protection Engineering*, **37**(1): 105–117.

[18] Cui J, Zhang L (2008) Metallurgical recovery of metals from electronic waste: A review. *Journal of Hazardous Materials*, **158**(2–3): 228–256.

[19] Ludwig C, Hellweg S, Stucki S (eds.) (2003). *Municipal Solid Waste Management: Strategies and Technologies for Sustainable Solutions* (Springer, Berlin), pp. 164–349.

[20] Havlik T, Orac D, Petranikova M, Miskufova A, Kukurugya F, Takacova Z (2010) Leaching of copper and tin from used printed circuit boards after thermal treatment. *Journal of Hazardous Materials*, **183**(1–3): 866–873.

[21] Ojeda MW, Perino E, Ruiz M (2009) Gold extraction by chlorination using a pyrometallurgical process. *Minerals Engineering*, **22**(4): 409–411.

[22] Jadhav UU, Hocheng H (2012) A review of recovery of metals from industrial waste. *Journal of Achievements in Materials and Manufacturing Engineering*, **54**(2): 159–167.

[23] Petrova TM, Tzaneva B, Fachikov L, Hristov J (2013) Silver recovery from spent photographic solutions by a magnetically assisted particle bed. *Chemical Engineering and Processing: Process Intensification*, **71**: 83–96.

[24] Marincovic J, Korac M, Kamperovic Z, Matic I (2006) Recycling of silver from exposed X-ray films. *Acta Metalurgica Slovaca*, **12**: 262–268.

[25] Kinoshita T, Akita S, Kobayashi N, Nii S, Kawaizumi F, Takahashi K (2003) Metal recovery from non-mounted printed wiring boards via hydrometallurgical processing. *Hydrometallurgy*, **69** (1–3): 73–79.

[26] Syed S (2006) A green technology for recovery of gold from non-metallic secondary sources. *Hydrometallurgy*, **82** (1–2): 48–53.

[27] Veglio F, Quaresima R, Fornari P, Ubaldini S (2003) Recovery of valuable metals from electronic and galvanic industrial wastes by leaching and electrowinning. *Waste Management*, **23**(3): 245–252.

[28] Silva JE, Paiva AP, Soares D, Labrincha A, Castro F (2005) Solvent extraction to the recovery of heavy metals from galvanic sludge. *Journal of Hazardous Materials*, **120**(1–3): 113–118.

[29] Ruiz O, Clemente C, Alonso M, Alguacil F, (2007) Recycling of an electric arc furnace flue dust to obtain high grade ZnO. *Journal of Hazardous Materials*, **141**(1): 33–36.

[30] Kentish SE, Stevens GW (2001) Innovations in separation technology for the recycling and re-use of liquid waste streams. *Chemical Engineering Journal*, **84**(2): 149–159.

[31] Xu K, Deng T, Liu J, Peng W (2007) Study on the recovery of gallium from phosphorus flue dust by leaching with spent sulfuric acid solution and precipitation. *Hydrometallurgy*, **86** (3–4): 172–177.

[32] Jha MK, Kumar V, Singh RJ (2001) Review of hydrometallurgical recovery of zinc from industrial wastes. *Resources Conservation and Recycling*, **33**(1): 1–22.

[33] Shankar S, More SV, Seeta Laxman R (2010) Recovery of silver from waste X-ray film by alkaline protease from *Conidiobolus coronatus*. *Kathmandu University Journal of Science Engineering and Technology*, **6**: 60–69.

[34] Klaus B (1997) Bioleaching: metal solubilization by microorganisms. *FEMS Microbiology reviews*, **20**(3–4): 591–604.

[35] Brandl H, Faramarzi MA (2006) Microbe-metal-interactions for the biotechnological treatment of metal-containing solid waste. *China Particuology*, **4**(2): 93–97.

[36] Hoque ME, Philip OJ (2011) Biotechnological recovery of heavy metals from secondary sources-An overview. *Material Science and Engineering C*, **31**(2): 57–66.

[37] Cerruti C, Curutchet G, Donati E (1998) Bio-dissolution of spent nickel–cadmium batteries using *Thiobacilus ferrooxidans*. *Journal of Biotechnology*, **62**(3): 209–219.

[38] Brierley CL (2008) How will biomining be applied in future. *Transactions of Nonferrous Metal Society of China*, **18**(6): 1302–1310.

[39] Brombacher CH, Bachofen R, Brandl H (1998) Development of a laboratory-scale leaching plant for metal extraction from fly ash by *Thiobacillus* strains. *Applied Environmental Microbiology*, **64**(4): 1237–1241.

[40] Mishra D, Kim DJ, Ralph DE, Ahn JH, Rhee YH (2008) Bioleaching of spent hydro-processing catalyst using acidophilic bacteria and its kinetics aspect. *Journal of Hazardous Materials*, **152**(3): 1082–1091.

[41] Zhou HB, Zeng WM, Yang ZF, Xie YJ, Qiu GZ (2009) Bioleaching of chalcopyrite concentrate by a moderately thermophilic culture in a stirred tank reactor. *Bioresource Technology*, **100**(2): 515–520.

[42] Gadd GM (2010) Metals, minerals and microbes: Geomicrobiology and bioremediation. *Microbiology*, **156** (Pt 3): 609–643.

[43] Singh J, Vohra RM, Sahoo DK (1999) Alkaline protease from a new obligate alkalophilic isolate of *Bacillus sphaericus*. *Biotechnology Letters*, **21** (10): 921–924.

[44] Nakiboglu N, Toscali D, Yasa I (2001) Silver recovery from waste photographic films by an enzymatic method. *Turkish Journal of Chemistry*, **25**: 349–353.

[45] Gupta R, Beg K, Lorenz P (2002) Bacterial Alkaline Protease: Molecular approaches and industrial application. *Applied Microbiology and Biotechnology*, **59**(1): 15–32.

[46] Fujiwara N, Yamamoto K, Masui A (1991) Utilization of a thermostable alkaline protease from an alkalophilic thermophile for the recovery of silver from used x-ray film. *Journal of Fermentation and Bioengineering*, **72**(4): 306–308.

[47] Masui A, Yasuda M, Fujiwara N, Ishikawa H (2004) Enzymatic hydrolysis of gelatin layers on used lith film using thermostable alkaline protease for recovery of silver and PET. *Biotechnology Progress*, **20**(4): 1267–1269.

[48] Nakiboglu N, Toscali D, Nisli G (2003) A novel silver recovery method from waste photographic films with NaOH stripping. *Turkish Journal of Chemistry*, **27**: 127–133.

[49] Ingale SS, Rele MV, Srinivasan MC (2002) Alkaline protease production by *Basidiobolus* (N.C.L. 97.1.1): effect of 'darmform' morphogenesis and cultural conditions on enzyme production and preliminary enzyme characterization. *World Journal of Microbiology and Biotechnology*, **18**(5): 403–408.

[50] Cavello IA, Hours RA, Cavalitto SF (2013) Enzymatic hydrolysis of gelatin layers of X-Ray films and release of silver particles using keratinolytic serine proteases from *Purpureocillium lilacinum* LPS # 876. *Journal of Microbiology and Biotechnology*, **23**(8): 1133–1139.

[51] Al-Abdalall AH, Al-Khaldi EM (2016) Recovery of silver from used X-ray film using alkaline protease from *Bacillus subtilis* sub sp. *Subtilis*. *African Journal of Biotechnology*, **15**(26): 1413–1416.

[52] Ford-Smith, MH (1964) *The Chemistry of Complex Cyanides* (Her Majesty's Stationery Office, London).

[53] Chadwick BM, Sharpe AG (1966) Transition metal cyanides and their complexes. *Advances in Inorganic Chemistry and Radiochemistry*, **8**: 83–176.

[54] Brandl H, Lehmann S, Faramarzi MA, Martinelli D (2008) Biomobilization of silver, gold, and platinum from solid waste materials by HCN-forming microorganisms. *Hydrometallurgy*, **94**(1–4): 14–17.

[55] Sathaiyan N, Nandakumar V, Ramachandran P (2006) Hydrometallurgical recovery of silver from waste silver oxide button cells. *Journal of Power Sources*, **161**(2): 1463–1468.

[56] Aktas S (2010) Silver recovery from spent silver oxide button cells. *Hydrometallurgy*, **104**(1): 106–111.

[57] Nan J, Han D, Yang M, Cui M, Hou X (2006) Recovery of metal values from a mixture of spent lithium-ion batteries and nickel-metal hydride batteries. *Hydrometallurgy*, **84**(1–2): 75–80.

[58] Jadhav U, Hocheng H (2013) Extraction of silver from spent silver oxide zinc button cells by using *Acidithiobacillus ferrooxidans* culture supernatant. *Journal of Cleaner Production*, **44**: 39–44.

[59] Aung K, Ting Y (2005) Bioleaching of spent fluid catalytic cracking catalyst using *Aspergillus niger*. *Journal of Biotechnology*, **116**(2): 159–170.

[60] Wu H, Ting Y (2006) Metal extraction from MSW incinerator fly ashechemical leaching and fungal bioleaching. *Enzyme Microbial Technology*, **38**(6): 839–847.

[61] Hocheng H, Chang J, Jadhav U (2012) Micromachining of various metals by using *Acidithiobacillus ferrooxidans* 13820 culture supernatant experiments. *Journal of Cleaner Production*, **20**(1): 180–185.

[62] Kim S, Bae J, Park H, Cha D (2005) Bioleaching of cadmium and nickel from synthetic sediments by *Acidithiobacillus ferrooxidans*. *Environmental Geochemistry and Health*, **27**(3): 229–235.

[63] Ma H, Suhling JC (2009) A review of mechanical properties of lead-free solders for electronic packaging. *Journal of Material Science*, **44**(5): 1141–1158.

[64] Hocheng H, Hong T, Jadhav U (2014) Microbial leaching of waste solder for recovery of metal. *Applied Biochemistry and Biotechnology*, **173**(1): 193–204.

[65] Fomina M, Gadd GM (2014) Biosorption: Current perspectives on concept, definition and application. *Bioresource Technology*, **160**: 3–14.

[66] Volesky B (1990) "Biosorption by Fungal Biomass." In *Biosorption of Heavy Metals*, B Volesky (ed.) (CRC Press, Boca Raton, FL), pp 139–172.

[67] Malik A (2004) Metal bioremediation through growing cells. *Environmetal International*, **30**(2): 261–278.

[68] Gadd GM (2009) Biosorption: Critical review of scientific rationale, environmental importance and significance for pollution treatment. *Journal of Chemical Technology and Biotechnology*, **84**(1): 13–28.

[69] Merroun ML, Ben Omar N, Alonso E, Arias MJM, Gonzalez-Munoz T (2001) Silver sorption to *Myxococcus xanthus* biomass. *Geomicrobiology Journal*, **18**(2): 183–192.

[70] Pethkar AV, Paknikar KM (2003) Thiosulphate biodegradation-silver biosorption process for the treatment of photofilm processing wastewater. *Process Biochemistry*, **38**(6): 855–860.

[71] Sneha K, Sathishkumar M, Mao J, Kwak IS, Yun YS (2010) *Corynebacterium glutamicum*-mediated crystallization of silver ions through sorption and reduction processes. *Chemical Engineering Journal*, **162**(3): 989–996.

[72] Das D, Das N, Mathew L (2010) Kinetics, equilibrium and thermodynamic studies on biosorption of Ag(I) from aqueous solution by macrofungus *Pleurotus platypus*. *Journal of Hazardous Materials*, **184**(1–3): 765–774.

[73] Li L, Hu Q, Zeng J, Qi H, Zhuang G (2011) Resistance and biosorption mechanism of silver ions by *Bacillus cereus* biomass. *Journal of Environmental Science*, **23**(1): 108–111.

[74] Holan ZR, Volesky B, Prasetyo I (1993) Biosorption of cadmium by biomass of marine algae. *Biotechnology and Bioengineering*, **41**(8): 819–825.

[75] Tsezos M, Volesky B (1982) The mechanism of uranium biosorption by *Rhizopus arrhizus*. *Biotechnology and Bioengineering*, **24**(2): 385–401.

[76] Zhao Y, Wang D, Xie H, Won SW, Cui L, Wu G (2015) Adsorption of Ag (I) from aqueous solution by waste yeast: Kinetic, equilibrium and mechanism studies. *Bioprocess and Biosystems Engineering*, **38**(1): 69–77.

Chapter 7
Silver Recovery Methods from Photographic Wastes

Nuri Nakiboğlu

Department of Chemistry
Balikesir University Science and Arts Faculty
Balıkesir-Türkiye

7.1 Introduction

The word "photography," derived from the Greek words *photos* (meaning "light") and *graphein* (meaning "drawing"), when literally translated, means "drawing with light". The effect of light on silver salts was known to early alchemists, but Thomas Wedgwood, a British inventor, was the first to successfully reproduce images as negatives on paper impregnated with silver salts at the beginning of the nineteenth century. In 1819, Herschel discovered the fixing properties of sodium thiosulfate, thus paving the way for permanent pictorial reproductions and making the first exhibit of so-called "photographs" in 1839 possible. Now photography plays an essential importance in modern civilization [1].

Photography is the process of producing a visible image upon a substance by the action of light or radiant energy. The light-sensitive substances employed are silver halides, diazo compounds, amorphous selenium, organic photoconductors, and zinc oxide. Silver halides are employed as the sensitive substance in most widely used photographic materials. Such materials have a natural sensitivity to ultraviolet and blue radiation. Their sensitivity to green, red, and infrared radiations is negligible unless special sensitization to these regions is induced by the adsorption of the so-called sensitizing dyes on their surfaces. Light-sensitive silver salts are prepared in the dark as aqueous gelatin emulsions, which are coated on a support such as glass, paper, plastic, or some other material, after proper additional treatment.

Exposure of the emulsion to radiant energy leads to a photochemical reaction occurring on the surface of the silver salt crystal. Photolytic silver, constituting a latent image, forms roughly in proportion to the amount of incident light of the color to which the crystal is sensitive. The latent image serves to catalyze the further conversion of the entire crystal to a metallic silver image upon the action of a developing agent. The density of the image depends upon the number of the crystals exposed and the extent to which development is permitted to occur. Removal of unexposed silver salts is accomplished by the formation of soluble complexes. The result of this process is an image, and the density of the image depends on the intensity of the initiating light energy.

All films consist essentially of an emulsion on a film support of cellulose acetate, polyester, or glass. The emulsion is composed of a suspension of minute silver halide crystals in gelatin, suitably sensitized by the addition of certain dyes or various classes of silver and gold compounds. In addition, anti-fogging agents, hardening agents, stabilizers, coating aids, and anti-halation baking are used. The back of the film is generally coated with a layer of hardened gelatin to prevent curling.

Silver is one of the natural precious metals known from the earliest times. The abundance of natural silver in the Earth's crust is about 0.080 mg/kg, and its annual production is about 24,000 tons, based on 2012 data [2]. It is found usually in the crust combined with sulfur, chlorine, antimony, and arsenic. The most widely used ores are argentite or "silver glance", Ag_2S, chlorargyrite or "horn silver", AgCl, and galena (lead(II) sulfide) containing silver. Silver can also be found in copper, gold, lead, and zinc ores and is obtained from them as a byproduct.

Figure 1 shows the global industrial use of silver based on 2010 data. Industrial silver demand in the world has been gradually increasing due to its useful properties, including high reflectivity, electrical conductivity, thermal conductivity, and high melting point (about 960°C). Silver is widely used in industrial and technological fields, such as in the manufacturing of batteries and electronics, catalysts for the production of ethylene oxide

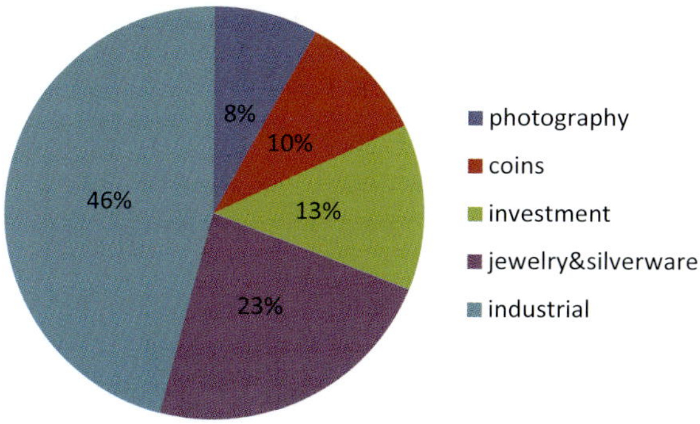

Fig. 1. Global industrial use of silver.

from ethylene, in the windows and the glass industry, in solar energy panels, in disinfection of water, an anti-bacterial in medicine, and in nanotechnology. Silver used in photography has tailed off as a result of digital technology. However, the classical X-ray technique using silver halide is still preferred in medical diagnosis by doctors because of its tremendous accuracy and cost-effectiveness compared to digital X-ray technology. For this reason, conventional X-ray techniques are preferred in developing countries. Considering that there is a fixed amount of silver present in the earth, and industrial and technological silver demand is increasing, we can say that the recovery of silver from photographic waste is still very important. Important photographic materials containing silver are waste emulsions and broken films, waste photographic films, waste X-ray films, and spent fixing solutions. Silver that can be recovered is found in various forms, such as metallic silver, silver ion, soluble silver thiosulfate complex, and insoluble silver halide, depending on the source used. Waste emulsion and broken films contain silver as silver halide. Waste photographic films, including waste X-ray films, contain metallic silver, while silver in the form of silver thiosulfate complex compound is found in the spent fixing solutions.

The methods used for silver recovery should have some advantages such as cost- and time-effectiveness, simplicity, efficiency, and being environmentally friendly. Therefore, this chapter focuses on the silver recovery methods from photographic waste and the methods for determination of purity and impurities of the silver recovered.

7.2 Silver Recovery from Photographic Processing Solutions

Generally, silver content in spent photographic fixing solutions varies from 5 g/L to 12 g/L. Various silver recovery methods such as electrolysis, metallic replacement, chemical precipitation, and ion exchange is used in photographic processing solutions. These methods can have different benefits according to the type, silver content, amount of waste, and budget of the user. Some techniques such as evaporation, distillation, and reverse osmosis are used for pre-concentration of the diluted silver solutions. Additionally, the combination of two methods can be preferred to reduce lost silver and to adapt to environmental regulations. For this reason, in this section, irrespective of the source of silver waste, the basic principles, advantages, and disadvantages of silver recovery methods will be discussed.

Electrolytic recovery of silver has the advantage of reducing the amount of silver lost and may allow chemical conservation in some situations. Metallic replacement is simple to use and relatively inexpensive, while refining costs and increased discharge of iron are the main disadvantages. Ion exchange, the most frequently used method for treating large volumes of wash water, has shown operational difficulties, such as biological growth, resin fouling, and the stripping of silver off the resin by thiosulfate. These problems are gradually being solved by process and equipment modifications.

7.2.1 *Silver recovery by electrolysis*

7.2.1.1 *Equipment for silver recovery by electrolysis*

Silver recovery by electrolysis is a process based on the application of low voltage causing the passage of direct current between two electrodes in the solution containing silver. There is a great variety of electrolytic cells depending on the size and type of application [3]. However, all cells have some features in common: electrodes, a power supply, some method of agitation, and various types of containers. The main components of an electrolysis unit are shown in Fig. 2. Graphite is usually used for anode, and graphite or stainless steel is used for cathode [4, 5].

The dc power supply is usually adjustable up to 3 or 4 V. The largest commercially available cells may have 200 A power supplies. The cells

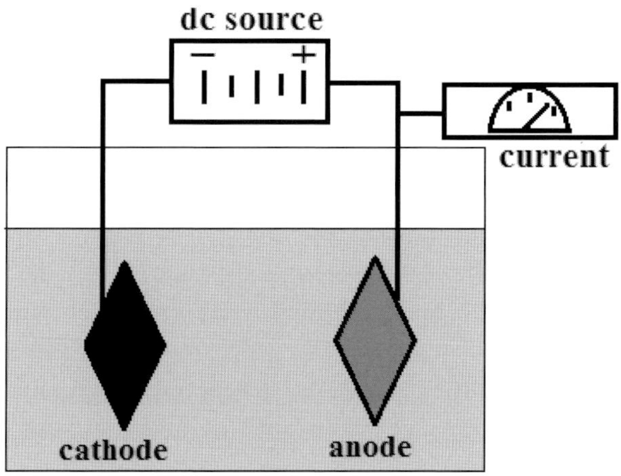

Fig. 2. Main components of an electrolysis unit.

may be turned on and off manually or with a timer, may be linked to the feed mechanism of a processor, or may sense silver concentration of the solution. Agitation is important to improve the mass transport of silver thiosulfate to the cathode surface.

Electrolytic cells may be operated in a batch mode recirculated to the processing machine or may treat the solution as it overflows from the processor. A fixed cathode or a rotating cathode can be used in the cell. Generally, a cell with a fixed cathode has been used for silver recovery from black-and-white fixers and color fixers, whereas a cell with a rotating cathode has been used for solutions containing high iron complexes. In such solutions, a competitive reduction reaction between iron and silver at cathode can occur. The iron may be more easily reduced at the cathode than silver, so the current efficiency for silver plating is decreased, sometimes to as low as 10% [6]. For this reason, it is necessary to improve the mass transportation of silver to the electrode surface by rotating the cathode, resulting in higher plating efficiency. Both features have been combined in a single unit by some companies. Each of the methods has its own advantages and limitations. However, the main factors taken into account for the selection of the electrolytic equipment for silver recovery from photographic solutions are the type of fixer or its ingredients and the volume of the solution being treated.

7.2.1.2 *Chemistry of photographic fixing and silver electroplating*

There are two kinds of photographic fixing solutions used in the photographic process. The solution used in development of black-and-white films, called "fix", contains sodium or ammonium thiosulfate ($Na_2S_2O_3$/$(NH_4)_2\ S_2O_3$) and sodium sulfite (Na_2SO_3), while the solution used in development of color films, called "bleach fix", contains sodium thiosulfate, sodium sulfite, and sodium iron (III) ethylenediaminetetraacetic (NaFeEDTA). After the development process of the film, silver halides (AgCl/AgBr) on the film dissolve to give a silver thiosulfate complex ($[Ag(S_2O_3)_2]^{3-}$).

$$\begin{aligned} AgCl + 2S_2O_3^{2-} &\rightarrow [Ag(S_2O_3)_2]^{3-} + Cl^- \\ AgBr + 2S_2O_3^{2-} &\rightarrow [Ag(S_2O_3)_2]^{3-} + Br \end{aligned} \qquad (1)$$

Thus, silver is found as a silver thiosulfate complex $[Ag_2S_2O_3]^{3-}$ in photographic processing solutions. For this reason, in an electrolytic cell, the silver thiosulfate complex is reduced to give metallic silver and thiosulfate ion at the negatively charged cathode, and usually sulfite is oxidized to give sulfate and hydrogen ions at the positively charged anode [7–9]. The electrode reactions as follows:

$$[Ag(S_2O_3)_2]^{3-} + e^- \rightarrow Ag_{(k)} + 2S_2O_3^{2-} \quad \text{(at cathode)} \qquad (2)$$

$$SO_3^{2-} + H_2O \rightarrow SO_4^{2-} + 2H^+ + 2\bar{e} \quad \text{(at anode)} \qquad (3)$$

On the other hand, the following reactions can occur in fixing solution:

$$S_2O_3^{2-} \rightarrow S + SO_3^{2-} \quad \text{(in fixing solution)} \qquad (4)$$

$$S_2O_3^{2-} + H_2O \rightarrow S^{2-} + SO_4^{2-} + 2H^+ \quad \text{(in fixing solution)} \qquad (5)$$

These reactions occur when the fixer solutions are allowed to stand without recovering the silver. Following this, the sulfide ion reacts with the silver ion in the solution, resulting in the formation of an insoluble silver sulfide by a process called sulfidization.

$$2[Ag(S_2O_3)_2]^{3-} + S^{2-} \rightarrow Ag_2S_{(k)} + 4S_2O_3^{2-} \quad \text{(sulfidization)} \qquad (6)$$

The silver deposited on the cathode is removed periodically and is stripped off for further usage.

Adjustment of pH is an important factor in the electrolytic silver recovery process. Electrolytic recovery of silver in the used black-and-white fixing solution may not require an adjustment of pH. However, pH adjustment is necessary for the electrolytic recovery of silver from the used bleach fix solutions since they contain iron (III) ethylenediaminetetraacetic complex, $[FeEDTA]^-$. Generally, the pH of solutions containing $[FeEDTA]^-$ is adjusted to 8 before electrolysis. $[FeEDTA]^-$ can oxidize the silver deposited, resulting in the dissolution of silver from the electrode surface at pH lower than 7. Maintaining the solution pH at about 8 increases the recovery efficiency of the silver as the oxidizing capability of $[FeEDTA]^-$ decreases. On the other hand, current density and deposition time are the other important factors to successfully conclude silver recovery. Increasing these two factors results in the precipitation of black silver sulfide at the cathode. This situation causes the recovery process to stop. Silver recovery efficiencies of about 95% and 90% have been reached for the used black-and-white fixing solution and the used bleach fix solutions, respectively, through proper equipment and pH level maintenance.

The main drawbacks of silver recovery by electrolysis are the high investment costs of the equipment, involving special electrical connections and installation, and high operational costs due to electricity consumption, maintenance, and monitoring. Additionally, the efficiency of electrolysis is limited if the silver concentration is low in the sample.

7.2.2 *Metallic replacement (cementation)*

Metallic replacement, sometimes called cementation, occurs when an active metal, such as iron, comes in contact with a solution containing ions or complexes of a less active metal such as silver. The more active metal goes into the solution as an ion or complex and the less active metal precipitates or is deposited on the surface of the more active metal. In the photographic processing industry, iron is the most used active metal. Steel wool is most often used, but steel window screens and various forms of powdered or finely divided iron are also used. The iron is usually packed into a bucket, cartridge, or column and used as a packed bed reactor.

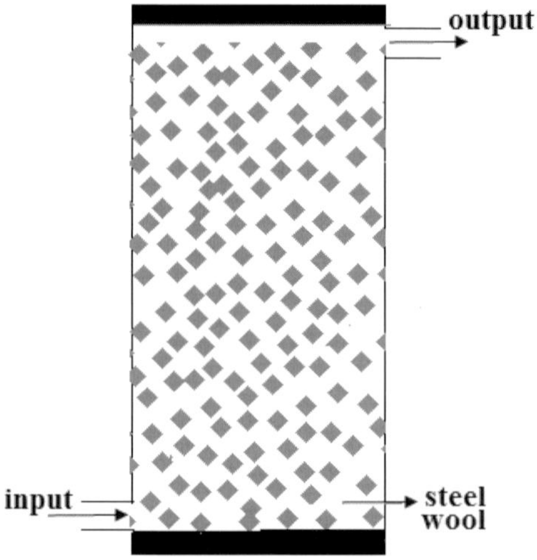

Fig. 3. Metallic replacement system.

A schematic representation of the metallic replacement system is shown in Fig. 3.

The reaction occurring in the metallic replacement system is a redox process involving the oxidation of the metallic iron and the reduction of the silver thiosulfate. The reaction is as follows:

$$Fe + 2[Ag(S_2O_3)_2]^{3-} + \rightarrow Fe^{2+} + Ag + 4S_2O_3^{2-} \quad (7)$$

The composition of the resulting solution from the cartridge may change over time with the following reactions:

$$\begin{aligned} &Fe^{2+} \rightarrow Fe^{3+} + e \\ &Fe^{3+} + 2S_2O_3^{2-} \rightarrow [Fe(S_2O_3)_2]^- \\ &Fe^{3+} + [Fe(S_2O_3)_2]^- \rightarrow Fe^{2+} + S_4O_6^{2-} \end{aligned} \quad (8)$$

Reacting with silver thiosulfate, the iron may be dissolved by the acidity of the fixer or by the ferric ion in bleach fix [10, 11]. For this reason, the pH of the solution is an important parameter affecting the silver recovery efficiency. The recommended pH range is usually between 5.5 and 6.5, as pH values lower than 5.5 may cause reduction in the capacity of the cartridge because of the rapid dissolution of the steel wool in the system. Additionally, lower pH conditions can cause the decomposition of the thiosulfate complex, resulting in the formation of elemental sulfur (S) and silver sulfide (Ag_2S) and release of the toxic hydrogen sulfide (H_2S) gas [12]. High pH values also decrease the silver recovery efficiency due to slow redox reaction between silver and iron. In practice, only 25% or less of the iron reacts to recover the silver. The rest of the iron either undergoes secondary reactions or is coated with silver so that the fresh solution cannot reach the surface to react. Sodium carbonate or acetic acid can be used for pH adjustment of the photographic fixing solution before the metallic replacement process. Besides pH, other factors such as flow rate and contact time of the photographic fixing solution and surface area of iron are important parameters affecting the recovery efficiency of silver. The increased contact time under air can cause the re-dissolution of the metallic silver into the solution according to the following reaction:

$$Ag^0 + 1/4O_{2(\text{from air})} + 2S_2O_3^{2-} + 1/2H_2O \rightarrow [Ag(S_2O_3)_2]^{3-} + OH^- \quad (9)$$

It is recommended to keep the solution under an argon atmosphere to avoid this unfavorable situation [12].

Metallic replacement system (cartridge or column) must be conditioned prior to the procedure. To do this, a weakly acidic fixing solution is placed inside the cartridge and allowed to stand for a few hours. Cartridges of various types and sizes produced by different manufacturers for the recovery of silver from the photographic fixing solution are commercially available in the market. Two cartridges or columns are usually used in series to ensure consistently low silver concentration.

Aluminum, copper, and zinc have also been tried for silver recovery. Aluminum has been found to be very erratic in performance, probably because of the tendency to form protective oxide coatings on its surface. Copper and zinc are more expensive, more environmentally harmful, and more strictly regulated than iron.

The greatest advantages of metallic replacement for silver recovery are the low cost of installation and ease of use. Although metallic replacement is a very effective technique, it has several disadvantages. It is not efficient, and the silver recovered contains more impurities due to the presence of other metals in the sludge compared to the electrolysis and precipitation methods. Additionally, further treatment is required as the recovered silver is obtained in the form of a sludge. Most of the iron that reacts with the solution cannot be recovered. In some countries with very low iron discharge requirements, metallic replacement cannot be used. Sometimes the iron precipitates as rust, causing the clogging of drain lines. The refining cost for silver mixed with the iron residue is more expensive than that for the relatively pure silver recovered by electrolytic cells. The mixture of silver and iron remaining in the cartridge may also be considered a hazardous substance unless it is rinsed to remove any unreacted soluble silver [13].

7.2.3 *Chemical precipitation*

Precipitation of silver as its insoluble compounds is the oldest technique capable of treating large volumes of the used photographic fixing solutions rapidly. For this purpose, a number of chemical reagents such as sodium or potassium sulfide (in the past) and trimercapto-s-triazine [14] (recently) have been used. On the other hand, some strong reducing agents such as sodium borohydride [15] and sodium dithionite [16] are able to precipitate silver as metallic silver sometimes mixed with silver sulfide.

Oxidizing reagents can also be used to precipitate silver from photoprocessing solutions [17]. However, a different chemical mechanism is involved. In almost every photographic process, the silver is complexed with thiosulfate that is in a molar excess of 10–100. When the thiosulfate is oxidized, usually to tetrathionate or sulfate, the silver precipitates as silver sulfide, a silver halide, or other insoluble salt. Usually all of the thiosulfate must be destroyed before a significant amount of silver will precipitate. The oxidants most often considered are sodium hypochlorite, hydrogen peroxide, and ozone.

If silver is precipitated as silver sulfide, the amount of silver in the solution must be determined to avoid overdosing. If excess sulfide is added to the solution, toxic hydrogen sulfide gas is released. Removing the precipitated silver from the solution is usually performed by filtration or a combination of settling, decanting, and filtering. For large-scale operations, the settle–decant–filter procedure has been found to work reasonably well [18]. The precipitate is allowed to settle and the supernatant decanted and filtered through a fine polishing filter placed at the drain.

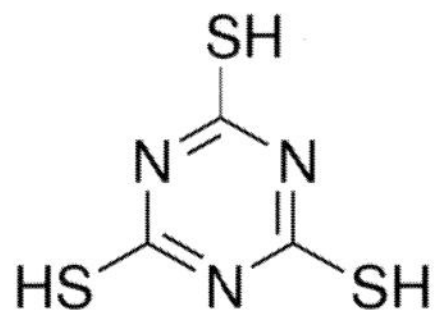

Fig. 4. Chemical structure of trimercapto-s-triazine.

The main advantage of the oldest chemical precipitation based on precipitation of silver sulfide is that large volumes of solution may be treated quickly and the silver lost is very low because of the less solubility of silver compound formed. The disadvantages are the possible contamination of the solution, the toxicity and difficulty in handling the chemical reagents, and the general problems of the liquid–solid separation step. In addition, sodium sulfide and other compounds containing sulfide or bisulfide can generate hydrogen sulfide, a very toxic gas.

Recently, trimercapto-s-triazine (see chemical structure in Fig. 4) has been used for the precipitation of silver from used photographic fixing solutions.

It is reported that the stochiometry of silver–trimercapto-s-triazine precipitate is dependent on solution pH and expressed as follows [14]:

$$3Ag^+ + (C_3N_3S_3)^{3-} \rightarrow Ag_3C_3N_3S_3.nH_2O \quad \text{(at pH} > 12.5) \qquad (10)$$

$$2Ag^+ + (C_3N_3S_3)^{3-} \rightarrow Ag_2C_3N_3S_3.nH_2O \quad \text{(at pH} = 8\text{–}10) \qquad (11)$$

$$Ag^+ + (C_3N_3S_3)^{3-} \rightarrow AgC_3N_3S_3.nH_2O \quad \text{(at pH} = 5\text{–}8) \qquad (12)$$

Compared with other methods that result in the precipitation of silver sulfide, precipitation of silver with trimercapto-s-triazine has important advantages such as simplicity, effectiveness, low laboratory cost, low toxicity, and formation of non-noxious products such as hydrogen sulfide. The only drawback is the high price of trimercapto-s-triazine.

7.2.4 *Ion exchange*

Ion exchange is used to recover silver from washes following fixers or bleach fixes. However, it is not effective for recovering silver from concentrated processing solutions [19]. In this procedure, silver thiosulfate, an anionic complex, is adsorbed on to an anion-adsorbing resin. The anion displaced may be thiosulfate, bisulfate, or chloride, depending on how the resin was regenerated. The resin may be regenerated by stripping the silver thiosulfate complex off the resin with a concentrated solution of ammonium thiosulfate [20] or ammonium chloride [21]. The silver is then electrolytically recovered, and the regenerating solution can be reused.

The capacity of an ion-exchange resin for adsorbing silver thiosulfate depends on both the silver concentration and the concentration of other salts, especially thiosulfate, in the feed stream or leakage concentration, and is also strongly affected by the thiosulfate concentration [22]. Silver in the treated stream is typically in the 0.5–1 mg/L range. With extreme care in monitoring and operation, it may be possible to reduce the silver concentration to 0.1–0.2 mg/L. However, if a fixer or bleach fix is added to the feed stream, the silver concentration in the treated stream will increase. It is very important to keep the thiosulfate concentration as low and as constant as possible. Conductivity controllers are often used to ensure that high concentrations of thiosulfate do not reach the resin columns. Biological growth in the wash water is difficult to control and can cause an accumulation of dirt in columns [23]. The organic components of film and paper are adsorbed on to the ion-exchange resin and may eventually decrease the exchange capacity of the resin. However, if the resin is changed yearly, this is not usually a major problem.

7.2.5 *Other silver recovery methods*

There are two other methods that have been used experimentally: biological treatment and ultraviolet radiation. In biological treatment, conventional activated sludge waste treatment has been used to treat photoprocessing effluents. In addition to reducing the biological oxygen demand (BOD) and chemical oxygen demand (COD), silver is precipitated as a mixture of silver and silver sulfide. This particulate silver was incorporated into the biomass of the treatment plant, and in one case, reached a concentration of $\sim$900 mg/L [24]. A few experiments have been conducted by the Eastman Kodak Company to evaluate the feasibility of using specialized microorganisms for silver recovery. The investigators at the company found a strain of green algae that would bioaccumulate silver from dilute thiosulfate solutions to 3–4% of the dry biomass weight [13].

Ultraviolet and X-ray radiation precipitate silver and silver sulfide from fixers and washes. Experiments with short-wave ultraviolet light showed that silver sulfide could be precipitated fairly rapidly, but that the solution soon became so opaque that the ultraviolet light can no longer penetrate far enough to be effective. Attempts at continuous filtration to remove the precipitate as rapidly as it formed were frustrated by the near-colloidal particle size of the silver.

Ultraviolet light is used to control biological growth in some water treatment equipment for photographic washes. The procedure is reported to be effective for its intended purpose, but some precipitation of silver or silver sulfide on the light element has also been reported [13]. So far there has not been enough experience to indicate whether ultraviolet treatment will be a useful method for silver recovery.

7.3 Silver Recovery Methods from Waste Photographic Films

A waste photographic film consists of three principal components; polyester film support material, gelatin, and metallic silver dispersed in gelatin. Therefore, mainly three ways are possible to recover silver from the film. One of these is burning the film directly in a proper medium. However, in this case, the support material cannot be recovered. The other two ways are to oxidize the metallic silver to silver ion with a chemical or to strip the gelatin–silver layer from the support material by hydrolysis of gelatin with a chemical or an enzyme. These possible methods are summarized in Fig. 5.

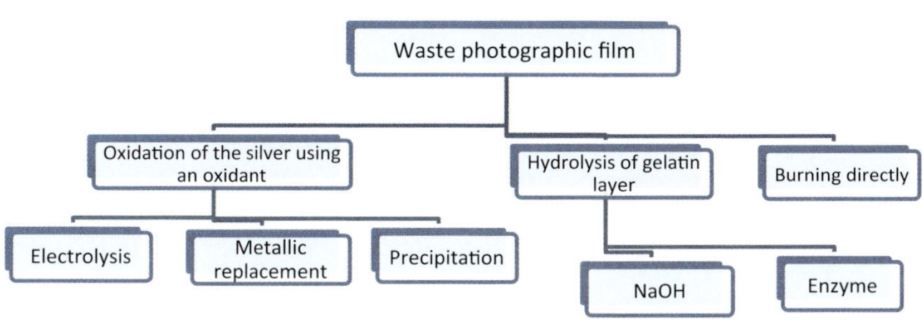

Fig. 5. Silver recovery methods from waste photographic films.

7.3.1 *Burning methods*

Burning methods used for silver recovery from the used photographic films, generally, are based on burning the used photographic films under controlled conditions and recovering the metallic silver by smelting [25] or by burning the films in a double chamber furnace and collecting the silver by using an electrostatic precipitator [26]. In the other studies, silver has been recovered by pyrolysis the films at 500–600°C [27] or at 500–800°C [28] in a non-combustible atmosphere and then by the separation of silver from the ash. It is reported that silver recovery can be performed without losing the silver in smoke and particules. Besides these methods, silver has been recovered by burning in the presence of $BaCO_3$ [29] or by smelting with Na_2CO_3 in the presence of C powder for 1 h at 1000°C after mechanically removing the emulsion layer [30]. In addition, some methods having small differences from each other have been reported for recovering silver from the waste photographic films by burning [31, 32]. However, burning methods used for silver recovery are old-style methods and generate a nasty smell. Additionally, these methods cause environmental pollution, and the polyester substrate, which is the other component of waste photographic film, cannot be recovered. For this reason, at present, new methods capable of recovering both silver and polyester substrate from waste photographic films are used.

7.3.2 *Oxidation of the silver on the film, then recovery by electrolysis*

Various chemicals have been used for silver recovery by oxidation of the silver on the used photographic film, such as aqua regia [33], $KMnO_4$, $K_2Cr_2O_7$ [34], and $Ce(SO_4)_2$ [35] in acidic medium; $FeCl_3$, $FeBr_3$, or $Fe_2(C_2O_4)_3$ in acidic or basic medium [36–39]; H_2O_2 in the presence of NaCN [40]; HNO_3 at room temperature [41]; and NaClO at room temperature [42–46]. After the pretreatment step, the silver ion, Ag^+, is recovered by a proper electrolysis system, as mentioned earlier.

7.3.3 *Stripping the gelatin–silver layer, then recovering by various ways*

The waste photographic films are a good source of silver containing about 1.5–2% metallic silver by weight in the gelatin layer. For this reason, various leaching methods using chemicals or enzymes for stripping the gelatin–metallic silver layer from the film base have been reported. In these methods, generally, metallic silver obtained has been recovered by smelting. There are also several methods to recover the silver by filtration or decanting.

7.3.3.1 *Chemical methods*

Silver has been recovered by using H_2O as a stripping agent to remove the gelatin–silver layer from waste film [47]. In this process, used films are cut into small chips and stirred in a warm water bath. Silver is recovered by electrolysis after separating the chips from the bath. A process involving rotating frames in hot water (90°C) for 15–20 min has been recommended by Mozgalov and Kuznetsov [48] to recover silver from waste films.

In another study [49], used films have been soaked in water for 0.5 h, acidified to pH 4 with diluted HCl, and heated to 80°C for 0.75 h. Ag–gelatin sol is coagulated at 60°C with $Al_2(SO_4)_3$. The Ag formed is separated by sedimentation and then washed and dried.

Silver has been also recovered from waste photographic films by immersing the films at 90–95°C in 15–20% urea nitrate, $(NH_2)_2CO.HNO_3$, solution for 3–5 min followed by transferring the films into clean water [50]. The silver precipitated at the bottom of the solution is filtered and fluxed in a muffle furnace at 1,100°C. It is reported that the purity of silver recovered was greater than 99%, and the method did not take a long time and has a low cost.

A process is described by Syed *et al.* [51] for the recovery of silver from processed radiographic films by heating the films with oxalic acid solution to boiling temperature. The silver is recovered by melting, after separating the silver from the polymer substrate.

Alkaline metal hydroxides have also been used for stripping the gelatin layer from the waste film. In some of the studies [52, 53] imaged photographic films are passed through $\geq$1 M NaOH baths (5–15% NaOH, at 80°C), and the gelatin layer is stripped as Na salts of amino acids and free metallic silver in the form of fine suspension. The metallic silver is then collected and electrolytically refined for photographic use. In another study [54], waste X-ray film chips were treated with an alkaline bath containing 0.2 L saturated KOH, 181.4 g KCl, and 1 L of water at 80°C and stirred for 1 h. The resulting slurry is filtered for separation of the silver.

Silver has been also recovered by using caustic (especially NaOH) solution to remove the gelatin–silver layer from the film base (especially polyester) by stirring at 60–95°C [55]. The sludge is then filtered to separate the film base and acidified to ~pH 3 to flocculate the metallic silver for settling and recovery.

The used photographic films with polyethylene terephthalate support and gelatin layer containing metallic silver have been heated in a boiling solution of NaOH in aqueous ethanol, with the recovery of silver having purity of greater than 99% [56].

In another study [57] using alkaline hydroxide, waste photographic films were treated for 1–2 min with a 10% NaOH or KOH solution in the presence of 5 g sodium dinaphthylmethane sulfonate/L at 70–80°C. Then the film support is removed from the solution and pH adjusted to 5.5–6.0 by the addition of 10–15% HCl, and silver is separated by filtration or centrifugation from the silver-containing sludge. After that, silver is recovered by dispersing the sludge in an aqueous 3% hydrazine solution, 5% hydrazine hydrate solution, 11% hydrazine hydrochloride solution, or 13% hydrazine sulfate solution with the precipitate/solution ratio of 2.1 at 40°C. The resulting silver powder is filtered, rinsed, and dried for 4 h at 70–80°C. Optionally, silver of 99% purity is recovered by calcining the sludge.

A silver recovery method [58] performing both stripping and recovering steps in the same NaOH solution at 70–80°C without any filtering, centrifuging, smelting or electrolysis has been reported. In this method, at first, silver stripped to the solution is observed as a colloid state. After heating the colloidal solution with effective stirring, coarse-grained and bright metallic silver particles with a purity level greater than 99% are obtained. The photograph of the recovered silver by this method is shown in Fig. 6, with the average diameter of the silver particles being about 0.6–0.7 mm. It is also reported that the method is simple, fast, cheap, and pollution-free because of less laboratory investment, low operation steps, and relatively low energy consumption. In another study [59], the spent fixing solution is used for recovering silver from waste X-ray films. The process is based on the reduction of silver in the used fixing solution by KBH4 at pH −9, and then the silver-free used fixing solution is used for silver recovery. It is reported that the purity of recovered silver is more than

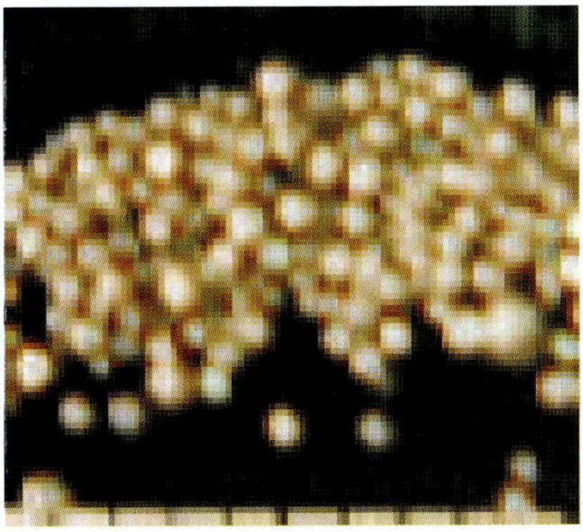

Fig. 6. A view of the silver obtained by NaOH stripping [58]. The real particle size is about 0.6–0.7 mm.

99.5%, and total silver recovery from the used fixing solution and waste X-ray films was 98.0% and 95.8%, respectively.

Some chemical methods in the literature for silver recovery from the used photographic films are summarized in Table 1.

7.3.3.2 *Enzymatic methods*

An alternative way to burning and chemical methods for stripping the gelatin–silver layer from the used photographic film surface is the use of gelatin-hydrolyzing enzymes. Various enzymes have been also used for stripping the gelatin–silver layer on waste films such as pancreatin [60], papain [61], trypsin, and alkaline proteases [62].

It has been suggested that microorganisms producing enzyme can be used for the decomposition of the gelatin layer from the waste film base. For example, in a study [63], 25 g of waste photographic films was immersed in 75 mL water and incubated with Pseudomonas B 132 at 45°C for 3 h to remove the gelatin layer. In the other study [64], the used photographic films were treated with an enzyme obtained from *Streptomyces rimosus* oxytetracycline for stripping the gelatin layer. Silver was recovered by smelting in both methods.

On the other hand, well-known enzymes used in silver recovery from waste films are proteases produced by the *Bacillus* species. *Bacillus* is a prokaryotic microorganism which forms an endospore (strong form against bad conditions in the cell) against destructive effects such as temperature,

Table 1. Chemical methods used for stripping silver from the support material.

Stripping medium	Stripping temperature (°C)	Recovery method used after stripping	References
aqua regia	At room temp.		[33]
$KMnO_4$, $K_2Cr_2O_7$	n.r	Electrolysis	[34]
$Ce(SO_4)_2$	n.r	Electrolysis	[35]
$FeCl_3$, $FeBr_3$ or $Fe_2(C_2O_4)_3$ in acidic or basic medium	n.r	Electrolysis	[36–39]
$NaCN + H_2O_2$	At room temp.	Electrolysis	[40]
HNO_3	At room temp.	Electrolysis in the presence of cyanide	[41]
NaClO	At room temp.	Electrolysis	[42–46]
H_2O (hot)	n.r	Electrolysis	[47, 48]
Diluted HCl (pH 4)	80°C for 45 min.	Coagulation in the presence of $Al_2(SO_4)_3$at 60°C	[49]
Urea nitrate (15–20%)	90–95	Smelting	[50]
$H_2C_2O_4$ (1% w/v)	at 97	Melting	[51]
NaOH (5–15%)	at 80	Electrolytic refining	[52, 53]
KOH (saturated) + KCl	at 80	n.r	[54]
NaOH with high shear	at 60–65		[55]
NaOH (in ethanol)	Boiling solution	Smelting	[56]
NaOH or KOH + Na dinaphtyl methane sulfonate (5 g/L)	70–80	Calcination	[57]
NaOH (1–1.5 M)	at 70–80	Decantation	[58]
Waste fixing bath $+KBH_4$ at pH 8 ∼ 9.	At room temp.	Melting	[59]

Note: n.r.: not reported.

detergent, and UV light. It can be isolated from air, water, soil, or compost. Many species are available in acidic, neutrophilic, and alkalophilic forms. Enzymes having industrial importance such as α-amylases and proteases have been produced from the *Bacillus* species [65, 66]. One of these species is neutrophilic *Bacillus subtilis*. For example, subtilisin BPM protease obtained from neutrophilic *Bacillus subtilis* takes 30 min at 50–60°C to decompose the gelatin layer. On the other hand, all alkaline proteases from neutrophilic *Bacillus* species take more than 20 min to act.

A method has been developed by using protease enzyme from the alkolophilic *Bacillus* species B21-2 to decompose the gelatin layer from the waste films at 40°C with pH 10.5 [67, 68]. It has been reported that a plot of the decomposition time and enzyme activity versus pH is hyperbolic, and the contribution of the enzyme activity to the decomposition time of the gelatin layer is less at increased temperature. To decrease the decomposition time and to increase the enzyme activity, thermophilic alkaline proteases such as protease B18 from *Bacillus* species, which can be stable at pH 13°C and 85°C, have been used for the same purpose [69]. Some enzyme methods in the literature for silver recovery from used photographic films are summarized in Table 2.

Silver with a purity of more than 99% can be recovered by conventional combustion of the obtained sludge or by using various electrolysis methods by dissolving the sludge in HNO_3 in all of the above-mentioned methods.

Primarily the waste photographic films are cut into small pieces. The pH and temperature of the previously prepared enzyme solution are set to optimal values at which the enzyme activity is the highest. The films cut into small pieces are submerged in this enzyme solution and allowed to stand until the gelatin–silver layer is stripped completely from the film base. The film base materials are removed from the enzyme solution, and a sludge containing metallic silver is obtained. If possible, the liquid part of the resulting sludge is decanted or filtered. The operator should avoid the loss of silver during this process. Generally, two options are used after the sludge containing metallic silver is obtained: smelting or fluxing and electrolysis. A general procedure for silver recovery from the used photographic films by the enzyme method is given in Fig. 7.

The enzymatic hydrolysis of the gelatin–silver layer on the used photographic film facilitates the recovery of not only silver but also the film base.The main drawback of the enzyme method that it always requires a microbiologist for the production, characterization, and safe elimination of the enzyme used.

7.4 Methods for Determination of Silver and Trace Metal Impurities in High-purity Silver

Determination of silver may be necessary before and after the recovery process to learn the silver content of the photographic waste and purity

Table 2. Enzyme methods used for stripping silver from the support material.

Enzyme	pH	Stripping temperature	Stripping time	Recovery method	References
Pseudomonas B 132	n.r	45°C	3 h	n.r	[63]
Alkalophilic isolate of *Bacillus sphaericus*	11.0	50°C	12 min	n.r	[70]
Bacillus subtilis ATCC 663	8.0	40°C	20 min	Smelting	[71]
Protease from *Bacillus* B21-2	10.5	50°C	8–10 min	n.r	[72]
Bromelain,	6.5	55°C	35 min	Fluxing with Na_2CO_3	[73]
Alkaline protease from *Conidiobolus coronatus* ATCC PTA-4132	10.0	40°C	6–7 min	n.r	[74]
Alkaline protease from *Aspergillus wenti*	8.0	50°C	20 min	n.r	[75]
Alkaline protease from *Bacillus tequilensis* P15	8.0	50°C	20 min	n.r	[76]
Aspergillus flavus MTCC 9952 mutant	11.0	37°C	20 min	n.r	[77]
Alkaline protease from *Bacillus subtilis* (NCIM 2724)	8	50°C	n.r	Smelting in the presence of borax	[78]
Bacillus subtilis sub sp. *Subtilis*	8	50°C	30 min	n.r	[79]

Note: n.r.: not reported.

of the obtained silver.Additionally, after the recovery process is completed, the impurity of the obtained silver should be determined. Various analytical techniques can be used for these purposes. In this section, silver determination methods and methods for determination of impurities in the recovered silver will be discussed.

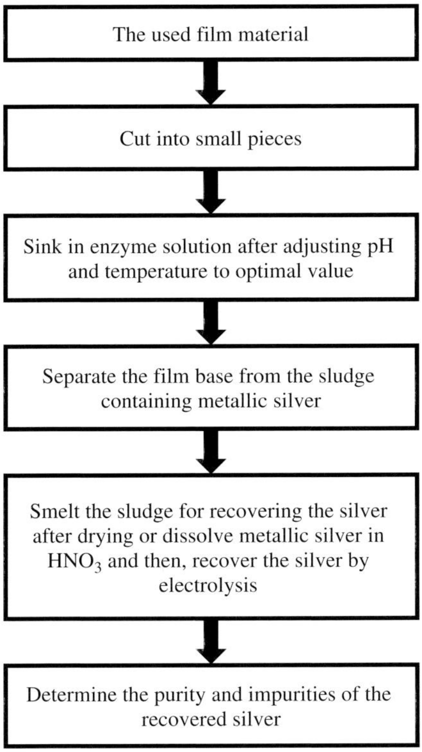

Fig. 7. A general procedure for silver recovery from used photographic films by the enzyme method.

7.4.1 *Methods for determination of silver*

Various techniques such as volumetry, potentiometry, voltammetry, atomic and molecular spectrometry, and neutron activation analysis are available for the determination of silver. Among these, volumetry and potentiometry are simple techniques and do not require any special expertise. The silver in a sample can be determined in gram level by using these methods. Conversely, voltammetry, atomic spectrometry, and neutron activation analysis techniques require special expertise, and these techniques can determine trace amounts of silver or any metal ion. Therefore, an analyst should know which technique is selected for the determination of silver by taking into account both silver content and the origin of the sample. If the purpose is to determine the purity of silver recovered, it is not necessarily

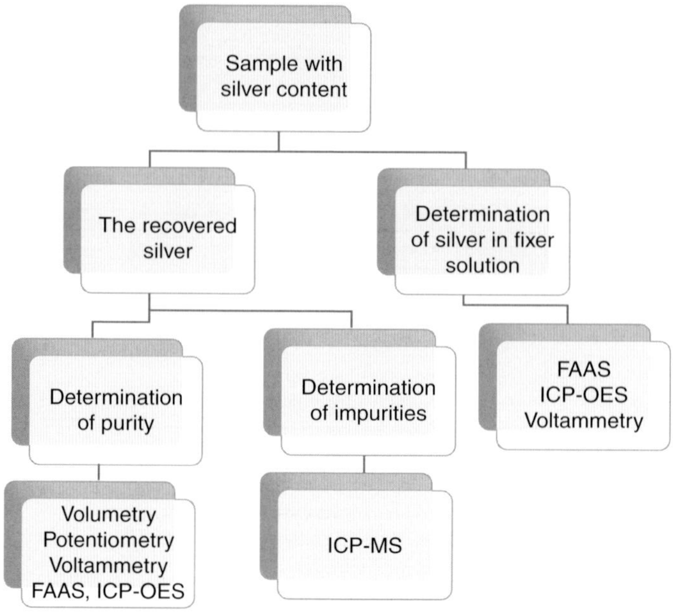

Fig. 8. Summary of different methods to be followed based on the requirements.

required to use a trace element analysis technique as the amount of silver is large enough in the sample. In this case, volumetric or potentiometric methods can be preferred. However, if the purpose is to determine the silver content of the solution after electrolysis or metallic replacement, voltammetric or atomic spectroscopic method should be used. Figure 8 provides a summary of the different methods to be followed based on the requirements. All of these methods will be described in detail in the subsequent sections.

7.4.1.1 *Volumetric method for determination of silver (argentometry, Volhard's Method)*

7.4.1.1.1 Principle of the method

Argentometry is the general name of the titration methods based on the precipitation of silver compounds. Although these methods are commonly used for the determination of chloride ions, they can also be used for the direct determination of bromide, iodide, and thiocyanate. Additionally, the ions which can form a precipitate with silver, such as arsenate, phosphate, and chromate, can be determined indirectly. This argentometric method is known as Volhard's Method. This method is based on the back titration of excess silver with standardized thiocyanate solution in the presence of iron (III) chloride indicator. On the other hand, Volhard's Method can also be used for the determination of silver ion. In this case, the silver ion in the sample is precipitated with excess standardized thiocyanate, and then the remaining thiocyanate is back titrated with standardized silver nitrate in the presence of iron(III) chloride indicator in an acidic medium. The titration end-point is determined by observing the disappearance of the red color. The difference between the numbers of millimoles gives the number of millimole of silver in the sample. Titration reactions used for the determination of silver by Volhard Method are as follows:

$$Ag^{+} + SCN\ (\text{excess}) \leftrightarrow AgSCN + SCN^{-}\ (\text{remaining}) \quad (13)$$

$$Fe^{3+} + SCN^{-}\ (\text{remaining}) \leftrightarrow [FeSCN]^{2+}\ (\text{red}) \quad (14)$$

$$[FeSCN]^{2+}\ (\text{red}) + Ag^{+} \leftrightarrow AgSCN + Fe^{3+} \quad (15)$$

7.4.1.1.2 Experimental procedure

Unless specified otherwise, all chemicals used should be analytical grade.

Silver nitrate solution: Standard silver nitrate solution can be prepared directly by weighing of the high-purity $AgNO_3$. The concentration of the silver solution often used is about 0.1 M. For this purpose:

(i) Dry high-purity $AgNO_3$ for 2 h at 150°C in an oven,
(ii) Weight the required amount of $AgNO_3$ and dissolve it in distilled water.

Silver nitrate solution should be kept in dark bottles as it decomposes (reduces to metallic silver) slowly when exposed to light.

Ammonium ferric sulfate solution: Prepare 10% w/w of ammonium ferric sulfate, $FeNH_4(SO_4)_2$, solution and acidify with nitric acid to avoid iron (III) hydroxide precipitation.

Potassium thiocyanate solution: Potassium thiocyanate is a hygroscopic salt and not pure enough. For this reason, it cannot be used as a standard substance and should be standardized against standard silver nitrate solution. The concentration of the potassium thiocyanate solution often used is about 0.1 M. The other steps of the standardization of KSCN are as follows:

(i) Prepare 500 mL of 0.1 N KSCN solution.
(ii) Take exactly known aliquot (min 20.0 ml) of standard $AgNO_3$ solution (about) 0.1 M into a 250 mL Erlenmeyer flask and add 50 mL of distilled water.
(iii) Add 1 mL of 10% $FeNH_4(SO_4)_2$ solution.
(iv) Fill 50 mL burette with 0.1 N solution of KSCN and start titration until a persistent red color is observed by adding KSCN drop by drop and note the volume.
(v) Repeat this titration at least three times under the same conditions.
(vi) Calculate the concentration of the potassium thiocyanate.
(vii) Calculate the average concentration of the potassium thiocyanate.

7.4.1.1.3 Calculation the concentration of the potassium thiocyanate

Assuming that a titration reaction is as follows:

$$aA + bT \rightarrow \text{Products} \tag{16}$$

where A is the analyte, T is titrant, a and b are the stoichiometric coefficients of analyte and titrant, respectively.

The number of millimoles of analyte and titrant equal each other at the equivalence point. Thus, the following equation can be used:

$$C_A \times V_A = C_T \times V_T \tag{17}$$

Assuming that 0.1058 M and 20.00 mL of standardized silver solution is titrated with unknown KSCN and the volume at titration end-point is 19.80 mL, the concentration of KSCN from Eq. (17) can be calculated as follows:

$$C_{\text{KSCN}} = \frac{C_{\text{Ag}} \times V_{\text{Ag}}}{V_{\text{KSCN}}} = \frac{0.1058 \times 20.00}{19.80} = 0.1068_7 M$$

7.4.1.1.4 Procedure for determination of silver purity

Weigh exactly the known amount of silver sample between 300 mg and 400 mg with an accuracy of 0.01 mg. Transfer it into a beaker and add 5 mL 1:1 nitric acid and heat gently until the evolution of nitrogen oxides stops by closing the mouth of the beaker with a watch glass. Rinse the watch glass into beakers after cooling and transfer the solution into a 250 mL Erlenmeyer flask. Rinse the beaker for several times with a minimum volume of distilled water and add about 20 mL distilled water. Add 50 mL (V_{SCN}) of 0.1 M (C_{SCN}) standard KSCN solution and 1 mL of 10% $FeNH_4(SO_4)_2$ solution. Titrate with of 0.1 M (C_{Ag}) standard $AgNO_3$ solution until the red color of the solution disappears and note the volume (V_{Ag}).

According to the titration reaction given above, the equation used for calculation the purity of silver in sample is given below:

$$\text{Purity\%} = \frac{10.7868 \times (C_{SCN} \times V_{SCN} - C_{Ag} \times V_{Ag})}{S_{Ag}} \tag{18}$$

where C_{SCN} and V_{SCN} are the molar concentration and volume of excess standard thiocyanate; C_{Ag} and V_{Ag} are the molar concentration and volume of standard silver solution used for back titration as titrant; and S_{Ag} is the weighed amount of silver sample.

7.4.1.2 *Potentiometric method*

7.4.1.2.1 Principle of the method

Silver is titrated with standard sodium chloride solution, and the measured potential values are used for potentiometric end-point (inflection point) detection.

7.4.1.2.2 Apparatus

A potentiometer, a magnetic stirrer, a magnetic bar, a burette (50 mL), a silver wire (1 mm in diameter), and Hg/Hg_2SO_4 reference electrode are the basic components of a potentiometric titration system. Other suitable reference electrodes such as Ag/AgCl saturated with KCl or saturated calomel electrode can also be used instead of Hg Hg/Hg_2SO_4. However, in this case, silver wire and the reference electrode used should be separated by using a salt bridge filled with saturated KNO_3.

Standard silver nitrate solution: Prepare and store as described in Sec. 3.1.1.2.

0.1 M NaCl solution: Weigh the required amount of dried NaCl (at 110°C for 2 h) and dissolve it in distilled water. Standardize against silver nitrate solution using the potentiometric method as follows:

(i) Set the potentiometric titration apparatus.
(ii) Transfer 10 mL of 0.1 M NaCl solution in a beaker.
(iii) Fill the salt bridge with saturated KNO_3.
(iv) Immerse silver wire and one end of the salt bridge into the solution.
(v) Immerse the other side of salt bridge and the reference electrode used in a bottle containing saturated KCl solution.
(vi) Complete the circuit using appropriate connection between electrodes and potentiometer.
(vii) Titrate with standard $AgNO_3$ solution by addition of 0.5 mL and record the potential for each addition.
(viii) Determine the volume range of the largest change in potential.
(ix) Transfer a second 10 mL of 0.1 M NaCl solution in a beaker.
(x) Repeat the titration by addition of 0.5 mL till the first significant change in potential is observed and then add 0.1 mL of $AgNO_3$ for subsequent additions.
(xi) Determine end-point volume by using first derivation method.
(xii) Calculate the exact concentration of NaCl using the equation given below:

$$C_{\mathrm{Ag}} \times V_{\mathrm{Ag}} = C_{\mathrm{Cl}} \times V_{\mathrm{Cl}} \tag{19}$$

Example: Assuming that 10.00 mL of unknown NaCl solution is titrated with 0.1088 M standard $AgNO_3$ and the volume at titration end-point is found as 9.8579 mL, the concentration of NaCl from Eq. (18) can be calculated as follows:

$$C_{\mathrm{NaCl}} = \frac{C_{\mathrm{Ag}} \times V_{\mathrm{Ag}}}{V_{\mathrm{NaCl}}} = \frac{0.1088 \times 9.8579}{10.00} = 0.1072_5 M$$

7.4.1.2.3 Procedure for determination of silver purity

Sample preparation procedure is the same as given in volumetric method. All of the titration steps are also the same as described in the standardization of NaCl, except that the place of analyte and titrant is changed. For this reason, a decreasing sigmoidal titration curve will be observed.

$$\text{Purity\%} = \frac{10.7868 \times C_{\text{Cl}} \times V_{\text{Cl}}}{S_{\text{Ag}}} \tag{20}$$

where C_{Cl} and V_{Cl} are the molar concentration and volume of NaCl solution and S_{Ag} is the weighed amount of silver sample.

The argentometric methods (e.g., Volhard's Method and potentiometric method) described above cannot be used for the determination of silver in fixer solutions containing thiosulfate as silver precipitates dissolve in this solution due to the formation of the silver thiosulfate complex.

7.4.1.3 *Voltammetric method*

The silver ion in the sample can be reduced to metallic silver when the potential is scanned to the negative direction. According to the potential scanning mode used, silver can be determined by several voltammetric techniques. One of these is anodic stripping voltammetry, in which silver is deposited as metallic silver on the electrode surface for a certain time (deposition time) at a constant potential (deposition potential) and then stripped off by scanning the potential to the positive direction. This technique is known as anodic stripping voltammetry. Generally, a three-electrode system consisting of a working electrode, an auxiliary electrode, and a reference electrode is used in the measurements. Figure 9 represents the general components of a voltammetric cell. Various electrodes such as dropping mercury electrode, hanging mercury electrode, platinum electrode, and pencil graphite electrode can be used as the working electrode. For general purpose, platinum wire is used as an auxiliary electrode, and saturated calomel electrode or silver/silver chloride (saturated with potassium chloride) electrode is used as the reference electrode. If the purpose is to determine the purity of the recovered silver by a voltammetric technique, it is important to separate the working electrode and the reference electrode to avoid the formation of silver chloride precipitate which causes

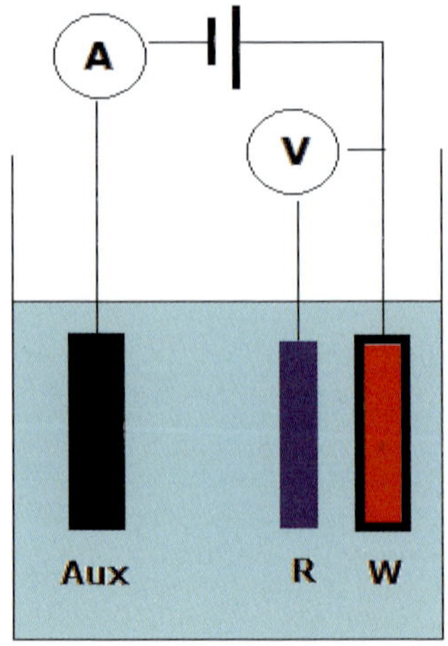

Fig. 9. A voltammetric cell comprising the three-electrode system; Aux: Auxiliary electrode, R: Reference electrode, and W: Working electrode.

erroneous results. In this case, a reference electrode jacket filled with potassium nitrate should be used. Mercury/mercury (I) sulfate electrode can be also used as an alternative reference electrode when the chloride problem is expected in measurements.

7.4.1.3.1 Purity determination of the recovered silver by Anodic Stripping Voltammetry

Principle of the method

Silver ions in the sample are deposited on the working electrode surface at a constant potential for a certain time, then the potential is scanned to positive direction by using differential pulse or square wave modulation. The measured peak current is linearly proportional to the concentration of silver.

Procedure for sample preparation

Weigh exactly a known amount of silver sample, about 1000 mg with an accuracy of 0.01 mg. Transfer it into a beaker and add 10 mL 1:1 nitric acid and heat gently until evolution of nitrogen oxides stops by closing the mouth of the beaker with a watch glass. Rinse the watch glass into beakers after cooling and transfer the solution into a 100 mL volumetric flask. Rinse the beaker several times with a minimum volume of distilled water and transfer it to a volumetric flask, then complete the volume up with distilled water. This solution contains about 10 g/L silver.

Procedure for voltammetric measurement

(i) Set up the voltammetric cell according to Fig. 9.
(ii) Prepare the working electrode to be used according to its type. For example, if the working electrode is platinum, polish it with an emery paper, then wash with acetone and finally distilled water. If the working electrode is a pencil graphite, activate it by applying a potential of 1.4 V for 60 s followed by the cycling of the potential between −1 and +1 V with a scan rate of 50 mV until a stable background current is observed.
(iii) Transfer a standard silver solution into the glass voltammetric cell and add KNO_3 (final concentration would be 0.1 M) and complete the volume up to 10 mL with distilled water.
(iv) Record the cyclic voltammogram (CV) by scanning the potential between 0 and −1000 mV (see the CVs for silver (I) at a pencil graphite electrode).
(v) Determine the cathodic and anodic peak potentials of the observed peaks.
(vi) Select deposition potential according to the peak potential values in CV. To do this, select more negative potential from the potential value at which a silver peak appears.
(vii) Allow the deposit of silver ions as metallic silver at selected deposition potential for 30 s, then strip it by scanning the potential to positive direction using differential or square wave modulation.
(viii) Repeat the same procedure described in step (vii) for the other standard silver solutions and sample and measure the current values (I_p) for each silver concentration and sample.
(ix) Construct a calibration curve ($\dot{I}_p = f(C_{Ag})$) and calculate silver concentration from the linear equation.

7.4.1.3.2 Voltammetric determination of silver content of fixer solutions

The voltammetric method can be recommended for the determination of silver in fixer solution. A voltammetric analyzer (potentiostat–galvanostat),

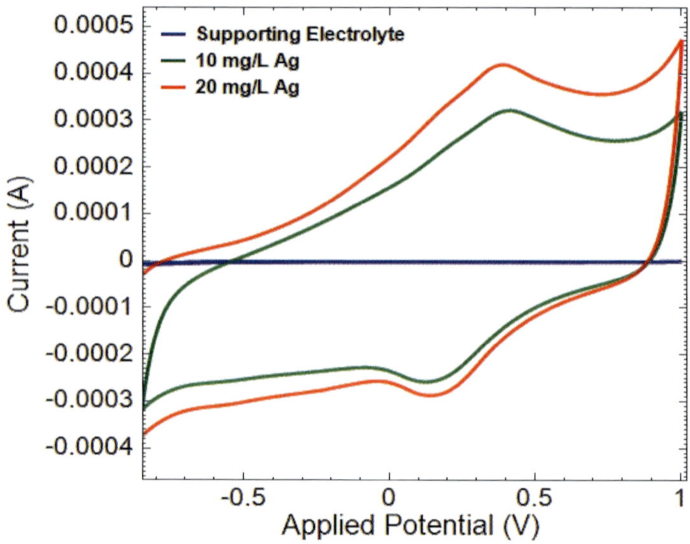

Fig. 10. Cyclic voltammograms obtained for silver (I) in 0.01 M HNO_3 at pencil graphite electrode.

a platinum disc or spiral wire electrode (or other suitable electrodes), a platinum black or platinum wire as auxiliary electrode, a saturated calomel electrode as a reference electrode, and a glass voltammetric cell are the basic component of the voltammetric system [80] (see Fig. 10).

Sample preparation

The pH of the fixer solution is adjusted to 5 using NaOH or HCl with a pH-meter.

Procedure for voltammetric measurement

(i) Set up the voltammetric cell according to Fig. 7.
(ii) Polish the platinum disc electrode with an emery paper, then wash with acetone and finally distilled water.
(iii) Transfer 10 mL of fixer sample into the glass voltammetric cell after adjusting the pH to 5 with NaOH or HCl by using a pH-meter.
(iv) Record the CV by scanning the potential between 0 and −1000 mV.
(v) Determine the two cathodic and two anodic peak potentials of the observed peaks.

(vi) The first cathodic peak belongs to the formation of metallic silver and the second belongs to sulfide formation.
(vii) Select the potential scanning range before the second peak potential to avoid the formation of silver sulfide for subsequent CV measurements.
(viii) After that, two options can be selected for silver determination in the fixer. One of these is CV measurements and the other is anodic stripping voltammetric measurements.

It is not required to provide more information here as these measurements require special expertise. Reference [80] is recommended for readers seeking more information.

7.4.1.4 *Atomic spectroscopic methods*

Atomic spectroscopic techniques can be used to determine the silver content of the photographic waste and the purity and the amount of impurities of the recovered silver.

As in voltammetric techniques, atomic spectroscopic techniques require also special expertise. Therefore, only the sample preparation method will be given here.

7.4.1.4.1 Sample preparation for purity determination of the recovered silver

Weigh exactly a known amount of silver sample, about 1000 mg with an accuracy of 0.01 mg. Transfer it into a beaker and add 10 mL 1:1 nitric acid and heat gently until the evolution of nitrogen oxides stops by closing the mouth of the beaker with a watch glass. Rinse the watch glass into beakers after cooling and transfer the solution into a 100 mL volumetric flask. Rinse the beaker several times with a minimum volume of distilled water and transfer it to a volumetric flask and then complete the volume up with distilled water. This solution contains about 10 g/L silver. Flame atomic absorption spectrometry (FAAS) or inductively coupled plasma optic emission spectrometry (ICP-OES) techniques can be used for purity determination of the recovered silver. These techniques can also be used for the determination of silver in fixer solutions without any sample pretreatment. The linear ranges for the determination of silver by FAAS and ICP-OES techniques are roughly between 0.1–5 and 0.01–300 mg/L, respectively. Therefore, dilution of the sample may be required with respect to the silver content of the sample. In order to obtain accurate results when using these devices, the manufacturer's guidelines should be followed.

7.4.2 *Methods for determination of impurities in the recovered silver*

It has been reported that spark source mass spectrometry and neutron activation analysis are widely used for the routine determination trace impurities in high-purity metals. However, owing to the unavailability of suitable standards and the difficulties connected with matrix interference, the quantitative applications of these methods are severely restricted. Atomic spectrometric methods, inductively coupled plasma optic emission spectrometric (ICP-OES) method, inductively coupled plasma mass spectrometric (ICP-MS) method, and electrochemical methods, sometimes combined with preconcentration and/or separation procedures, are also employed for the precise determination of trace impurities. All of these methods require special expertise. Among these, the ICP-MS technique is a "gold standard" and is usually preferred for the determination of trace impurities.

A rapid and simple ICP-MS method for the determination of some trace element impurities in high-purity silver has been reported in [81]. By using this method, the impurities in silver recovered from the waste photographic films have been determined by Nakiboglu *et al.* [71]. In these studies, the silver matrix is separated completely by the addition of HCl after dissolving the silver in HNO_3, and the presence of 11 elements (Al, Au, Cd, Co, Cu, Fe, Mg, Mn, Ni, Pb, and Sn) have been determined by the ICP-MS method in high-purity silver samples (EM9465, EM9343, and FS 14). The results obtained are shown in Table 3 with a comparison of the impurity levels for high-purity silver samples given in the literature.

7.4.2.1 *Sample preparation for determination of impurities in the recovered silver*

(i) Transfer the recovered silver to a 100 mL PTFE beaker.
(ii) Add 2.5 mL of cold 0.1 M HNO_3 and shake for 10 min.
(iii) Rinse several times with distilled water, then dry by nitrogen purging and store in a desiccator.

Table 3. Trace impurities (μg/g) in the recovered silver with a comparison of some results for high purity silver given in [58].

Element	Sample 1	Sample 2	EM9465	EM9343	FS 14
Fe	2.3	20.01	1.85	0.26	47.6
Cu	0.047	1.784	0.132	0.078	61.8
Mg	0.011	0.033	0.087	0.0640	—
Cr	0.499	0.671	—	—	—
Pb	<0.001	0.010	0.597	0.011	33.8
Al	0.024	4.620	0.082	0.024	—
Mn	0.003	0.021	0.011	0.007	—
Co	1.16	20.89	0.0024	—	—
Ni	0.362	9.962	0.011	0.007	53.9
Sn	0.044	0.021	0.006	0.039	44.0
Cd	<0.001	<0.001	0.018	0.012	—

Note: Sample 1: the recovered silver by NaOH method; Sample 2: the recovered silver by enzyme method; EM9465: high-purity silver having 5’9 grade silver; EM9343: high-purity silver having 6’9 grade silver; FS 14: Fine Silver.

(iv) Weigh exactly about 1.00 g of the cleaned and dried silver and transfer it to a 100 mL PTFE beaker.
(v) Add 3 mL ultrapure water and 3 mL of concentrated HNO_3.
(vi) Heat the solution on a hotplate below the boiling point of HNO_3 until it completely dissolves the silver.
(vii) Add expeditiously 4 mL of 2.3 M hydrochloric acid to obtain fine AgCl precipitate.
(viii) Add 2 mL of concentrated hydrochloric acid and stir for 20 h.
(ix) Filter by using G4 crucible under the vacuum and rinse several times with ultrapure water.
(x) Transfer the filtrate to a 25 mL calibrated flask and complete the volume up with ultrapure water.
(xi) Determine the impurities in the filtrate by the ICP-MS technique.

References*

[1] Austin GT (1984) *Shreve’s Chemical Process Industries*, Fifth Edition (McGraw-Hill), New York, pp. 409–410.

*Prepresents the patent studies published in *Chemical Abstracts Service*, and the abstracts of these studies were used for information.

[2] See: https://en.wikipedia.org/wiki/Abundance_of_elements_in_Earth's_crust. Accessed 5 March 2017.

[3] Dagenkolb DJ (1988) Current silver recovery techniques. *Journal of the Society of Motion Picture Engineers*, **97**(8): 630–637.

[4] Grau JM, Bisang JM (1992) Silver electrodeposition from photographic processing solutions. *Journal of Chemical Technology and Biotechnology*, **53**: 105–110.

[5] Chatelut M, Gobert E, Vittori O (2000) Silver electrowinning from photographic fixing solutions using zirconium cathode. *Hydrometallurgy*, **54**(2–3): 79–90.

[6] Krauss SR (1990) The electrolytic desilvering of bleach-fix. Paper Summaries. In *43rd SPSE Annual Conference*, Rochester, NY, 283 p.

[7] Hickman KC, Weyerts SW (1931) The electrolytic regeneration of fixing baths. *Journal of the Society of Motion Picture Engineers*, **17**: 568–590.

[8] Crabtree JI, Rasch AA (1955) Electrolytic silver recovery from spent fixing bathes at low current density. *Photographic Science and Technology Series II*, **2**: 15–33.

[9] Chang JC, Mina R (1982) Electrolytic silver recovery from spent fixing solutions-An electroanalytical study. *Photographic Science and Engineering*, **26**: 223–227.

[10] Levenson GIP (1981) Silver recovery by metal exchange. *Journal of Photographic. Science*, **29**: 31–38.

[11] Cooley AC (1988) Silver recovery using steel wool metallic replacement cartridges. *Journal of Imaging Technology*, **14**: 167–173.

[12] Aktas S (2008) Silver recovery from silver-rich photographic processing solutions by copper. *Canadian Metallurgical Quarterly*, **47**(1): 37–44.

[13] Cooley AC (1993) Silver recovery for environmental compliance in photographic processing. *Journal of Imaging Science and Technology*, **37**(4): 374–379.

[14] Yazıcı EY, Deveci H, Yazıcı R (2011) Recovery of silver from X-ray film processing effluents using trimercapto-s-triazine (TMT) *Separation Science and Technology*, **46**: 2231–2238.

[15] Cook MM, Lander JA (1970) Use of sodium borohdride to control heavy metal fischarge in the photographic industry. *Journal of Applied Photographic Engineering*, **5**: 144–147.

[16] Rivera I, Roca A, Cruells M, Patiño F, Salinas E (2007) Study of silver precipitation in thiosulfate solutions using sodium dithionite: Application to an industrial effluent. *Hydrometallurgy*, **89**: 89–98.

[17] Knorre H, Maening D, Stvesel K (1989) Chemical treatment of effluent from photofinishing plants. *Journal of Imaging Technology*, **14**: 154–156.
[18] Spears NR, Sentell B (1992) Silver recovery from Photographic waste processing solutions by using the trisodium salt of 2.4.6 trimercapto-s-triazine. *International Symposium on Photofinishing Technology*, Las Vegas, pp. 45–49.
[19] Marsh DG (1978) Removal of residual silver from processing washwater by ion exchange. *Journal of Applied Photographic Engineering*, **4**: 17–21.
[20] Cooley AC (1981) Ion exchange silver recovery for process EP-2 with nonregenerated bleach-fix. *Journal of Applied Photographic Engineering*, **7**: 106–110.
[21] Kreiman RT (1984) Photo wash water recycling system utilized ion exchange technology. *Journal of Imaging Technology*, **10**: 244–246.
[22] Lorenzo GA (1988) In-situ ion exchange silver recovery for population control. *Journal of Imaging Technology*, **14**: 174–178.
[23] Barbo BC (1987) Ion exchange silver recovery: Field experience. *Journal of Imaging Technology*, **13**: 89–92.
[24] Dagon TJ (1972) Biological treatment of photoprocessing effluents. *Journal of Water Pollution Control Federation*, **45**: 2123–2135.
[25] Fisher EW, O'Brian, JE (1960) Silver recovery from photographic wastes. *Chemical Abstract*, **54**:*P 20598b.
[26] Ewell TW, Piper J (1970) Silver-recovery incinerator. *Industrial Water*, **7**(11): 30–32.
[27] Cameron EC (1971) Recovery of silver from used photographic materials. *Chemical Abstract*, **74**:*P 89788b.
[28] Van Slyke WJ (1971) Silver recovery from photographic films. *Chemical Abstract*, **75**: *P 143104w.
[29] Trojan M, Kuhn . (1989) Silver recovery in thermal treatment of photographic wastes. *Chemical Abstract*, **110**(2): *P 99371w.
[30] Liu W (1991) Simple and easy method for recovery of silver from waste spectral photophalates. *Lihua Jianyan Huaxue Fence*, **27**(1): 19.
[31] Kunda W, Etsell TH (1986) Recovery of silver from X-ray film. *Proceedings of the 9th Precious Metals Institute Conference*, New York, June 10–13, pp. 289–304.
[32] Ramachandran V, Govindoswami K (1972) Recovery of silver from photographic wastes. *Transactions of the Society for Advancement of Electrochemical Science and Technology*, **7**(2): 59–60.
[33] Martinez MA, Criado AJ (1985) Recovery of purification of silver and gold. *Quimica Industrial*, **31**(11): 725–728.

[34] Ocker H (1959) Removal of silver containing residues from plastic parts. *Gottlob epple Mineraölwerk*, **68**: 840.
[35] Tereshkova SG, Pradon EA (1988). Alkali-free processing of cine photographic film wastes by extraction of silver and regeneration of the polimer supports. *Tekhnika Kino i Televideniya*, **5**: 14–18.
[36] Fabian J (1964). Recovering silver from photographic emulsions. *Chemical Abstract*, **60**: *P 8822f.
[37] Anderson R (1972) Recovery of silver from photographic material. *Chemical Abstract*, **60**: *P 8822f.
[38] Hein H (1986) Procedure and equipment for silver recovery. *Chemical Abstract*, **104**(22): *P 190303n.
[39] Messerschnidt H (1989) Recovery of silver from photographic emulsion by oxidation with ferric chloride. *Chemical Abstract*, **110**(8): *P 66820p.
[40] Moreno Garcia R (1986) The recovery of silver from photographic film: A study of the leaching reaction with cyanide solution for industrial use. *Hydrometalurgy*, **16**(3): 395–400.
[41] Ajiwe VIE, Anyadiegwu IE (2000) Recovery of silver from industrial wastes, cassava solution effects. *Separation and Purification Technology*, **18**: 89–92.
[42] Hilton DB (1992) Nontoxic process for recovery of silver from photographic wastes and emulsions. *Chemical Abstract*, **116**(14): *P 133343z.
[43] Nişli G, Toscalı D, Yenigül B, Duran C (1992) An investigation on the reaction of hypochlorite with waste films: The use of products on preparing new photographic emulsions and films. *VIIIth National Symposium on Chemistry and Chemical Engineering*, Istanbul, **1**: 341–345.
[44] Kuya K (1993) Recovery of silver from radiographs: An experiment using household materials. *Quimica Nova*, **16**(5): 474–476.
[45] Senecal G (1994) Recovery of silver and plastics from photographic materials. *Chemical Abstract*, **120**(4): *P 35118t.
[46] Mc Loren RH (1995) Method for recovering photographic and industrial wastes such as exposed films. *Chemical Abstract*, **122**(18): *P 221985g.
[47] Lindau ES (1970) Recovery of silver from film. *Chemical Abstract*, **73**: *P 40453p.
[48] Mozgalov IA, Kuznetsov SF (1994) Recovery of a silver containing layer with regenation of the base. *Chemical Abstract*, **120**(16): *P 204523d.
[49] Srivastava SP, Laxmi S, Krisha PV, Lyanger MS (1972) Processing photographic waste. *Chemical Abstract*, **77**: *P 130423.
[50] Guo J, An W (1987) New simple method for removing silver from waste photographic films. *Huaxue Shijie*, **28**(4): 174–175.

[51] Syed S, Suresha SL, Sharma M, Syed AA (2002) Clean technology for the recovery of silver from processed radiographic films. *Hydrometallurgy,* **63**: 277–280.

[52] Wainer E (1972). Recovery of silver, polyester and amino acids from processed photographic film. *Chemical Abstract*, **76**: *P 119902e.

[53] Nişli G, Yenigül B, Yıldıran H (1994) Plastik malzeme üzerine electrokaplamada rontgen film atıklarının kullanılmasına ilişkin bir inceleme. X. *Ulusal Kimya Kongresi*, Bursa-Turkey, p. 48.

[54] Schoenhard JD (1989) Filter process for silver recovery from polymeric films. *Chemical Abstract*, **110**(10): *P 79930z.

[55] Hochberg J (1989) Recovery of silver from photographic film by high -shear mixing with caustic solution. *Chemical Abstract*, **110**(20): *P 117153t.

[56] Chwojnowski A, Lada WA (1990) Recovery of silver and monomer components from used photographic films. *Chemical Abstract*, **112**(4): *P 28007n.

[57] Sandu I, Panait SV, Chifor OF, Ichim V (1996) Recovery of silver and supports from used photographic films. *Chemical Abstract*, **124**(18): *P 237980m.

[58] Nakiboğlu N, Toscalı D, Nişli G (2003) A novel silver recovery method from waste photographic films with NaOH stripping. *Turkish Journal of Chemistry*, **27**(1): 127–133.

[59] Zhouxiang H, Jianying W, Ma Z, Jifan H (2008) A method to recover silver from waste X-ray films with spent fixing bath. *Hydrometallurgy,* **92**: 148–151.

[60] Sramek M (1963) Silver recovery from photographic film. *Chemical Abstract*, **59**: *P 4715g.

[61] Bakker L (1973) Silver recovery from photographic film. *Chemical Abstract*, **79**: *P 99189j.

[62] Krusteva M, Peev G, Iotova L (1987) Study on the enzyme hydrolysis of waste photoemulsions. *Acta Biotechnology*, **7**(1): 93–96.

[63] Gehre C, Iske U, Glombitza F, Klein I (1989) Recovery of silver and carrier from waste film by use gelatin degrading microbes. *Chemical Abstract*, **110**(22): *P 198681d.

[64] Ozegowski JH, Mueller PJ, Koehler W, Hess W, Knabe K, Schmidt D,Schorning D, Roessler E, Kehlert F, Mueller G (1990) Method for treating photographic film for silver recovery of support material and silver. *Chemical Abstract*, **113**(2): *P 14702u.

[65] Priest FG (1984) *Extracelluer Enzymes: Commercial Enzymes* (American Society for Microbiology, USA), pp. 32–38.

[66] Frost G, Moss DA (1987) *Production of Enzymes by Fermentation* In Biotechnology: HJ Rhem, G Reed and JF Kennedy (eds.), VCH, Weinheim, **7a**: pp. 95–180.

[67] Fujiwara N, Tsumiya T, Katada T, Hosobushi T, Yamamoto K (1989) Continous recovery of silver from used x-ray films using a proteolytic enzyme. *Process Biochemistry*, **24**(4): 155–156.

[68] Ishikawa H, Ishimi K, Sugiura M, Sowa A, Fujiwara N (1993) Kinetics and mechanism of enzymatic hydrolysis of gelatin layers of X-ray film and release of silver particles. *Journal of Fermentation and Bioengineering*, **76**(4): 300–305.

[69] Fujiwara N, Yamamoto K, Masuri A (1991) Utilization of thermostable alkaline protease from an alkolophilic thermophile for the recovery of silver. *Journal of Fermentation and Bioengineering*, **72**(4): 306–308.

[70] Singh J, Vohra RM, Sahoo DK (1999) Alkaline protease from a new obligate alkalophilic isolate of Bacillus sphaericus. *Biotechnology Letters*, **21**: 921–924.

[71] Nakiboglu N, Toscalı D, Yaşa İ (2001) Silver recovery from waste photographic films by an enzymatic method. *Turkish Journal of Chemistry*, **25**(3): 349–353.

[72] Masui A, Yasuda M, Fujiwara N, Ishikawa H (2004) Enzymatic hydrolysis of gelatin layer on used lith film using thermostable alkaline protease for the recovery of silver and PET film. *Biotechnology Progress*, **20**: 1267–1269.

[73] Radha KV, Arun C (2010) Recycling of exposed photographic X-ray films and recovery of silver using Bromelain. *Transactions on Ecology and the Environment*, **142**: 421–430.

[74] Shankar S, More SV, Laxman RS (2010) Recovery of silver from waste x-ray film by alkaline protease from *Conidiobolus coronatus. Kathmandu University Journal of Science, Engineering and Technology*, **6**(1): 60–69.

[75] Prema K, Vandana R (2012) Silver recovery from waste x-ray films using alkaline protease from *Aspergillus Wentii* and its antibacterial studies. *Research Journal of Biotechnology,* **7**: 19–24.

[76] Bose A, Chawdhary V, Keharia H, Subramanian RB (2014) Production and characterization of a solvent-tolerant protease from a novel marine isolate Bacillus tequilensis P15. *Annual Microbiology*, **64**: 343–354.

[77] Yadav SK, Bisht D, Tiwari S, Darmwal NS (2015) Purification, biochemical characterization and performance evaluation of an alkaline serine protease from Aspergillus flavus MTCC 9952 mutant. *Biocatalysis and Agricultural Biotechnology*, **4**: 667–677.

[78] Parpalliwar JP, Patil PS, Patil ID, Deshannavar UB (2015) Extraction of silver from waste X-ray films using protease enzyme. *International Journal of Advanced Biotechnology and Research*, **6**(2): 220–226.

[79] Al-Abdalall AH, Al-Khaldi E (2016) Recovery of silver from used X-ray film using alkaline protease from *Bacillus subtilis* sub sp. *Subtilis.African Journal of Biotechnology*, **15**(26): 1413–1416.

[80] Sathaiyan N, Adaikkalam P, Venkateswaran KV (2002) Estimation of silver in the photographic fixer solution by voltammetric technique. *Transactions of the Society for Advancement of Electrochemical Science and Technology*, **37**(3–4): 83–85.

[81] Sun Y, Mierzwa J, Lin C, Yeh TI, Yang M (1997) Selective precipitation separation and inductively coupled plasma mass spectrometric determination of trace metal impurities in high purity silver. *Analyst*, **122**: 43–440.

Chapter 8
Recovery of Silver from E-wastes Using Acidothiourea

Katsutoshi Inoue*, Biplob Kumar Biswas*,‡, Manju Gurung*, Hidetaka Kawakita*, Keisuke Ohto* and Shafiq Alam†

**Department of Applied Chemistry*
Faculty of Science & Engineering, Saga University
Honjo-machi, Saga 840-8502, Japan

†Department of Chemical and Biological Engineering
College of Engineering, University of Saskatchewan
57 Campus Drive, SK S7N 5A9, Canada

8.1 Introduction

Due to rapid technological advancement in recent years, electronics are continuously being replaced by newer, cheaper, and more advanced products. Consequently, their lifespans and lifecycles are rapidly changing due to people's changing lifestyles and aspirations. Among such electronic products, the use of mobile phones has rapidly grown in recent decades because people are upgrading their mobile phones more frequently than before. On the other hand, such frequent replacement of mobile phones has brought about a big problem, and now spent mobile phones constitute the fastest growing component among solid municipal wastes. Mobile phones are similar in composition to other electronic devices: they consist of plastics, metals, ceramics, and glass [1]. The most useful parts of the mobile phones as a resource are printed circuit boards (PCBs) that contain many

‡Present address: Department of Chemical Engineering, Jessore University of Science and Technology, Jessore 7408, Bangladesh.

kinds of valuable metals [2]. With the advancement of recycling technology for e-wastes, different technological options are currently being explored for the reuse, recycling, and disposal of spent mobile phones. From an economical point of view, recovery of precious metals from spent mobile phones is highly attractive as the contents of precious metals in waste PCBs are about ten times higher than those of rich-content ores [3–5]. Therefore, waste PCBs are considered as "urban mine" of precious metals from which these valuable metals can be economically and conveniently recovered.

Another representative e-waste is spent TV monitors. For example, scrap plasma TV monitors also create an environmental problem due to their massive disposal. Owing to rapid developments in technology in the field of TV, they are being replaced by newer models at a much faster rate, for example, the continuous replacement of cathode ray tube (CRT) monitors by LCD displays and/or plasma monitors and the recent introduction of high-definition TV sets. Plasma TV monitor scrap is a potential source of silver, meaning there is an urgent need to develop efficient technologies to recover this precious metal in an environmentally benign way.

In the past two decades, the most active research on the recovery of valuable metals from e-waste such as spent PCBs has focused on hydrometallurgical techniques, which are more suitable for complicated feed materials such as waste PCBs in comparison to pyrometallurgical techniques [6]. Hydrometallurgical processes consist of leaching, purification or concentration of the leach liquor, and recovery of valuable metals from the leach liquor. In general, leaching is a starting step for the recycling of e-wastes in hydrometallurgical processes and involves dissolution of valuable metals into aqueous solution. Cyanide leaching has been in extensive use for more than 100 years since the early twentieth century for extracting gold and silver from their ores. The advantages of the cyanidation process over other extraction methods include its simplicity, fast leaching kinetics, and ease of recovery [7]. Despite its great success in hydrometallurgical processes, there has been increasing pressure against the use of cyanide because of its acute toxicity. Taking into account the deadly environmental impact of using cyanide, it seems to be essential to establish more economical and environmentally benign options for the recovery of gold and silver by using non-toxic lixiviants. Consequently, in recent years, there has been an increase in research activity about alternative non-cyanide lixiviants such as halogens (Cl^-, Br^-, I^-) [8], ammonia (NH_3) [9], thiocyanate (SCN^-) [10], thiosulfate ($S_2O_3^{2-}$) [11, 12], and thiourea (NH_2CSNH_2) [13–18]. Some of these alternatives not only offer a safer and more environmentally benign method but it is also expected that the use of these lixiviants can also increase the extent of the extraction of gold and silver [13, 19, 20]. Among these alternative lixiviants, thiourea (abbreviated as TU hereafter) mixed in acidic solutions is the most promising with regard to its very low toxicity and rapid kinetics of gold and silver leaching because it can readily dissolve gold and silver in acidic media as a stable complex by forming co-ordination bonds with these metals through the lone pairs of sulfur and nitrogen atoms, which eventually facilitates the leaching from complex ores, concentrates, and other potential sources of gold and silver. Further, the co-ordination with sulfur and nitrogen atoms suggests the high selectivity to these precious metals over base metals such as iron, copper, and zinc. Consequently, effects of leaching parameters such as concentrations of TU and acid, temperature, and pH on the leaching from ores of gold and silver have been studied by many authors to determine the optimum leaching conditions [13–20]. Nevertheless, significant strides are necessary to optimize the efficient leaching conditions to achieve satisfactory extraction of precious metal values from e-wastes such as spent mobile phones.

Pre-concentration and recovery of valuable metals from leach liquor is another essential step in hydrometallurgical processes. Several processes have been proposed for the recovery of gold and silver from acidothiourea solutions, which include cementation using powders of some base metals such as zinc [21–24], adsorption on activated carbon [25, 26], direct electrowinning [27], solvent extraction, and ion exchange [28–30]. Activated carbon and cation exchange resins have been reported to exhibit excellent adsorption behaviors for gold–TU complex and actually have been employed as commercially available adsorbents to concentrate gold [25, 26]. However, these commercially available adsorbents suffer from some disadvantages because they are expensive. In recent years, interest has been focused on gold and silver adsorption from acidiothiourea solutions on biomass as adsorbent. The rice husk carbonized at 573 K and *Chlorella vulgaris* heated at 673 K, containing specific silanol (–Si(OH)) and oxygen functional groups (>C=O, $-COO^-$, etc.) at carbon surface, were employed to adsorb the gold–TU complex [31, 32].

We investigated the hydrometallurgical method in order to fully understand the optimum conditions necessary for the extraction of gold and silver from PCBs of mobile waste [33] as well as for that of silver from the scrap of plasma TV monitors using acidothiourea solution [34]. Further, the potentiality of biomass-based adsorption material for the recovery of gold and silver from pregnant leach liquor was investigated as an alternative to conventional adsorbents.

8.2 Recovery of Silver from the Scrap of Plasma TV Monitors

The scrap sample of silver electrode of plasma TV monitors, which consists of glass scraps containing trace concentrations of silver, was kindly provided in powdered form by Nishinihon Kaden Recycle Corporation (NKRC), Kitakyushu, Japan. The metal contents in this sample was measured by means of the total dissolution of the sample using boiling aqua regia followed by the spectrochemical analysis using a Shimadzu model ICPS-8100 inductively coupled plasma atomic emission spectrometer (ICP/AES) and evaluated as follows (unit: mg/g). Na 2.96, K 6.12, Ca 6.63, Ba 49.98, Fe 1.43, Ni 0.16, Cu 0.37, Zn 7.14, Ag 4.03, Al 0.27, Si 3.21, and Au 0.18.

As shown in Fig. 1, the particle size distribution of the scrap showed that 43.2% and 30.7% of the scrap corresponded to particle size fractions

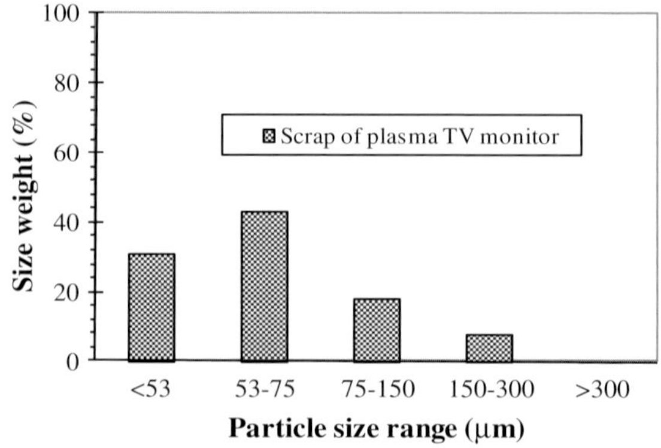

Fig. 1. Particle size distribution of the scrap of plasma TV monitors.

of 53–75 μm and <53 μm, respectively, while the amount of the largest particle size fraction (>300 μm) was very insignificant. The 53–75 μm and <53 μm particle size fractions of the scrap were mixed together to be used in the study of silver recovery as follows.

8.2.1 *Screening test for the most suitable non-cyanide lixiviant*

8.2.1.1 *Leaching using thiosulfate, hydrochloric acid, and acetic acid solutions*

Figure 2 shows the effect of pH on the extraction efficiency of silver at three different concentrations, from 0.25 to 1.0 M (M = mol/dm^3) of sodium thiosulfate at constant liquid/solid (L/S) ratio of 150 mL/g at 303 K. It is seen from Fig. 2 that silver extraction is only slightly influenced by pH and concentration of thiosulfate and did not reach even 50% at pH ranging from 1 to 5.

Figure 3 shows the effect of L/S ratio on the extraction efficiency of silver at three different concentrations of thiosulfate. It is seen from this figure that the extraction of silver is not higher than 50% even at increasing thiosulfate concentration.

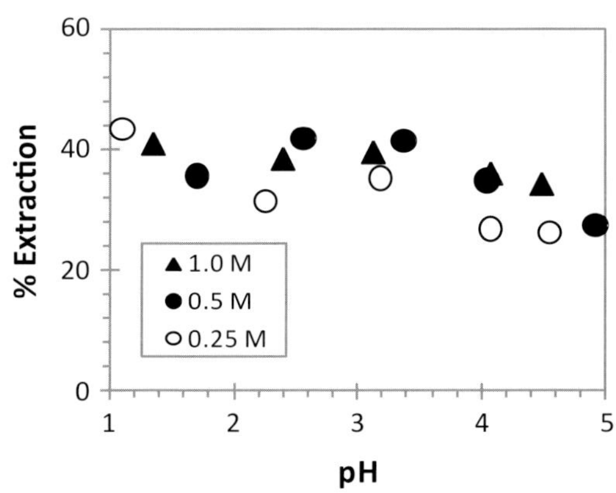

Fig. 2. Effect of pH on extraction efficiency of silver from scrap sample of plasma TV monitors using three different concentrations of sodium thiosulfate at constant L/S ratio (150 mL/g) at 303 K [34].

Source: With permission for reuse from the Minerals, Metals & Materials Society.

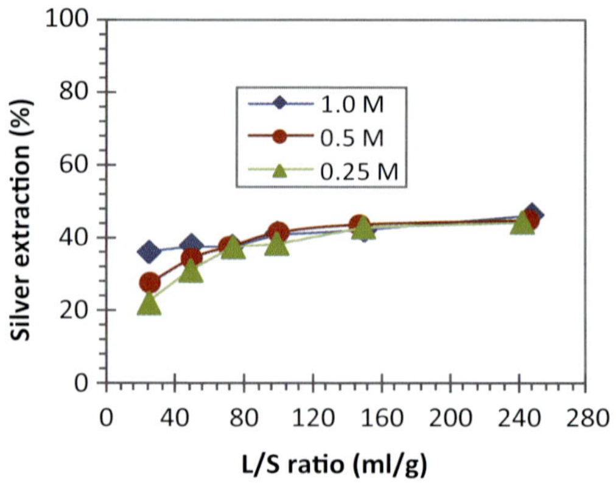

Fig. 3. Effect of L/S ratio on silver extraction from scrap sample of plasma TV monitors using three different concentrations sodium thiosulfate at 303 K at pH 3 [34].

Source: With permission for reuse from the Minerals, Metals & Materials Society.

Figures 4 and 5 show the effect of L/S ratio on silver extraction from scrap sample of plasma TV monitors using varying concentrations of hydrochloric acid and acetic acid, respectively.

As seen from Fig. 4, although the silver extraction increases with increasing concentration of hydrochloric acid, it is still lower than 50%. On the other hand, it is only negligible in the extraction using acetic acid as shown in Fig. 5.

8.2.1.2 *Acidothiourea leaching*

Since TU gives rise to a strong cationic complex with silver(I), ($Ag[CS(NH_2)_2]_3^+$), it represents an attractive lixiviant for silver(I) alternative [11]. Although, as shown in Fig. 6, dilute TU cannot extract silver from the scrap of plasma TV monitors by itself, an improved silver extraction can be achieved by using TU together with a dilute acidic solution such as a dilute sulfuric acid solution. This mixture is referred to as acidothiourea solution hereafter. In some literatures, the addition of ferric ion is reported to be required to enhance the extraction of silver [12].

Figure 7 shows the effects of concentrations of TU and sulfuric acid on % silver extraction at 303 K. From this figure, it is obvious that silver extraction is more effective by using a combination of low-concentration acid and low-concentration TU. That is, although dilute sulfuric acid alone

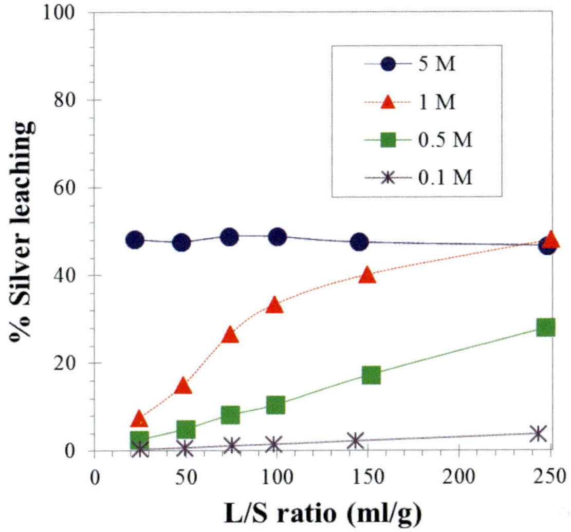

Fig. 4. Effect of L/S ratio on extraction efficiency of silver from scrap sample of plasma TV monitors using varying concentrations of hydrochloric acid at 303 K [34].

Source: With permission for reuse from the Minerals, Metals & Materials Society.

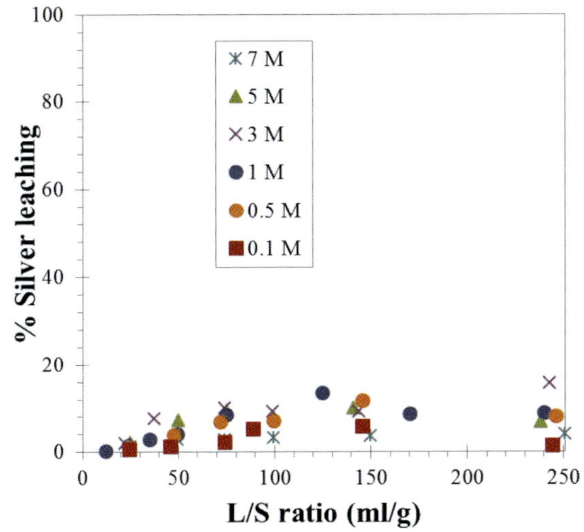

Fig. 5. Effect of L/S ratio on extraction efficiency of silver from scrap sample of plasma TV monitors using varying concentrations of acetic acid at 303 K.

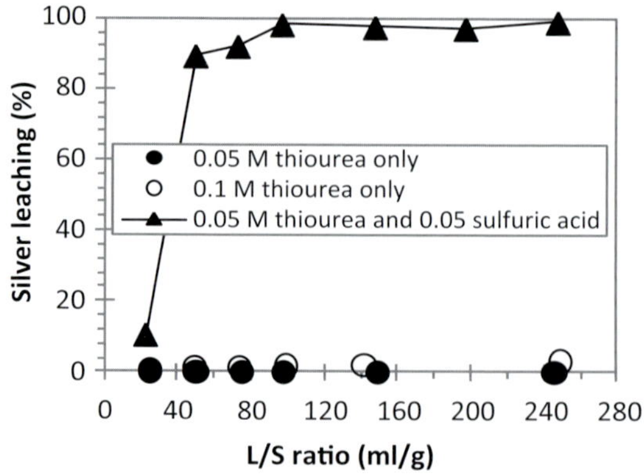

Fig. 6. Effect of L/S ratio on silver extraction using TU solution together or except for dilute sulfuric acid [34].

Source: With permission for reuse from the Minerals, Metals & Materials Society.

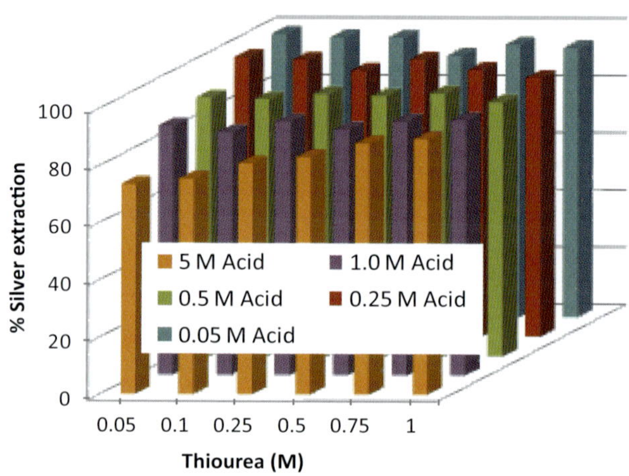

Fig. 7. Silver extraction using varying concentrations of TU solution mixed with varying concentrations of sulfuric acid. M = mol/dm^3 [34].

Source: With permission for reuse from the Minerals, Metals & Materials Society.

has only very little effect on silver extraction, a mixture of sulfuric acid and TU offers an excellent leaching performance. This is because low-concentration sulfuric acid facilitates the complexation of silver with TU, while high-concentration acid impedes the complexation by protonating TU, which results in the low extraction of silver.

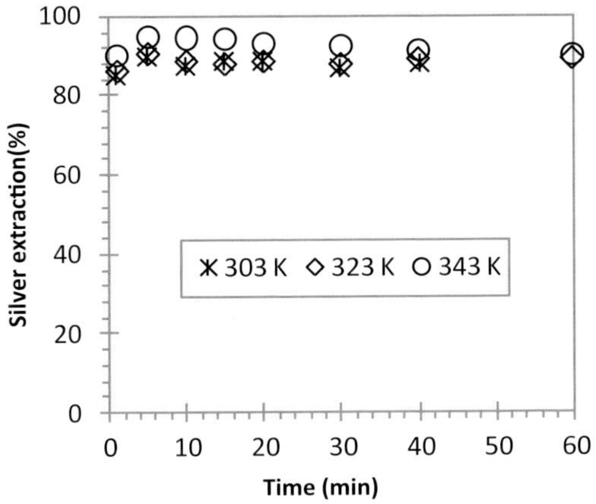

Fig. 8. Time variation of silver extraction using mixture of 0.05 M TU and 0.05 M sulfuric acid at three different temperatures [34].

Source: With permission for reuse from the Minerals, Metals & Materials Society.

Figure 8 shows the effect of shaking time on silver extraction at three different temperatures. As seen from this figure, silver extraction was extremely fast and nearly quantitative over the whole period of shaking and was nearly independent of temperature.

From the above-mentioned results, acidothiourea can be concluded to be the most effective lixiviant for the leaching of silver.

Silver can be recovered from the leach liquor of acidothiourea by various conventional methods, among which cementation using zinc powder appears to be the simplest and most economical method.

Figure 9 shows the effect of solid/liquid ratio on the recovery of silver from acidothiourea leach liquor containing around 7–8 mg/dm^3 silver by means of cementation using zinc powder. As seen from this figure, by adding around 6 g zinc powder to 1 dm^3 leach liquor, quantitative recovery can be achieved.

8.3 Leaching of Silver and Gold from PCBs of Spent Mobile Phones Using Acidothiourea

As mentioned earlier, acidothiourea was found to be the most suitable lixiviant for silver extraction, and the recovery of silver from PCBs of

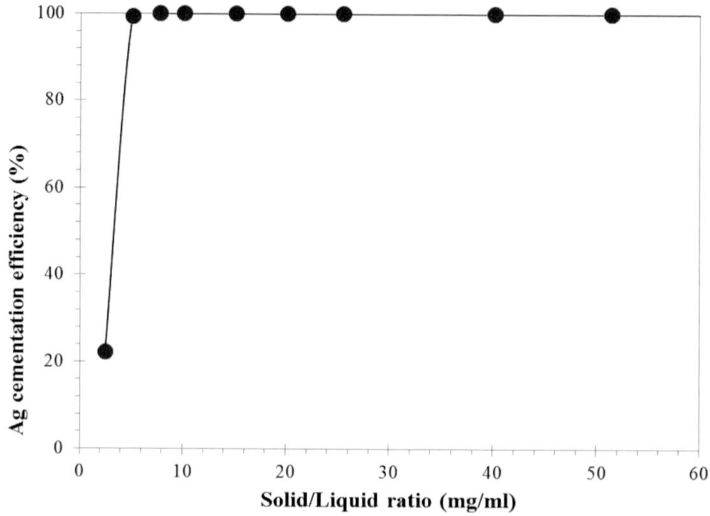

Fig. 9. Effect of solid/liquid ratio on the recovery of silver from acidothiourea leach liquor (silver concentration = 7–8 mg/dm^3) by means of cementation using zinc powder [34].

Source: With permission for reuse from the Minerals, Metals & Materials Society.

spent mobile phones was attempted to be achieved by using this lixiviant and also compared with that of gold.

8.3.1 *Pre-treatment of PCBs of spent mobile phones*

In the present study, as shown in Fig. 10, samples of waste PCBs from approximately 200 spent mobile phones were dismantled into some parts, including PCBs by hand at the factory of Shibata Industrial Co. Ltd., Omuta, Japan. Among them, the samples of PCBs were sent to the Shonan factory of Tanaka Kikinzoku Kogyo K.K., Hiratsuka, Japan, where they were mechanically crushed using a two-axis crusher in order to effectively separate the rare earth magnets. Such crushed samples of PCBs were then calcined at 750° for 6 h to remove epoxy resins, the main body of the PCBs, on which various devices are attached, by incineration, leaving metals among the ash.

Figure 11 shows the particle size distribution of the ash of the calcined PCBs sample, which suggests that around 25% of the ash has a particle size smaller than 75 μm while the majority of the ash has that greater than 300 μm.

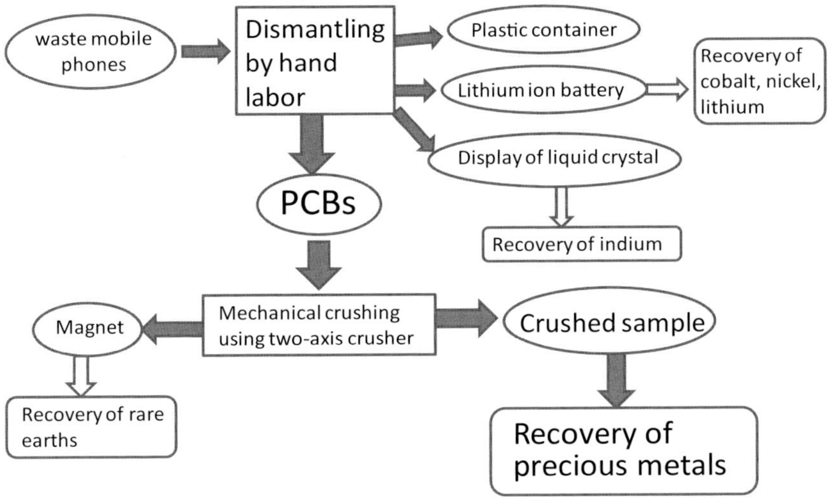

Fig. 10. Flow-sheet of the dismantling of waste mobile phones into some parts.

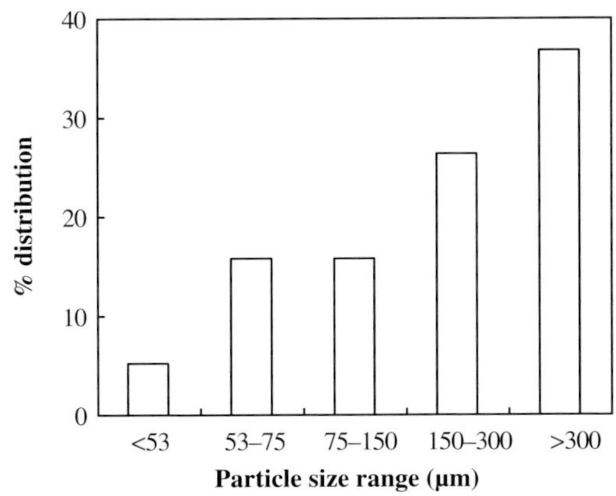

Fig. 11. Particle size distribution of the ash of the calcined sample of PCBs of spent mobile phones.

Figure 12 shows the scanning electron microscopy (SEM) image of the ash, which reveals the extruded shape consisting of aggregated small particles with very irregular shapes.

Figure 13 shows the spectra obtained using energy-dispersive X-ray (EDX) spectroscopy of the ash of the calcined sample of PCBs of spent

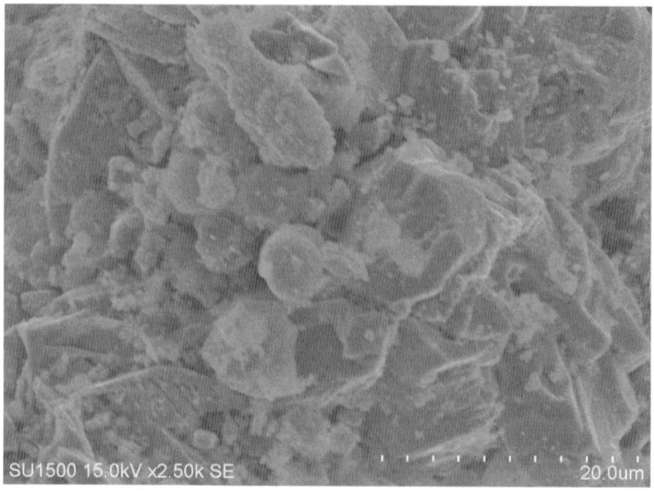

Fig. 12. SEM image of the ash of the calcined sample of PCBs of spent mobile phones.

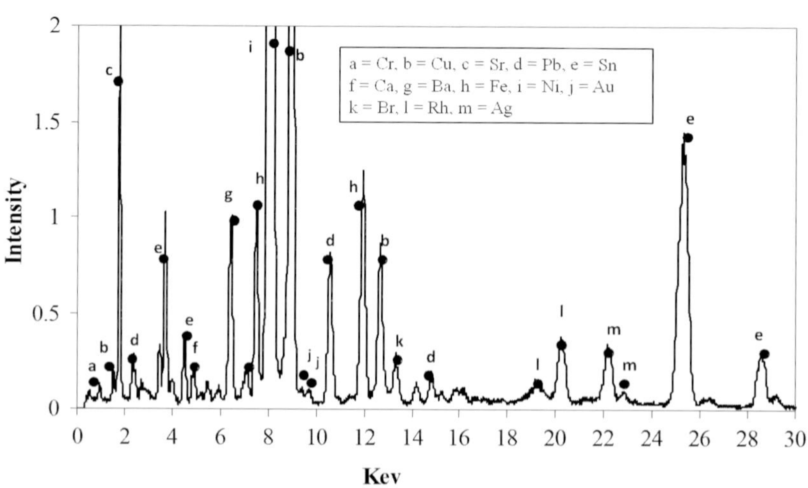

Fig. 13. EDX spectra of the ash of the calcined sample of PCBs of spent mobile phones.

mobile phones, while Table 1 shows the result of the quantitative analysis of the same sample, which was measured using an inductively coupled plasma atomic emission spectrometer (ICP-AES) after totally dissolving the sample in boiling aqua regia. Both of these results indicate that copper is the main element existing in PCBs together with iron, lead, and nickel,

Table 1. Quantitative analysis of metals contained in the ash of the calcined sample of PCBs of spent mobile phones.

Metals	Au	Pt	Pd	Ag	Cr	Fe	Co	Ni	Zn	Pb	Cu	Al
Content (mg/g)	1.63	0.01	0.39	0.52	0.11	10.5	0.20	19.1	1.69	13.2	349	0.69

Table 2. Effect of particle size on the dissolution of gold and silver from the calcined sample of PCBs of spent mobile phones.

	Amount of metal dissolved (mg/g)	
Particle size (μm)	Gold	Silver
>300	1.98	5.94
150–300	2.28	6.15
75–150	2.50	6.05
53–75	2.51	6.16

contributing to the significant proportion of the metals in a relatively small portion of scrap. Nevertheless, it can be confirmed that the PCBs of spent mobile phones are an outstanding secondary resource of these valuable metals.

The calcined sample was brought to the next step of leaching, in which valuable metals were extracted into aqueous solutions.

8.3.2 *Leaching of silver and gold using acidothiourea*

8.3.2.1 *Effect of particle size*

The effect of particle size on the leaching of gold and silver was investigated by using different particle size (>300, 150–300, 75–150, and 53–75 μm) of pulverized PCBs kept under the constant concentrations of TU (0.5 M) and sulfuric acid (0.05 M), constant agitation speed (500 rpm), and constant temperature (303 K), as presented in Table 2. It is evident from Table 2 that although the fraction with the smallest particle size yields the highest quantity of dissolved gold and silver, the effect is not so large. But particle size samples that were less than 75 μm were employed in further experiments.

8.3.2.2 *Effect of TU concentration*

Figures 14 and 15 show the effect of L/S ratio at varying TU concentration on the leaching of gold and silver, respectively, at constant acid concentration.

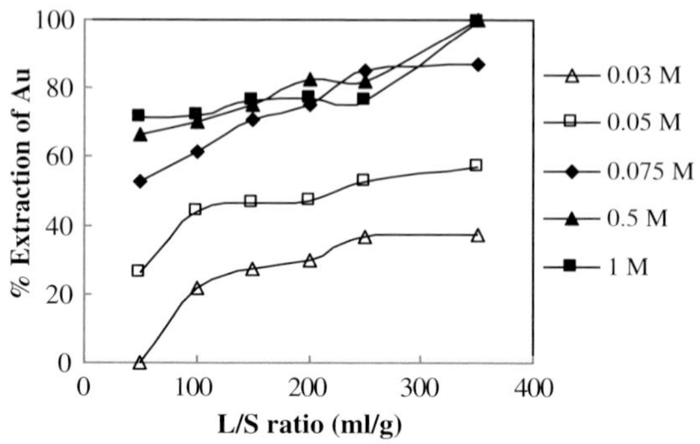

Fig. 14. Effect of TU concentration and L/S ratio on the leaching of gold from the calcined PCBs sample under the following constant conditions. Concentration of sulfuric acid = 0.05 M, contact time = 24 h, temperature = 303 K where, in the case of 100% leaching, the leached amount of gold was 2.55 mg/g.

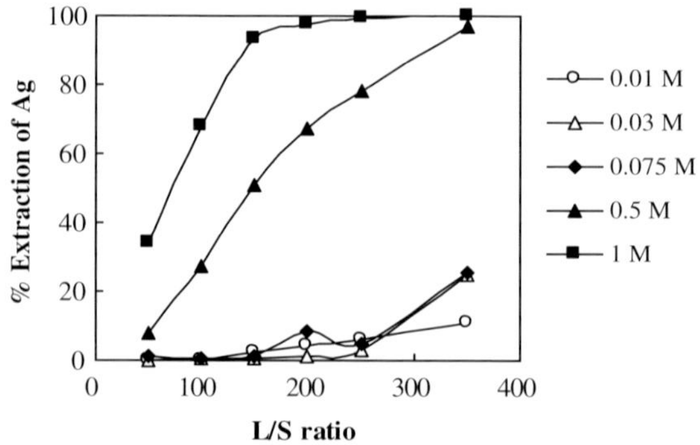

Fig. 15. Effect of TU concentration and L/S ratio on the leaching of silver from the calcined PCBs sample under the following constant conditions. Concentration of sulfuric acid = 0.05 M, contact time = 24 h, temperature = 303 K where, in case of 100% leaching, the leached amount of silver was 6.36 mg/g.

It is apparent from the comparison of these figures that the effects are much different between gold and silver though the extraction efficiency increases with increasing L/S ratio and TU concentration for both metals. However, taking into account the cost-effective consumption of TU and achievement of complete leaching, 0.5 M TU at L/S ratio of 350 mL/g is considered to be recommended as the most suitable condition for leaching of both metals.

8.3.2.3 *Effect of sulfuric acid concentration*

Figures 16 and 17 show the effect of L/S ratio at varying sulfuric acid concentrations on the leaching of gold and silver, respectively, at constant TU concentration. Also in these figures, it is apparent that the effects are much different between gold and silver.

It is seen from Figure 16 that there is no beneficial effect of increasing sulfuric acid concentration higher than 0.05 M for the dissolution of gold. Also as seen from this figure, the gold leaching was increased nearly linearly with increasing L/S ratio, and quantitative leaching was achieved at L/S ratio of 350 mL/g regardless of the sulfuric acid concentration. Taking the consumption of acid into account, 0.05 M sulfuric acid at L/S ratio

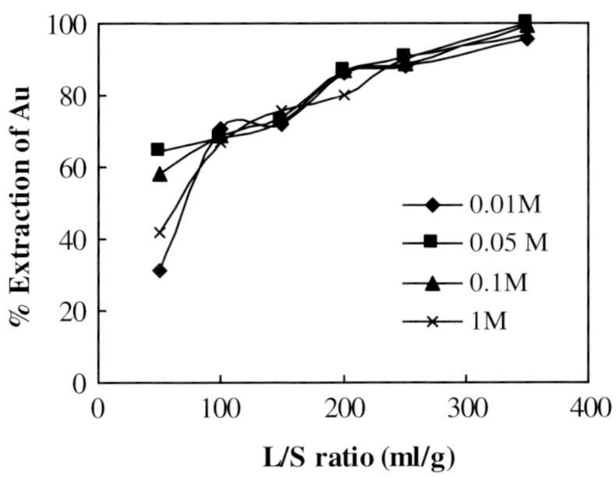

Fig. 16. Effect of sulfuric acid concentration and L/S ratio on gold leaching from the calcined PCBs sample under the following constant conditions: TU concentration = 0.5 M, temperature = 303 K, contact time = 24 h where, in the case of 100% leaching, the amount of gold leached was 2.63 mg/g.

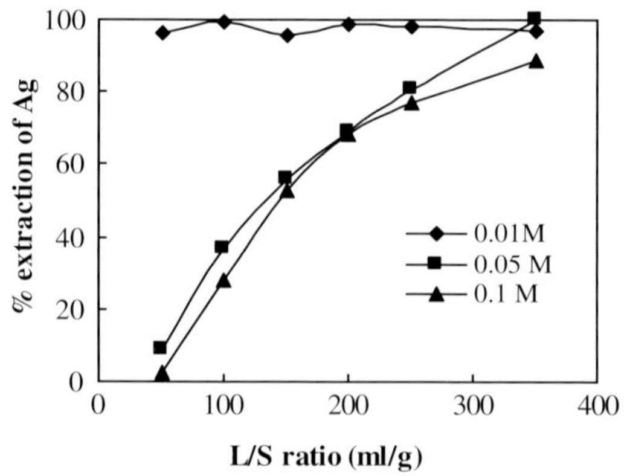

Fig. 17. Effect of sulfuric acid concentration and L/S ratio on silver leaching from the calcined PCBs sample under the following constant conditions. TU concentration = 0.5 M, temperature = 303 K, contact time = 24 h where, in the case of 100% leaching, the amount of silver leached was 6.56 mg/g.

of 350 mL/g seems to be the best-suited condition for complete leaching of gold.

On the other hand, it is seen from Fig. 17 that 0.01 M sulfuric acid is adequate to achieve a high dissolution of silver even at low L/S ratios. Concentration higher than 0.01 M was impotent for silver dissolution, which can be explained by the fact that the low concentration of sulfuric acid activates the lone-pair electrons on TU to facilitate the complexation with silver while, contrarily, high concentration of sulfuric acid protonates TU and impedes complexation, resulting in low silver extraction [23].

8.3.2.4 *Effect of the oxidant used and contact time*

It has been reported that gold and silver dissolution in acidothiourea is catalyzed by ferric ion [9, 10]. The role of ferric ion in the complexation process is to facilitate the oxidation of metallic gold and silver to aurous (Au^+) and argentous (Ag^+) ions, respectively, which is shown in the following reactions:

$$Au + 2CS(NH_2)_2 + Fe^{3+} \rightarrow Au[CS(NH_2)_2]_2^+ + Fe^{2+}$$

$$Ag + 3CS(NH_2)_2 + Fe^{3+} \rightarrow Ag[CS(NH_2)_2]_3^+ + Fe^{2+}$$

Huyhua *et al.* [35] reported that leaching of gold with TU in presence of ferric sulfate as an oxidant were up to four times faster than TU alone in air.

Table 1 indicates that the PCBs of mobile waste contain appreciable amounts of iron. Thus, its reasonable that the iron present in the waste serves as a partial supply of ferric ion oxidant for TU–gold interaction, and a small additional amount of ferric ion gives better results of gold dissolution.

Figure 18 shows the effect of the addition of ferric ion on the leaching rate of gold by acidothiourea. It can be seen that ferric ion slightly enhances the leaching rate of gold. Because, as shown in Table 1, the sample contains appreciable amount of iron, it is considered to serve as a partial supply of ferric ion oxidant for TU–gold interaction.

On the other hand, Fig. 19 shows the time variation of the percentage leaching of silver in the presence as well as absence of ferric ions. It is seen that there was no beneficial effect of ferric ions on leaching of silver. Rather, the dissolution efficiency was slightly decreased. These results suggest that the proper amount of ferric ion oxidant necessary for the oxidation of silver is supplied by the feed material itself. The difference in leaching behavior

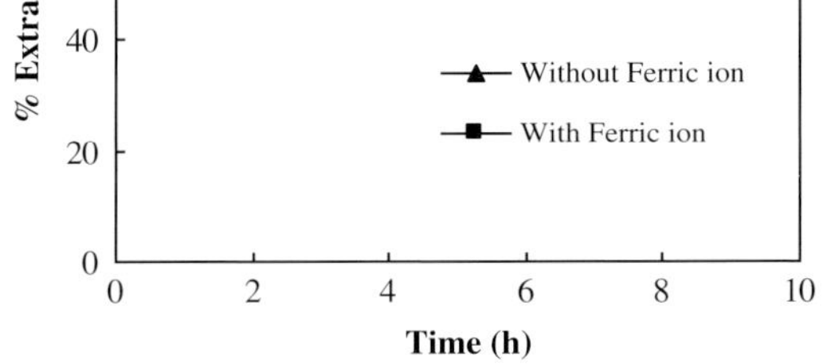

Fig. 18. Effect of shaking time in the presence or absence of ferric ion on the leaching of gold from the calcined PCBs of spent mobile phones under the following conditions: Leachant = 0.5 M TU in 0.05 M sulfuric acid, L/S ratio = 350 mL/g, temperature = 303 K, added ferric ion concentration = 0.01 M [33].

Source: With permission for reuse from Elsevier B.V.

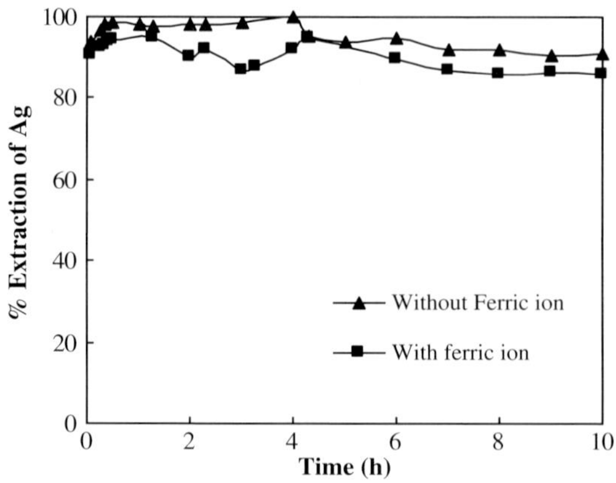

Fig. 19. Effect of shaking time in the presence or absence of ferric ion on the leaching of silver from the calcined PCBs of spent mobile phones under the following conditions: Leachant = 0.5 M TU in 0.01 M sulfuric acid, L/S ratio = 350 mL/g, temperature = 303 K, added ferric ion concentration = 0.01 M [33].

Source: With permission for reuse from Elsevier B.V.

of gold and silver in the presence of the oxidizing agent is attributable to the respective oxidation–reduction potential (ORP) values of these metals. Since gold has lower ORP values than silver, it seems reasonable that the presence of external oxidizing agent enhances the gold dissolution.

8.3.2.5 *Effect of temperature*

The effect of temperature on the extraction rate of gold and silver are presented in Figs. 20 and 21, respectively. Leaching of gold was examined in presence of 0.01 M ferric ions while that of silver was investigated in absence of ferric ions as there was no observed beneficial effect for the leaching of silver in presence of ferric ions as mentioned earlier. The results indicate that the extent of gold dissolution slightly increased with increasing temperature from 303 up to 318 K but decreased with further increase in temperature to 333 K. It may be considered that increasing temperature activates the reactant species to promote the reaction, while it was suppressed at further elevated temperature. Since the gold dissolution was examined in the presence of an external oxidant, this may be attributable to the rapid decomposition of TU, resulting in the formation of colloidal sulfur [28] and retardation of the reaction through surface passivation.

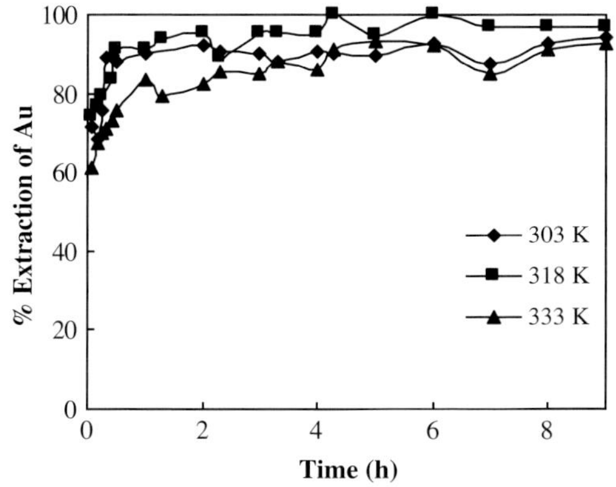

Fig. 20. Effect of temperature on gold leaching from the calcined sample of PCBs of spent mobile phones under the following conditions. Leachant = 0.5 M TU in 0.05 M sulfuric acid in presence of 0.01 M ferric ion, L/S ratio = 350 mL/g [33].

Source: With permission for reuse from Elsevier B.V.

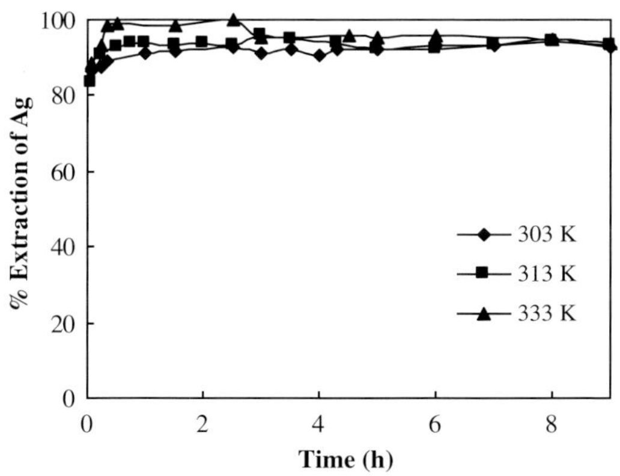

Fig. 21. Effect of temperature on silver leaching from the calcined sample of PCBs of spent mobile phones under the following conditions. Leachant = 0.5 M TU in 0.01 M sulfuric acid in presence of 0.01 M ferric ion, L/S ratio = 350 mL/g [33].

Source: With permission for reuse from Elsevier B.V.

Thus, under the authors' experimental conditions, 318 K appears to be the optimum temperature for the leaching of gold. Nevertheless, the results support the fact that substantial dissolution of gold from PCBs sample can be achieved with acidothiourea solution in presence of ferric ions even at ambient temperature. However, as it is obvious from Fig. 21, increasing temperature up to 333 K only slightly enhanced the silver dissolution.

8.3.3 *EDX and SEM study after leaching*

The changes in the chemical composition of the PCBs sample after gold and silver leaching were examined by means of EDX measurement. Figures 22 and 23 show the EDX spectrum of the sample after the leaching of gold and silver, respectively, under their optimum leaching conditions mentioned in the figure captions. The absence of a gold peak in EDX spectrum of Fig. 22 confirms the complete leaching of gold from the sample. It is interesting to note the absence of both gold and silver peaks in the EDX spectrum of silver leached sample shown in Fig. 22. These results indicate that both gold and silver were quantitatively leached out from the PCBs sample.

The morphological changes, which occurred in PCBs sample after leaching of metal values, were also observed in the SEM image of the sample after leaching the metal values, and are presented in Fig. 24. It is seen from the comparison of SEM images presented in Fig. 12 with this figure that the surface roughness decreased after leaching.

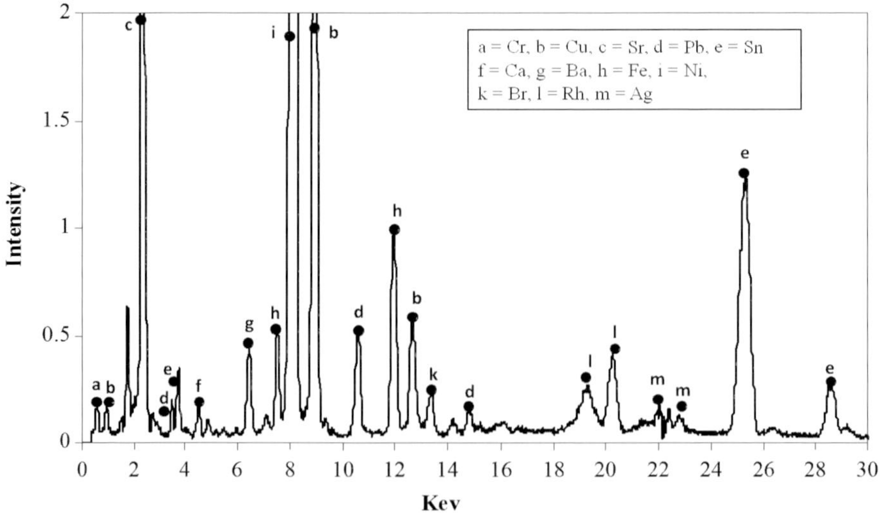

Fig. 22. EDX spectra of the residue after leaching gold using 0.5 M TU in 0.05 M sulfuric acid.

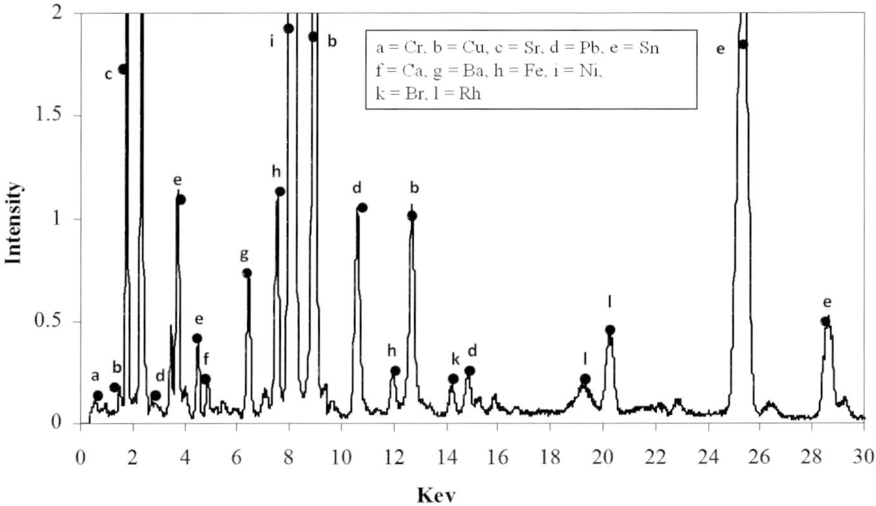

Fig. 23. EDX spectra of the residue after leaching silver using 0.5 M TU in 0.01 M sulfuric acid.

Fig. 24. SEM image of gold and silver leached residue.

8.4 Recovery of Gold and Silver from the Leach Liquor of Acidothiourea

From the above-mentioned leach liquor of acidothiourea, it was attempted to recover gold and silver by means of cementation using copper powder, a typical conventional method, and through reductive adsorption using a bio-adsorbent developed by the authors.

8.4.1 *Recovery of gold and silver by means of cementation using copper powder*

It has been reported that Cu, Pb, Zn, Ni, and Fe have the potential to form metal–TU complexes such as $Cu[CS(NH_2)_2]_4^{2+}$, $Pb[CS(NH_2)_2]_4^{2+}$, $Zn[CS(NH_2)_2]_2^{2+}$, $Ni[CS(NH_2]_4^{2+}$, and $Fe[CS(NH_2)_2]_2^{3+}$, respectively [29–31], although acidiothiourea solution is less sensitive to base metals compared with gold and silver. In the above-described leaching tests using acidothiourea, it was found that 12% Zn, 15% Fe, 24% Ni, and 34% Pb (as compared to the content of total dissolution with aqua regia) were leached along with gold and silver. It is desirable that the leached gold and silver should be selectively recovered from acidothiourea solutions containing such base metals. Reductive precipitation of gold by treating with metal hydride [32], electrowinning, and cementation are typical conventional methods for gold recovery. The most popular reductant is sodium borohydride [32, 33], which has been employed to reduce gold and silver from leach liquors at a commercial scale. However, in the authors' case, reduction with sodium borohydride was found to be ineffective for gold and silver recovery from acidothiourea solution since it precipitated base metals along with gold and silver. In addition, due to the environmental toxicity of boron, the use of boron compounds is not recommended.

Cementation is one of the most effective and economical techniques extensively employed in hydrometallurgical processes of solution streams, including the recovery of precious metal values from pregnant leach liquors. The advantages of the cementation process include its relative simplicity, ease of control, and its ability to recover valuable metals.

For gold and silver cementation, there are two simultaneous reactions. The cationic gold and silver existing as the complexes $Au[SC(NH_2)_2]_2^+$ and $Ag[SC(NH_2)_2]_3^+$ are reduced to elemental gold and silver while the oxidation of less noble metals such as copper and zinc supplies the electrons necessary for the reduction reaction. A number of research works have been conducted on cementation of noble metals [11–13]. Iron, zinc, and copper have been studied as the reducing metals for gold and silver recovery from acidothiourea solution. Since the standard reduction potential of zinc (−0.76 V) and iron (−0.44 V) are lower than that of nickel (−0.26 V), lead (−0.13 V), copper (+0.34 V), silver (+0.80 V), and gold (+1.50 V), zinc and iron were considered unsuitable for the cementation process as they would reduce and precipitate some base metals co-existing along with gold and silver. In fact, lead, copper, and nickel were also precipitated when the zinc cementation process was attempted. Copper appeared more preferable because it is relatively inexpensive and has lower reduction potential than those of gold and silver and higher reduction potential than those of other base metals; consequently, it is expected to selectively reduce gold and silver over base metals.

Figure 25 shows the results of cementation test for gold and silver using copper powder. It is clear from this figure that complete recovery of silver was achieved by copper cementation. Quantitative recovery of silver from the leach liquor was achieved even at lower solid/liquid ratio; that is, 4 g

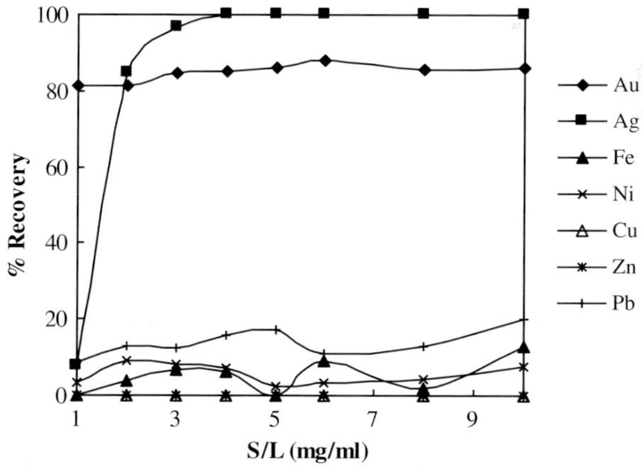

Fig. 25. Effect of solid/liquid ratio on % recovery of gold and silver by means of cementation with copper powder from the acidothiourea leach liquor by shaking for 10 h at 303 K [33].

Source: With permission for reuse from Elsevier B.V.

of copper powder per liter of leach liquor was effective for the complete precipitation of silver. On the other hand, although the recovery of gold was worthy of consideration at lower solid/liquid ratio (more than 80% recovery was achieved even at solid/liquid ratio of 1 g/l), it did not increase with increasing solid/liquid ratio up to 10 g/l. The difference in the ORP values of gold (+1.50 V), silver (+0.80 V) and copper (+0.34 V) would certainly play a major role to achieve this feat. In the cementation process, the copper powder reduces silver(I) and gold(I) to their respective elemental forms according to the following reactions.

$$2Au[SC(NH_2)_2]_2^+ + Cu \rightarrow Cu^{2+} + 2Au + 4SC(NH_2)_2$$

$$2Ag[SC(NH_2)_2]_3^+ + Cu \rightarrow Cu^{2+} + 2Ag + 6SC(NH_2)_2$$

Although quantitative precipitation of silver from the leach liquor was achieved by using a relatively lower amount of copper, a much higher amount of copper was required for the quantitative precipitation of gold. Hence, it is necessary to investigate the more economically viabale method for pre-concentration and recovery of leached gold from the pregnant leach liquor.

8.4.2 *Adsorptive recovery of gold and silver from acidothiourea leach liquor using bio-adsorbent*

The bio-adsorbent employed for the recovery of gold and silver in the present work was prepared from powdered persimmon extract rich in persimmon tannin, a kind of polyphenol compound, the chemical structure of which is shown in Fig. 26.

Figure 27 shows the schematic flowsheet for the preparation of the present bio-adsorbent, in which powdered persimmon extract was treated in boiling concentrated sulfuric acid for the purpose of cross-linking among tannin compounds [36], thus preparing the cross-linked persimmon tannin (CPT) gel.

The adsorptive recovery of gold and silver was attempted from acidothiourea leach liquor using CPT gel and commercially available activated charcoal for comparison. Figures 28 and 29 show the plot of % recovery of gold and silver versus the dry weight of the adsorbent added to the unit volume of the leach liquor (solid/liquid ratio) in the cases of the recovery using CPT gel and activated charcoal, respectively. As seen from

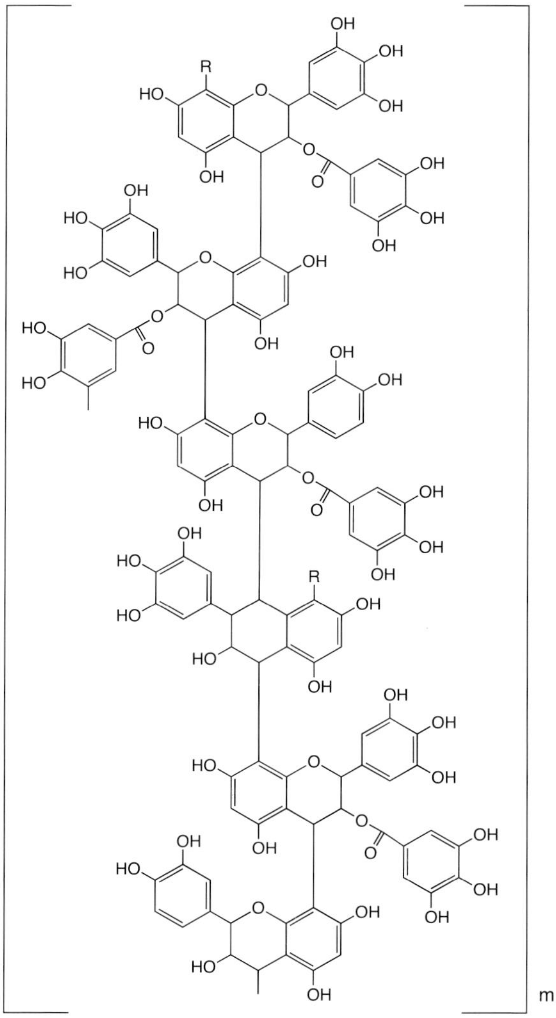

Fig. 26. Chemical structure of persimmon tannin.

these figures, the adsorptive recovery of both gold and silver increased with the increase in adsorbent dose of CPT gel as well as the activated charcoal. Although the adsorption behaviors for gold appear nearly the same for both the adsorbents, CPT gel was found to be more effective than activated charcoal for silver adsorption. Quantitative adsorption of silver on CPT gel was achieved at solid/liquid ratio of 15 mg/mL and that of gold was achieved at solid/liquid ratio of 40 mg/mL. More interestingly, the CPT gel

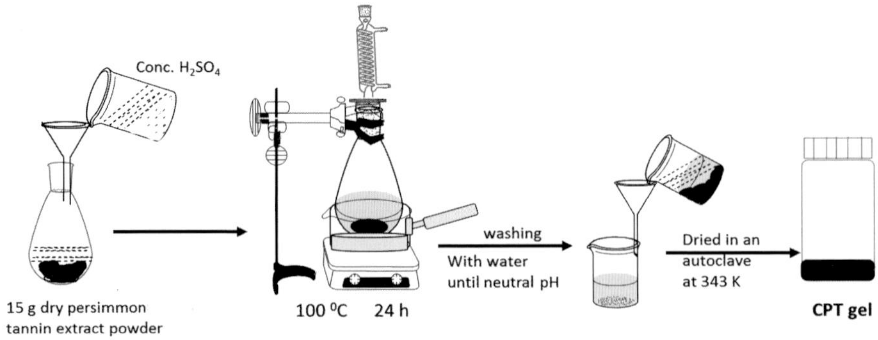

Fig. 27. Schematic flowsheet for the preparation of CPT gel.

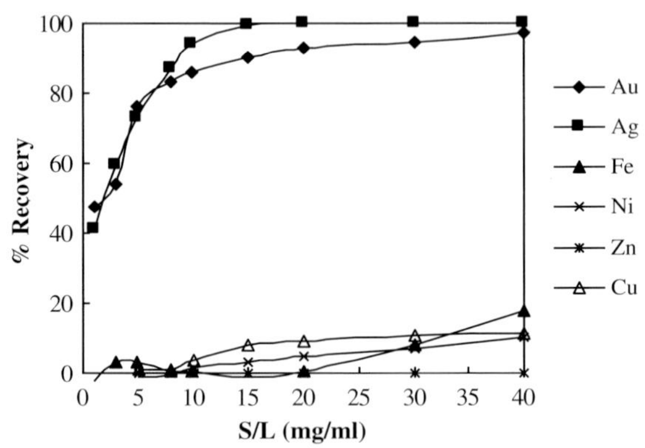

Fig. 28. % adsorptive recovery of gold and silver as well as co-existing base metals from acidothiourea leach liquor using CPT gel at varying solid/liquid ratios [33].

Source: With permission for reuse from Elsevier B.V.

exhibited negligible adsorption for base metals, suggesting the possibility of selective recovery of gold and silver from the leach liquor of acidothiourea. The loaded gold and silver can be easily recovered as metallic gold and silver by incinerating the loaded gel at around 500°.

On the basis of the above-described experimental results, a flow-sheet for the recovery of gold and silver from PCBs of spent mobile phones can be proposed as shown in Fig. 30 through the combination of acidothiourea leaching followed by adsorption using CPT gel.

Thus recovered gold and silver can be easily separated from each other by conventional methods, for example, by selective dissolution using nitric acid.

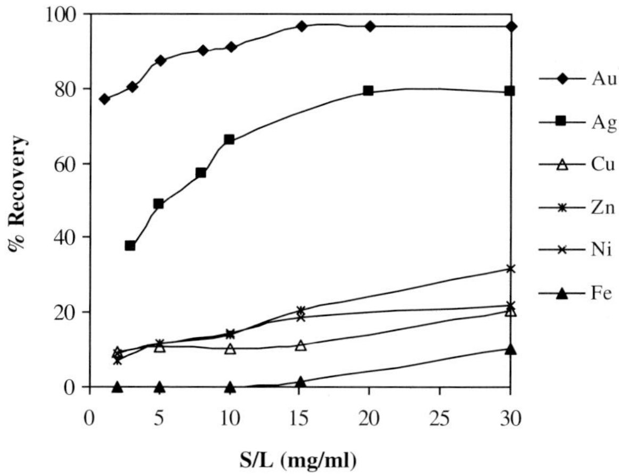

Fig. 29. % adsorptive recovery of gold and silver as well as co-existing base metals from acidothiourea leach liquor using commercially available charcoal at varying solid/liquid ratios [33].

Source: With permission for reuse from Elsevier B.V.

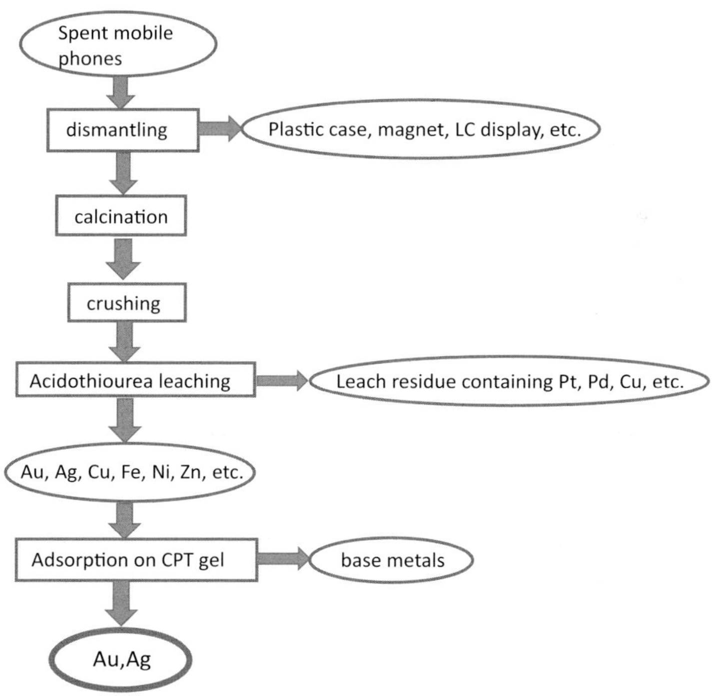

Fig. 30. Flowsheet for the recovery of gold and silver from spent mobile phones.

8.5 Conclusions

For the purpose of recovering silver existing in trace concentrations in scrap plasma TV monitors, a kind of glass waste, leaching tests were carried out using lixiviants such as thiosulfate and acidothiourea, a mixture of TU in sulfuric acid solution, to screen the most suitable lixiviant. As the result, acidothiourea was found to be the most suitable. The effects of the concentrations of TU and sulfuric acid, temperature, shaking time, and addition of ferric ion were investigated to find the most optimum leaching condition. The most optimum mixture was found to be 0.05 mol/dm^3 for both sulfuric acid and TU, where quantitative silver extraction was achieved at the L/S ratio of 100 cm^3/g within 5 min at 303 K. Effects of temperature and the addition of ferric ion were found to be negligible. From such impregnated leach liquor, silver was quantitatively recovered by means of cementation using zinc powder.

Referring to such results, the recovery of silver and gold was also attempted from PCBs of spent mobile phones by means of leaching using acidothiourea followed by adsorption on persimmon tannin gel prepared by treating the persimmon extract powder with boiling concentrated sulfuric acid to generate cross-linking. The quantitative recovery of both silver and gold was achieved at the liquid/solid ratio (pregnant acidothiourea leach liquor/weight of added adsorption gel) of 40 mg/cm^3 at 30°.

References

[1] Osibanjo O, Nnorom IC (2008) Material flows of mobile phones and accessories in Nigeria: Environmental implications and sound end-of-life management options. *Environmental Impact Assessment Review*, **28**: 198–213.

[2] Park YJ, Fray DJ (2009) Recovery of high purity precious metals from printed circuit boards. *Journal of Hazardous Materials*, **164**: 1152–1158.

[3] Guo J, Rao Q, Xu Z (2008) Application of glass-nonmetals of waste printed circuit boards to produce phenolic moulding compound. *Journal of Hazardous Materials*, **153**: 728–734.

[4] He W, Li G, Ma X, Wang H, Huang J, Xu M, Huang C (2006) WEEE recovery strategies and the WEEE treatment status in China. *Journal of Hazardous Materials*, **B136**: 502–512.

[5] Li J, Lu H, Guo J, Xu Z, Zhou Y (2007) Recycle technology for recovering resources and products from waste printed circuit boards. *Environmental Science and Technology*, **41**: 1995–2000.

[6] Cui J, Zhang L (2008) Metallurgical recovery of metals from electronic waste: A review. *Journal of Hazardous Materials*, **158**: 228–256.
[7] Annon (2010) Cyanide management in the gold industry. *Mining Environmental Management*, **July**: 26–27.
[8] Hilson G, Monhemius AJ (2006) Alternative to cyanide in the gold mining industry: What prospects for the future. *Journal of Cleaner Production*, **14**: 1158–1167.
[9] Umetsu Y, Tozawa K, Sasaki K (1971) Dissolution of silver in ammonia solution under oxygen pressure. *Nippon Kogyokaishi*, **87**: 29–32 [in Japanese].
[10] Kononova ON, Kholmogorov AG, Danilenko NV, Goryaeva NG, Shatnykh KA, Kachin VS (2007) Recovery of silver from thiosulfate and thiocyanate leach solution by adsorption on anion exchange resins and activated carbon. *Hydrometallurgy*, **88**: 189–195.
[11] Abbruzzese C, Fornari P, Mesidda R, Veglio F, Ubaldini S (1995) Thiosulphate leaching for gold hydrometallurgy. *Hydrometallurgy*, **39**: 265–276.
[12] Jeffery MI (2001) Kinetic aspects of gold and silver leaching in ammonia-thiosulphate solutions. *Hydrometallurgy*, **60**: 7–16.
[13] Chen CK, Lung TN, Wan CC (1980) A study of the leaching of gold and silver by acidothioureation. *Hydrometallurgy*, **5**: 207–212.
[14] Deschenes G, Ghali E (1988) Leaching of gold from a chalcopyrite concentrate by thiourea. *Hydrometallurgy*, **20**: 179–202.
[15] Lacoste-Bouchet P, Deschenes G, Ghali E (1998) Thiourea leaching of copper-gold ore using statistical design, *Hydrometallurgy*, **47**: 189–203.
[16] Li J, Miller JD (2007) Reaction kinetics of gold dissolution in acid thiourea solution using ferric sulfate as oxidant. *Hydrometallurgy*, **89**: 279–288.
[17] Murthy DSR, Prasad PM (1996) Leaching of gold silver from miller process dross through non-cyanide leachants. *Hydrometallurgy*, **42**: 27–33.
[18] Murthy DSR, Kumar V, Rao KVS (2003) Extraction of gold from an Indian low-grade refractory gold ore through physical beneficiation and thiourea leaching. *Hydrometallurgy*, **68**: 125–130.
[19] Ficeriova J, Balaz P, Dutkova E, Gock E (2008) Leaching of gold and silver from crushed Au-Ag wastes. *Open Chemistry Journal*, **2**: 6–9.
[20] Groenewald T (1976) The dissolution of gold in acidic solution of thiourea. *Hydrometallurgy*, **1**: 277–290.
[21] Lee HY, Kim SG, Oh JK (1997) Cementation behavior of gold and silver onto Zn, Al, and Fe powders from acid thiourea solutions. *Canadian Metallurgical Quaternary*, **36**: 149–155.
[22] Nguyen HH, Tran T, Wong PIM (1997) A kinetic study of the cementation of gold from cyanide solutions onto copper. *Hydrometallurgy*, **46**: 55–69.

[23] Sulka GD, Jaskula M (2002) Study of the kinetics of the cementation of silver ions onto copper in a rotating cylinder system from acidic sulphate solutions. *Hydrometallurgy*, **64**: 13–33.

[24] Zhang HG, Doyle JA, Kenna CC, La Brooy SG, Hefter GT, Ritchie IM (1996) A kinetic and electrochemical study of the cementation of gold onto mild steel from acidic thiourea solutions. *Electrochimica Acta*, **41**: 389–395.

[25] Schmidt R, Moya SA, Foerster JE, Varela AR (1993) Carbon adsorption and leaching of gold by acid thiourea. *International Journal of Mineral Processing*, **38**: 257–265.

[26] Zhang H, Ritchie IM, La Brooy SR (2004) The adsorption of gold thiourea complex onto activated carbon. *Hydrometallurgy*, **72**: 291–301.

[27] Juarez CM, Dutra AJB (2000) Gold electrowinning from thiourea solutions. *Mineral Engineering*, **13**: 1083–1096.

[28] Becker E, Knothe M, Loebel J (1983) Gold recovery from non-metallic secondary raw materials by leaching with thiourea and adsorption on ion exchangers. *Hydrometallurgy*, **11**: 265–275.

[29] Conradie PJ, Johns MW, Fowles RJ (1994) Elution and electrowinning of gold from gold-selective strong-base resins. *Hydrometallurgy*, **37**: 349–366.

[30] Nakahiro Y, Horio U, Niinae M, Kusaka E, Wakamatsu T (1992) Recovery of gold with ion exchange resin from leaching solution by acidothioureation. *Mineral Engineering*, **5**: 1389–1400.

[31] Nakbanpote W, Thiravetyan P, Kalambaheti C (2000) Preconcentration of gold by rice husk ash. *Mineral Engineering*, **13**: 391–400.

[32] Nakbanpote W, Thiravetyan P, Kalambaheti C (2002) Comparison of gold adsorption by *Chlorella vulgaris*, rice husk and activated carbon. *Mineral Engineering*, **15**: 549–552.

[33] Gurung M, Adhikari BB, Kawakita H, Ohto KK, Alam S (2013) Recovery of gold and silver from spent mobile phones by means of acidothiourea leaching followed by adsorption using biosorbent prepared from persimmon tannin. *Hydrometallurgy*, **133**: 84–93.

[34] Inoue K, Biswas BK, Kawakita H, Ohto K, Hoshino A (2011) Recovery of silver from spent plasma TV monitors. *Late News Proceedings TMS 2011*, San Diego, USA, pp. 229–237.

[35] Huyhua JC, Zegarra CR, Gundiler JH (1989) A comparative study of oxidants on gold and silver dissolution in acidic thiourea solutions. In *Precious Metals '89*, MC Jha, SD Hill (eds.) (TMS, The Minerals, Metals and Materials Society, Warrendale, USA), pp. 287–303.

[36] Gurung M, Adhikari BB, Kawakita H, Ohto K, Inoue K, Alam S (2011) Recovery of Au(III) by using low cost adsorbent prepared from persimmon tannin extract. *Chemical Engineering Journal*, 174: 556–563.

Chapter 9
Silver Extraction and Recovery with Macrocyclic and Tripodal Compounds

Keisuke Ohto*, **Yuki Ueda, Ramachandra Rao Sathuluri, Hidetaka Kawakita, Shitaro Morisada and Katsutoshi Inoue**

Department of Chemistry and Applied Chemistry
Saga University, 1-Honjo, Saga 840-8502, Japan
**ohtok@cc.saga-u.ac.jp*

9.1 Introduction

Silver has been extensively used in a variety of common goods, including money, jewelry, mirrors, and tableware and in advanced materials such as photographically sensitive materials, dental materials, relay contacts, anti-bacterial materials, battery cells, and solar cells, all due to its excellent physicochemical property. Although the total amount of silver reserves in mines has been decreasing year on year, it has been accumulating rapidly in a so-called "urban mine" — in spent small home appliances. The need for the recycling and recovering of silver from these appliances is increasingly important to meet the growing demand for silver.

Solvent extraction is one of the primary techniques used to recover a target ion from the co-existing ions due to its versatility, possibility for mass disposal, and efficiency [1]. The efficiency of solvent extraction is predominantly dependent on the performance of the extraction reagents [2–6]. Since the mutual separation of some elements is still complicated, the preparation of more effective reagents is required.

Calixarenes are macrocyclic oligomers prepared by a condensation reaction with *p*-alkylphenol and formaldehyde. They provide a

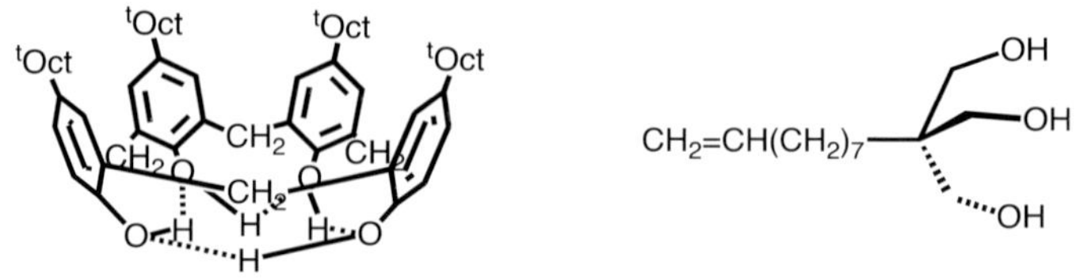

p-t-Octylcalix[4]areneTrimethylolnonene

(^{t}***Oct[4]H***) (^{8}***None{3}CH$_2$OH***)

Fig. 1. Structures and abbreviations of two frameworks as calixarene and alkyltrimethylol.

three-dimensionally (3D) arranged co-ordination site and discriminate a certain guest molecules and ions with their remarkable features and structural effects [7, 8]. Alkyltrimethylols are linear tripodal compounds prepared by a condensation reaction with α-saturated aldehyde and formaldehyde. They can also provide a 3D arranged co-ordination site and discriminate a certain guest [9, 10] and was termed the "Trident" molecule [10], which was derived from a weapon of Poseidon in Greek myths. Both frameworks provide attractive properties for metal uptake. There are some different properties between calix [4]arene and tripodal frameworks, because the former provides reactive phenolic hydroxyl groups and a relatively larger co-ordination site template by the sodium ion, while the latter provides less reactive alcoholic hydroxyl groups and a relatively narrower site. For high lipophilicity and suppressed hydrophobicity, *p*-t-octylcalixarene as calixarene framework and 1,1,1-tris(hydroxymethyl)-9-decene as the tripodal one have been usually employed in our research works. Figure 1 shows the structures and the abbreviations of the two frameworks.

In this review, our research works on silver extraction is introduced.

9.2 Silver Extraction with Calixarene Derivatives

In this section, the research works on silver extraction with calixarene derivatives together with the extraction using micro-reactor system are introduced.

9.2.1 *Methylketonic derivative of p-t-octylcalix[4]arene*

First, methylketonic derivatives of *p*-t-octylcalixarene and the corresponding monomer were employed [11]. The methylketonic derivatives were

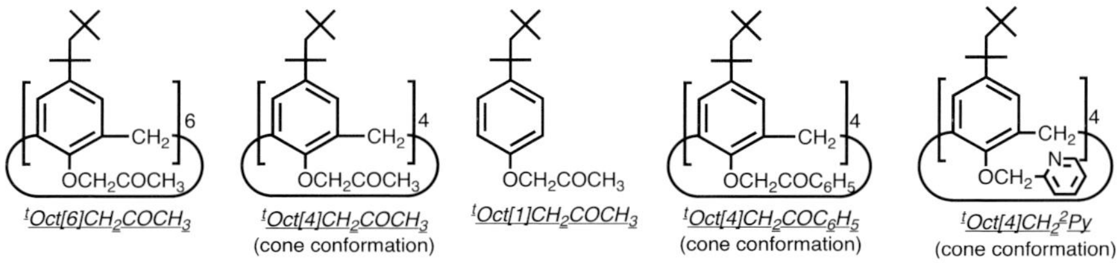

Fig. 2. Structures and abbreviations of calixarene and monomeric derivatives.

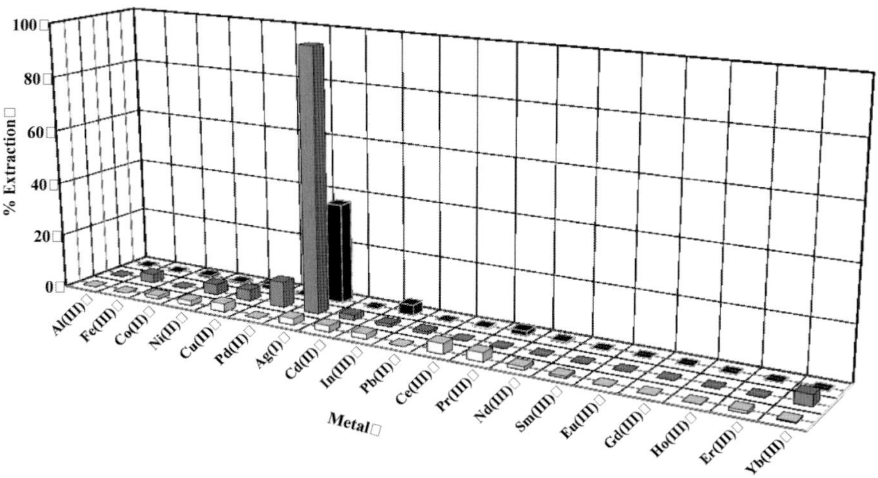

Fig. 3. Extraction percentages of various metal ions in 1 M (M = mold m^{-3}) HNO_3 with three ketonic derivatives in an individual system. [Metal ion] = 0.1 mM each in 1 M HNO_3, [^{t}Oct[6]CH_2COCH_3], [^{t}Oct[4]CH_2COCH_3] = 5 mM, and [^{t}Oct[1]CH_2COCH_3] = 20 mM in $CHCl_3$. The white, gray and black color bars represent ^{t}Oct[1]CH_2COCH_3, ^{t}Oct[4]CH_2COCH_3, and ^{t}Oct[6]CH_2COCH_3 at 30°C.

prepared by the parent reaction of the *p*-t-octylcalixarenes or corresponding monomer with chloroacetone in the presence of sodium carbonate, sodium iodide, and acetone. Figure 2 shows the structures and abbreviations of ketonic derivatives used in the work, together with the 2-pyridyl one. Calix[4]arene derivatives assume the cone conformation, all of whose functional groups face the same direction.

Figure 3 shows the extraction percentages of various metal ions in 1 M (M = mold m^{-3}) nitric acid with three ketonic derivatives. Silver was the only metal remarkably extracted with calixarene derivatives among the metal ions examined. It is attributed to lower metal valency because the ketonic derivatives are neutral and the cation extraction takes place by a solvation mechanism. Since the calixarene derivatives exhibit a converging effect of functional groups [7, 8], the extraction ability of calixarene derivatives is significantly higher than that of the corresponding monomer. Particularly, the extraction ability of ^{t}Oct[4]CH_2COCH_3 is higher than that of ^{t}Oct[6]CH_2COCH_3 due to size-fitting and conformational effects.

Silver and divalent palladium are classified as soft acid types based on the Please change to hard and soft acids and bases (HSAB) theory [12], and both get effectively extracted with soft-base–type extraction reagents. Amines and sulfur-containing reagents exhibit palladium selectivity over silver. Methylketonic derivatives are classified into a hard base type, which has an advantage for monovalent cation selective extraction. Figure 4 shows the effect of nitric acid concentration on silver and divalent palladium extraction percentage with three methylketonic derivatives. As expected from the result in Fig. 3, ^{t}Oct[4]CH_2COCH_3 exhibited the highest selectivity for silver over palladium in nitric acid media around 1 M [11]. (Silver and palladium distribute to chloroform in nitric acid media over 6 M.)

Based on the result of Fig. 4, the removal of a trace amount of silver from a large excess of palladium using ^{t}Oct[4]CH_2COCH_3 was also investigated. The result is shown in Figure 5. The concentration of silver selectively extracted was over 100 times higher than that of palladium [11], although

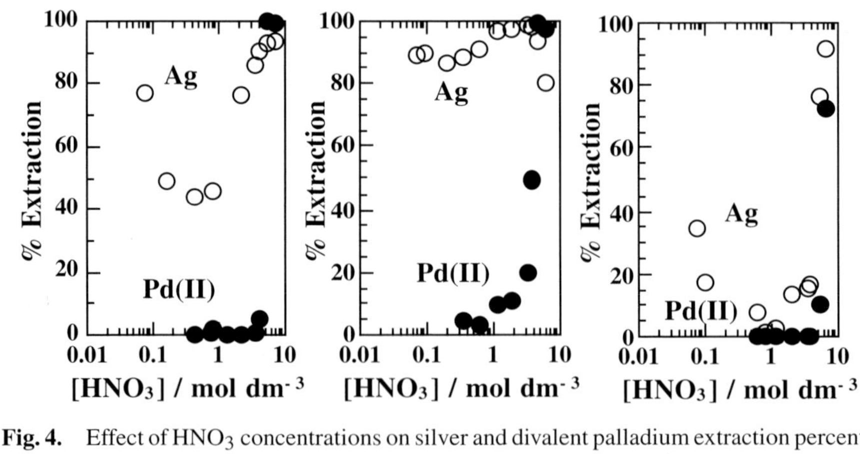

Fig. 4. Effect of HNO_3 concentrations on silver and divalent palladium extraction percentage with three methylketonic derivatives. [Metal ion] = 0.1 mM, [^{t}Oct[6]CH_2COCH_3], [^{t}Oct[4]CH_2COCH_3] = 5 mM and [^{t}Oct[1]CH_2COCH_3] = 20 mM in $CHCl_3$ at 30°C.

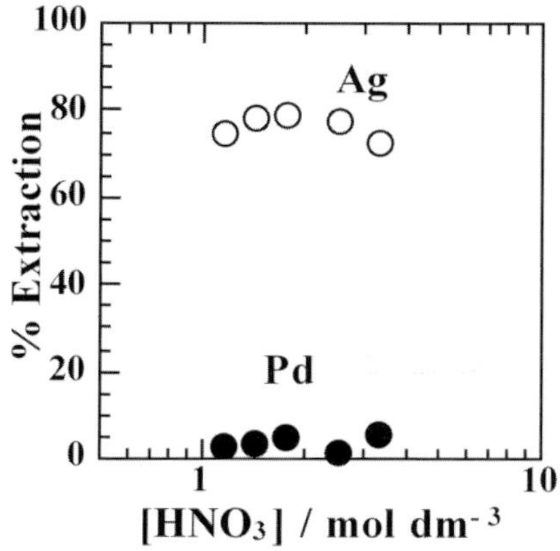

Fig. 5. Effect of HNO_3 concentrations on extraction percentage of silver and palladium with ^{t}Oct[4]CH_2COCH_3. $[Ag^+] = 2.5 \times 10^{-5}$ mM, $[Pd^{2+}] = 2.5 \times 10^{-3}$ mM, [^{t}Oct[4]CH_2COCH_3] = 3.3 mM in $CHCl_3$ at 30°C.

the extraction percentage of silver was slightly decreased compared with the result in Fig. 4. Such a higher silver selectivity is effective for recovery from a large excess of coexisting ions. For confirming the co-ordination site of ^{t}Oct[4]CH_2COCH_3 for silver uptake, ^{1}H-NMR spectra of before and after silver loading were investigated. The result showed that phenoxy and carbonyl oxygen atoms co-ordinate to silver ion, as shown in Fig. 6. Since all peaks in the spectra before and after silver loading appear as singlets, eight oxygen atoms equivalently co-ordinate to silver under the equilibrium [11], although the fact that the silver ion takes 2 or 4 as co-ordination number is well known.

Methylketonic ^{t}Oct[4]CH_2COCH_3 exhibited excellent silver selectivity, while the extraction rate was extremely slow. Figure 7 shows the effect of shaking time on an extraction percentage of silver and palladium with ^{t}Oct[4]CH_2COCH_3. The time to reach extraction equilibrium of silver was 72 h (3 d) or more for silver, while that for palladium was less than 30 min [13, 14] The silver extraction rate using other reagents is generally fast. The slow extraction may relate to the less polar structure of the extracting reagent, which leads to a less adsorption property on the interface. The activity of ^{t}Oct[4]CH_2COCH_3 on the interface between the organic and aqueous phases was investigated. Figure 8 shows the effect of extractant concentration on the interfacial tension of ^{t}Oct[4]CH_2COCH_3

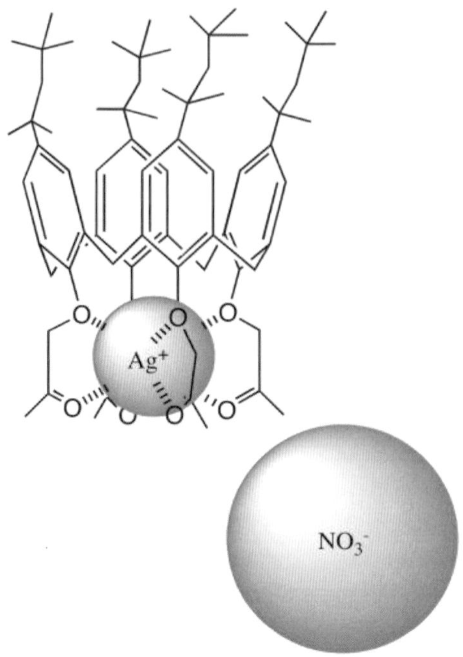

Fig. 6. Proposal structure of a silver complex with ^{t}Oct[4]CH_2COCH_3.

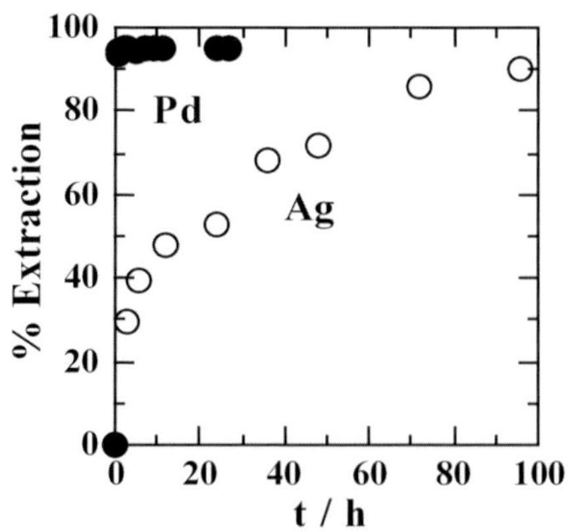

Fig. 7. Effect of shaking time on the extraction percentage of silver with ^{t}Oct[4]CH_2COCH_3.[HNO_3] = 1 M, [^{t}Oct[4]CH_2COCH_3]. = 5 mM, [Metal ion] = 0.1 mM each at 30°C.

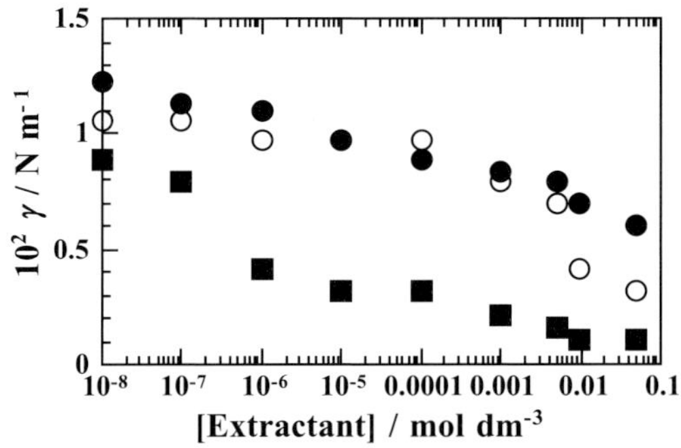

Fig. 8. Effect of extractant concentrations on the interfacial tension of ^{t}Oct[4]CH_2COCH_3 (●), ^{t}Oct[4]$CH_2COC_6H_5$ (○), ^{t}Oct[4]CH_2^2Py (■) at chloroform–1 M nitric acid interface at 30°C.

at the chloroform–nitric acid interface, together with the results for ^{t}Oct[4]$CH_2COC_6H_5$ and ^{t}Oct[4]$CH_2$2Py shown in Fig. 2. The fact that the interfacial tension of ^{t}Oct[4]CH_2COCH_3 is much higher than those of ^{t}Oct[4]$CH_2COC_6H_5$ and ^{t}Oct[4]$CH_2$2Py means ^{t}Oct[4]CH_2COCH_3 adsorbs less on the interface than the others, consequently it exhibits a slower extraction rate compared with the others [14].

In order to improve the slow extraction rate of ^{t}Oct[4]CH_2COCH_3 with high silver selectivity, the use of a micro-reactor system was investigated [15, 16]. The micro-reactor system can contribute to reduced amounts of the extraction reagents, solvents, and energy; reduced space; fast extraction; and smooth and lossless operation. The illustrations of parallel two-phase flow (PTPF) and slug flow (SF) types micro-reactors are shown in Fig. 9.

The effect of space–time on silver extraction percentage with ^{t}Oct[4]CH_2COCH_3 using PTPF- and SF-type micro-reactor systems together with a batch-wise method is shown in Fig. 10. The times to reach an equilibrium level of the extraction percentage using PTPF- and SF-type micro-reactor systems together with a batch-wise method were 16s, 4s and 72 h, respectively. Both micro-reactor systems remarkably improved the silver extraction rate due to the increased interfacial area together with a short diffusion distance. Particularly, the SF-type micro-reactor system contributed to significantly a rapid extraction rate. Although the micro-reactor system has drawbacks from the viewpoint of mass productivity, it is a possible candidate to solve the metal resource reservation.

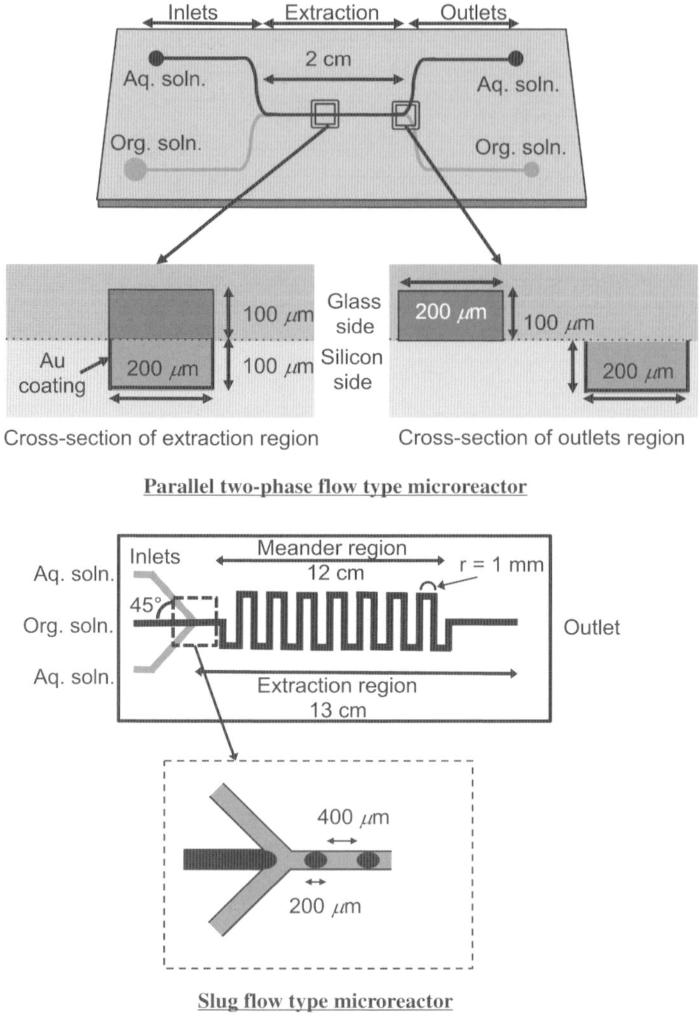

Fig. 9. Illustration of parallel two-phase flow and slug-flow micro-reactors types.

9.2.2 *N,N-Diethylamido derivative of p-t-octylcalix[4]arene*

Second, *N*,*N*-diethylamido derivatives of *p*-t-octylcalix[6]arene, *p*-t-octylcalix[4]arene and the corresponding monomer were employed [17]. For the preparation of *N*,*N*-diethylamido derivatives, acetic acid derivatives of *p*-t-octylcalixarene or the corresponding monomer were first prepared

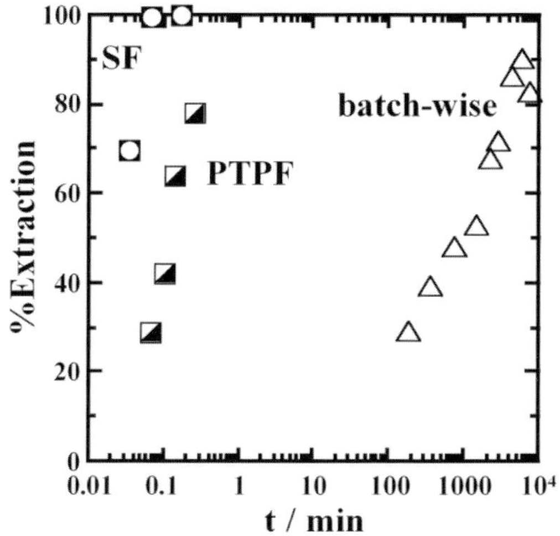

Fig. 10. Effect of space–time on silver extraction percentage with ^{t}Oct[4]CH_2COCH_3 using PTPF- and SF-type micro-reactor systems together with a batch-wise method.

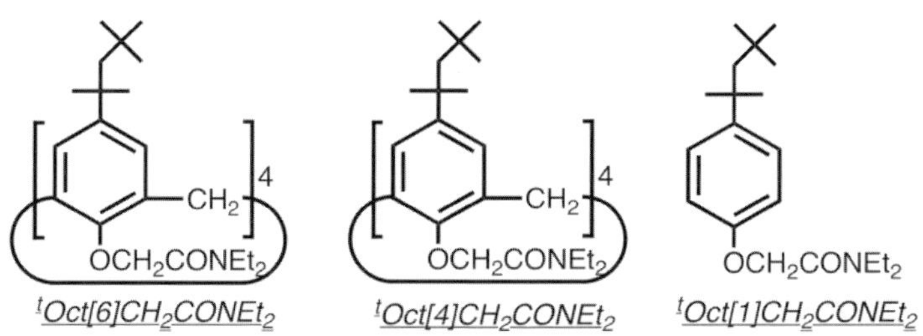

Fig. 11. Structures and abbreviations of amido derivatives of *p*-t-octylcalix[6]arene, calix[4]arene and the corresponding monomer.

by the reaction reported previously [18]. The acetic acid derivatives were reacted with thionyl chloride in toluene to prepare acid chloride compounds, then the acid chlorides were reacted with *N*,*N*-diethylamine in toluene to obtain *N*,*N*-diethylamido derivatives. Figure 11 shows the structures and abbreviations of *N*, *N*-diethylamido derivatives in the work. Calix[4]arene derivative (^{t}Oct[4]CH_2CONEt_2) assumes the cone conformation, all of whose amido groups face the same direction.

Since ^{t}Oct[4]CH_2COCH_3 was effective to selectively extract silver by co-ordinating with carbonyl and phenoxy oxygen atoms, extraction of some precious metals with amido derivatives, ^{t}Oct[4]CH_2CONEt_2 and ^{t}Oct[1]CH_2CONEt_2 in nitric acid media were also investigated. Figure 12

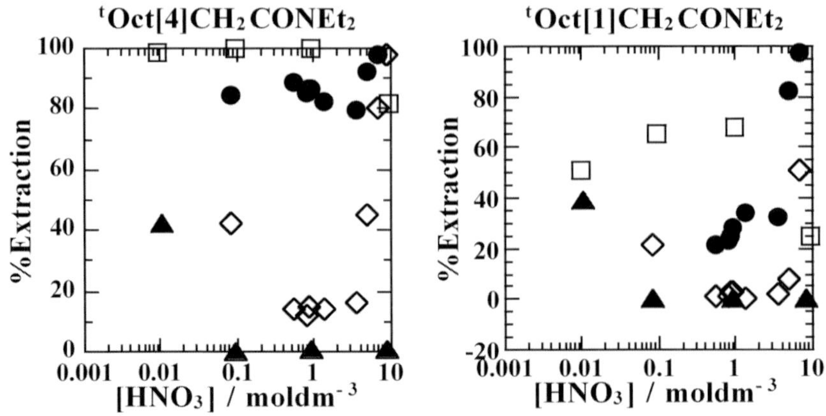

Fig. 12. Effect of HNO_3 concentrations on precious metals extraction percentage with amide derivatives. Au(III) (□), Ag (■), Pd(II) (◇), Pt(IV) (▲), [Metal ion] = 0.1 mM, [^{t}Oct[4]CH_2CONEt_2] = 5 mM and [^{t}Oct[1]CH_2CONEt_2] = 20 mM in $CHCl_3$ at 30°C.

shows the effect of nitric acid concentration on precious metals extraction percentage with amido derivatives. As anticipated, ^{t}Oct[4]CH_2CONEt_2 exhibited silver extractability with less silver selectivity over divalent palladium in nitric acid media (Fig. 12) [17]. The selectivity order is as follows: Au(III) > Ag > Pd(II) > Pt(IV). ^{t}Oct[4]CH_2CONEt_2 also exhibited higher extractability compared with ^{t}Oct[1]CH_2CONEt_2. It was attributed to the size-fitting effect and the converging effect of the amido group functionality [8].

Figure 13 shows the effect of hydrochloric acid concentration on silver extraction percentage with three amido derivatives. Lithium chloride (it was preliminarily investigated that lithium does not affect silver extraction) was added not to suppress silver chloride precipitation so as to adjust the chloride concentration to be 5 and 3 M, respectively. The silver extraction with ^{t}Oct[6]CH_2CONEt_2 was not much different at both the LiCl concentrations tested, while those with ^{t}Oct[4]CH_2CONEt_2 and ^{t}Oct[1]CH_2CONEt_2 were significantly different at both the chloride concentrations. A higher chloride concentration was preferable for silver extraction with ^{t}Oct[1]CH_2CONEt_2, but not with ^{t}Oct[4]CH_2CONEt_2. It is attributed to the different extraction mechanisms. That is, silver is extracted as a cation with ^{t}Oct[4]CH_2CONEt_2, while it was obtained as an anion with ^{t}Oct[1]CH_2CONEt_2. Figure 14 shows proposed structures of silver complexes with ^{t}Oct[4]CH_2CONEt_2 and ^{t}Oct[1]CH_2CONEt_2.

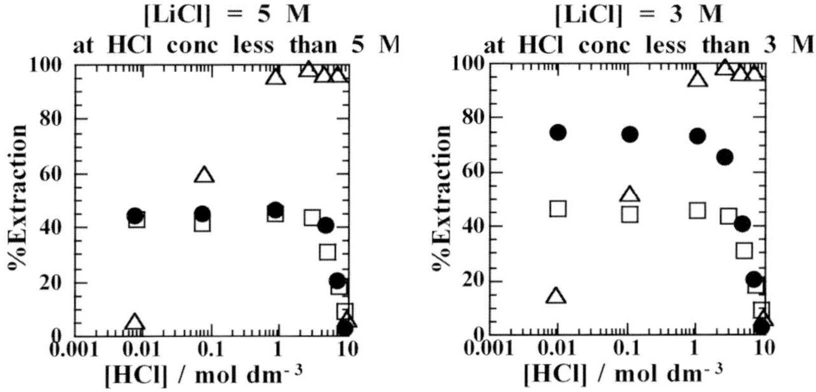

Fig. 13. Effect of HCl concentrations on silver extraction percentage with amide derivatives. ^{t}Oct[6]CH_2CONEt_2 (□), ^{t}Oct[4]CH_2CONEt_2 (■), ^{t}Oct[1]CH_2CONEt_2 (△), [Metal ion] = 0.1 mM, [^{t}Oct[6]CH_2CONEt_2] and [^{t}Oct[4]CH_2CONEt_2] = 5 mM; and [^{t}Oct[1]CH_2CONEt_2] = 20 mM in $CHCl_3$, LiCl was added at low HCl concentration to adjust the chloride concentration at 30°C.

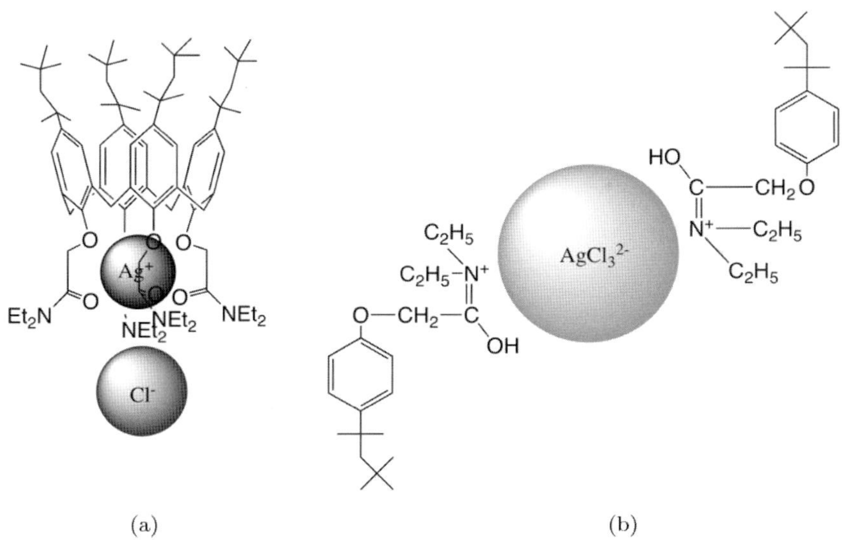

Fig. 14. Proposed structures of silver complexes with (a)^{t}Oct[4]CH_2CONEt_2 and (b) ^{t}Oct[1]CH_2CONEt_2 in chloride media.

For confirming the co-ordination site of ^{t}Oct[4]CH_2CONEt_2 for silver uptake, ^{1}H-NMR spectra of before and after silver loading were also investigated. The result showed phenoxy and carbonyl oxygen atoms co-ordinate to silver ion, as shown in Fig. 14(a). Since all peaks in the spectra before and after silver loading appear as singlets, eight oxygen atoms equivalently co-ordinate to silver under the equilibrium [17]. Different from the silver complex with ^{t}Oct[4]CH_2CONEt_2, two molecules of protonated ^{t}Oct[1]CH_2CONEt_2 may extract the anionic species of the silver ion.

9.2.3 *2-Pyridyl derivative of p-t-octylcalix[4]arene*

Next, 2-, 3-, and 4-pyridylmethyl derivatives of *p*-t-octylcalix[4]arene and the corresponding monomer were employed [19]. For the preparation of pyridyl derivatives, the original *p*-t-octylcalix[4]arene was reacted with n-chloromethyl pyridines or the corresponding monomer Figure 15 shows the structures and abbreviations of the pyridyl derivatives. All calix[4]arene derivatives assume cone conformation, all of whose pyridyl groups face to the same direction.

Figure 16 shows the effect of pH on silver extraction percentage with 2-pyridyl derivatives of calix[4]arene and the monomer. Comparison of the results apparently proves the higher extraction ability of calix[4]arene,

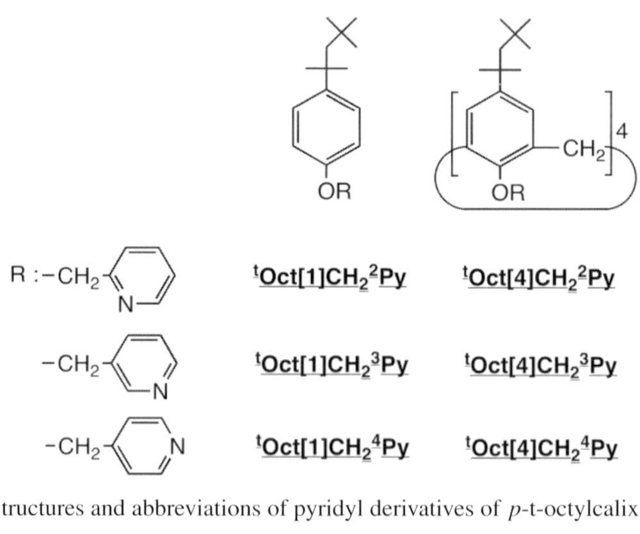

Fig. 15. Structures and abbreviations of pyridyl derivatives of *p*-t-octylcalix[4]arene and monomer.

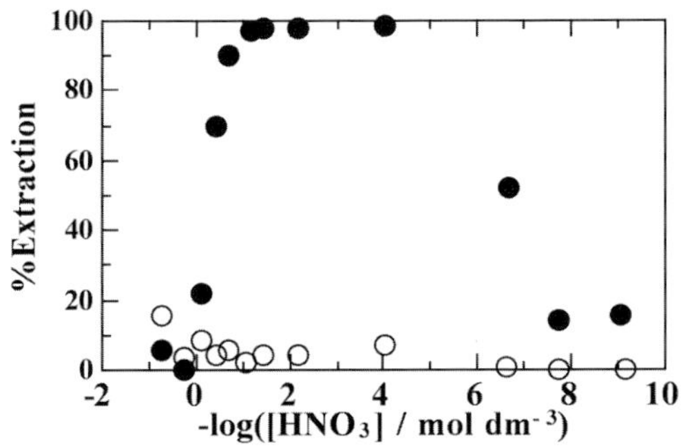

Fig. 16. Effect of nitric acid concentration on silver extraction percentage with ^{t}Oct[4]CH$_2^2$Py (●, 5 mM in $CHCl_3$) and ^{t}Oct[1]CH$_2^2$Py (○, 20 mM in $CHCl_3$). $[Ag^+]$ = 0.1 mM, 30°C.

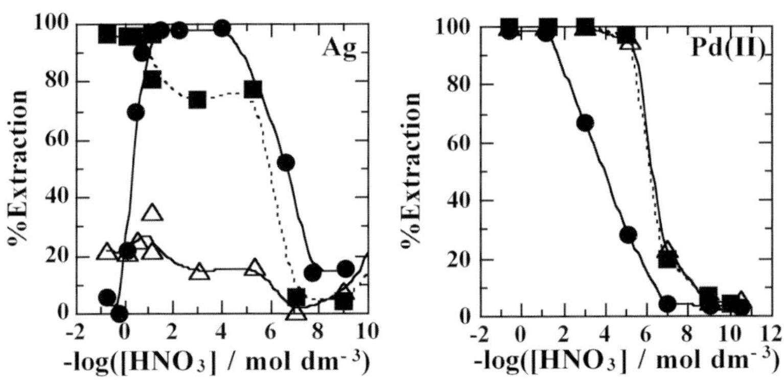

(a) (b)

Fig. 17. Effect of nitric acid concentration on (a) Ag and (b) Pd(II) extraction percentage with ^{t}Oct[4]CH$_2^2$Py(●), ^{t}Oct[4]CH$_2^3$Py(△), ^{t}Oct[4]CH$_2^4$Py (■). [^{t}Oct[4]CH$_2^n$Py] = 5 mM in $CHCl_3$,[Metal] = 0.1 mM at 30°C.

and consequently, the structural effect of calix[4]arene. Figure 17 shows the effect of pH on silver and divalent palladium extraction with three pyridyl derivatives of calix[4]arene with pyridyl nitrogen atoms at different positions. ^{t}Oct[4]CH$_2^2$Py and ^{t}Oct[4]CH$_2^4$Py exhibited higher extraction ability than ^{t}Oct[4]CH$_2^3$Py. The silver extraction ability of ^{t}Oct[4]CH$_2^3$Py is slightly higher than that of ^{t}Oct[1]CH$_2^2$Py. The results of Figs. 16 and 17 suggest that a single pyridyl group hardly works for silver extraction, and the position of nitrogen atom was related to higher extraction ability. That is, the higher extraction ability of ^{t}Oct[4]CH_2^2Py and ^{t}Oct[4]CH_2^4Py is probably attributed to the converging effect of the functional groups. On the contrary, the position of the nitrogen atom was slightly related to the extraction ability for palladium, although the extraction ability of ^{t}Oct[4]CH_2^2Py is slightly lower than the others probably due to stable protonation. It means that functional group effect dominantly contributes to palladium extraction rather than the structural effect of calix[4]arene. Figure 18 shows the effect of pH on other metal ion extractions with three pyridyl derivatives of calix[4]arene with pyridyl nitrogen atoms at different positions. All metal ions investigated were hardly extracted with any extractants. ^{t}Oct[4]CH_2^2Py and ^{t}Oct[4]CH_2^4Py exhibited high silver selectivity over other metal ions except for palladium, which was extracted based on the functional group effect.

For confirming the co-ordination site of ^{t}Oct[4]CH_2^2Py for silver uptake, ^{1}H-NMR spectra before and after silver loading were also investigated. Clear peak shifts were observed before and after silver loading for pyridyl aromatic protons but were vague for methylene protons next to phenoxy oxygens atom. The result showed carbonyl oxygen atoms co-ordinate to the silver ion, as shown in Fig. 19. It is different from the result with the amido derivative, where the phenoxy oxygen atoms did not co-ordinate to the silver ion due to the principle of HSAB theory [12].

Next, impregnating resins of 2-, 3-, and 4-pyrydinomethyl derivatives *p*-t-butylcalix[4]arene into macroporous matrix Amberlite XAD-7 shown in Fig. 20 were prepared and were used for silver adsorption [20]. Three pyridyl derivatives were prepared from the original *p*-t-butylcalix[4]arene by a similar method as described for *p*-t-butylcalix[4]arene derivatives. In this work, the *t*-butyl branch was selected for its lower molecular weight than the *t*-octyl one, consequently resulting in higher loading on XAD-7. Each of the extractants was dissolved in chloroform and the preconditioned and dried XAD-7 was added to the mixture and kept overnight. The solvent was evaporated. The procedure was repeated until the extractant was not seen at the flask wall, after which the impregnated resins ^{t}Bu[4]CH_2^nPy−IR were obtained. The impregnation percentage of each extractant was 33 wt%. The leakage of the extractant into the aqueous phase and adsorption rate were preliminarily checked. The leakage of ^{t}Bu[4]CH_2^2Py was found to

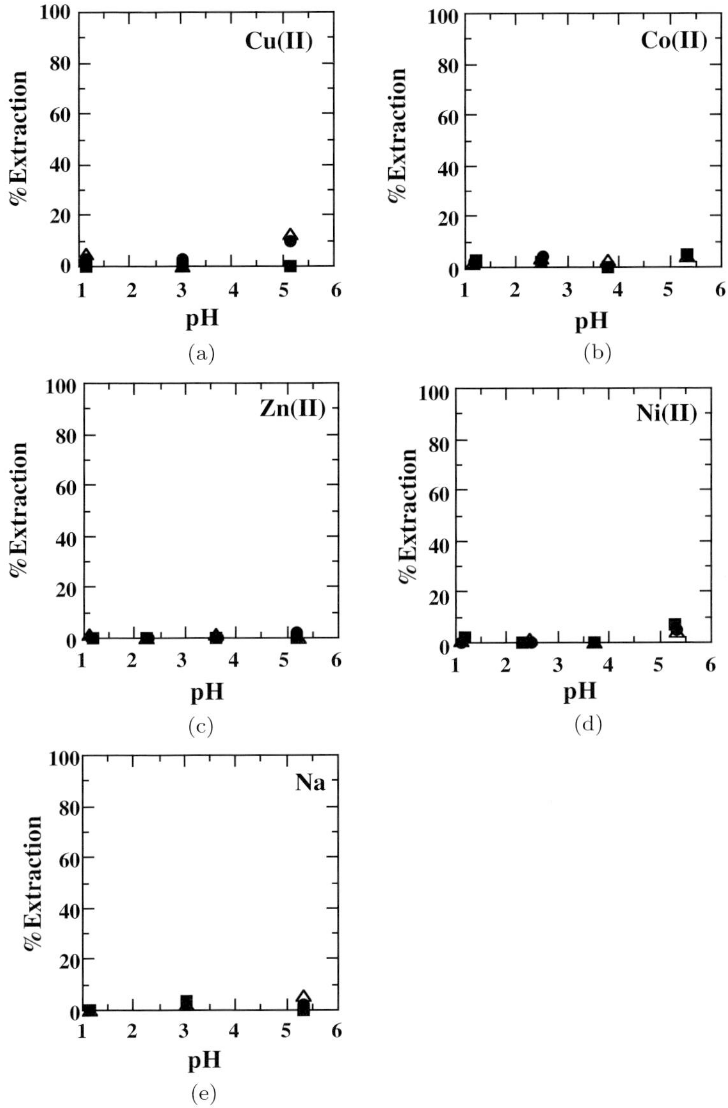

Fig. 18. Effect of pH on (a) Cu(II), (b) Co(II), (c) Zn(II), (d) Ni(II), (e) Na extraction percentage with ^{t}Oct[4]CH$_2^2$Py (●), ^{t}Oct[4]CH$_2^3$Py (△), ^{t}Oct[4]CH$_2^4$Py(■). [^{t}Oct[4]CH$_2^n$Py] = 5 mM in $CHCl_3$, [Metal] = 0.1 mM at 30°C.

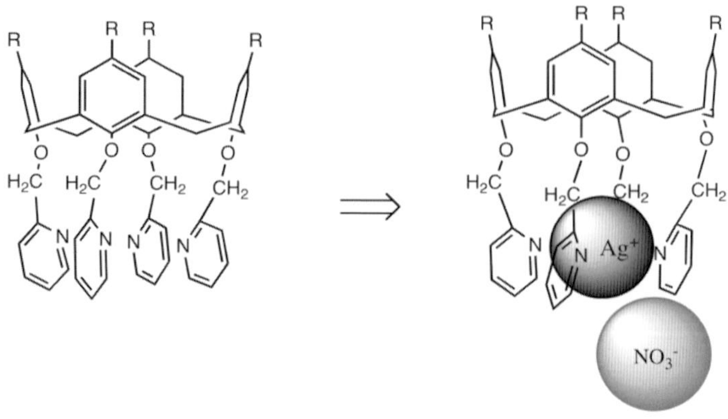

Fig. 19. Silver extraction reaction with ^{t}Oct[4]CH_2^2Py and the structure of the silver complex.

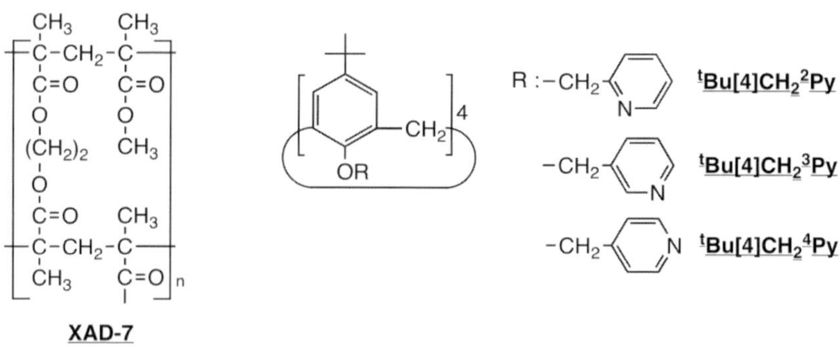

Fig. 20. Structures and abbreviations of Amberlite XAD-7 and pyridyl derivatives of *p*-t-butylcalix[4]arene at different nitrogen positions.

be negligible at 0.38% at pH 1.01 and 0.19% at pH 5.02 at pH 1.01–5.02. Silver adsorption on ^{t}Bu[4]CH_2^nPy−IR resins was found to take 5 h to reach equilibrium.

Figure 21 shows the effect of pH on the adsorption percentage of various metal ions on three impregnated resins, ^{t}Bu[4]CH_2^nPy−IR and XAD-7 itself ^{t}Bu[4]CH_2^2Py−IR exhibited silver selectivity among metal ions examined, although XAD-7 itself also adsorbed some metal ions at pH 3–5 due to low metal concentration. The extraction property of the originally impregnated extractant, ^{t}Bu[4]CH_2^2Py is reflected the adsorption property. Figure 22 shows the silver adsorption

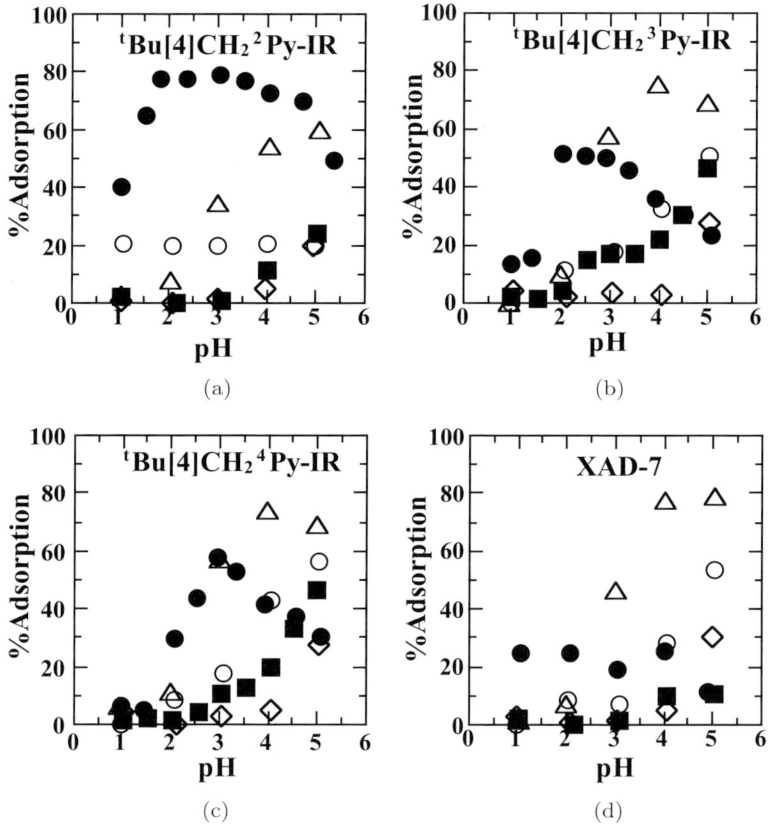

Fig. 21. Effect of pH on adsorption percentage of various metal ions on three impregnated resins and XAD-7. (a) ^{t}Bu[4]CH$_2^2$Py–IR, (b) ^{t}Bu[4]CH$_2^3$Py–IR, (c) ^{t}Bu[4]CH$_2^4$Py–IR, (d) XAD-7, ●: Ag, △: Fe(III), ■: Cu(II), ○: Pb(II), ◇: Zn(II), amount of IR and XAD-7 = 20 mg, [metal] = 0.1 mM, adjustment of pH = 0.1 M HNO_3 –0.1 M HEPES, 30°C.

isotherm on three ^{t}Bu[4]CH$_2^n$Py–IR resins. The maximum loading capacity for ^{t}Bu[4]CH$_2^2$Py–IR was 0.39 mol kg^{-1}, which corresponds to the ^{t}Bu[4]CH$_2^2$Py amount impregnated into XAD-7, while those for ^{t}Bu[4]CH$_2^3$Py–IR and ^{t}Bu[4]CH$_2^4$Py–IR were 0.24 mol kg^{-1} and 0.25 mol kg^{-1}, which are roughly half of the impregnated amount. The results verify that the stoichiometries of ^{t}Bu[4]CH$_2^2$Py, ^{t}Bu[4]CH$_2^3$Py and ^{t}Bu[4]CH$_2^4$Py with the silver ion are 1:1, 2:1, and 2:1, respectively. These correspond to those of ^{t}Oct[4]CH$_2^2$Py and ^{t}Oct[4]CH$_2^3$Py with silver ion obtained by slope analysis.

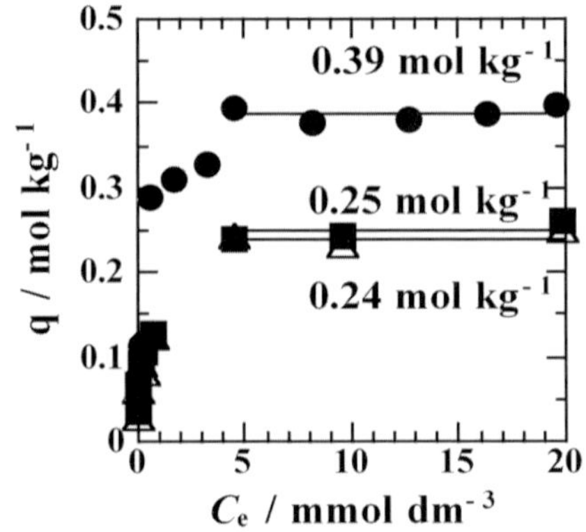

Fig. 22. Adsorption isotherm of silver on ^{t}Bu[4]CH$_2^2$Py$-$IR (●), ^{t}Bu[4]CH$_2^3$Py$-$IR (△), ^{t}Bu[4]CH$_2^4$Py$-$IR (■). [^{t}Bu[4]CH$_2^n$Py$-$IR = 20 mg pH$_i$ = 3.00 at 30°C.

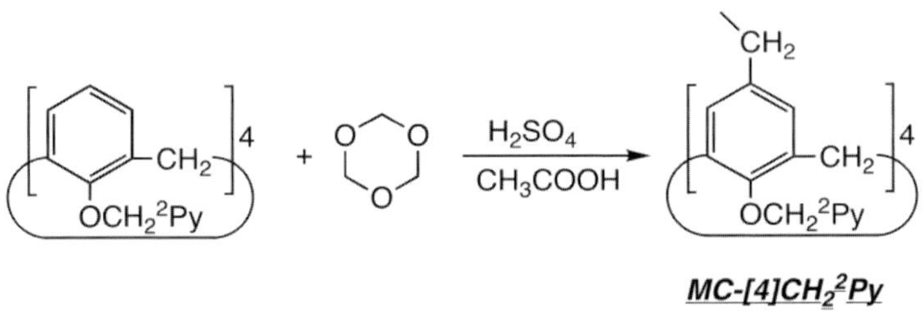

Fig. 23. Preparation of methylene cross-linked resin containing 2-pyrydino type calix[4]arene.

Methylene cross-linked resin, MC$-$[4]CH$_2^2$Py, was also prepared by the reaction of *s*-trioxane with 2-methylpyridyl type of debutylated calix[4]arene under acidic condition shown in Fig. 23 [21].

The prepared resins were not dissolved in any solvents and exhibited high silver selectivity and loading capacity. The resin prepared by 3:1 ratio of calix[4]arene and trioxane, in particular, exhibited excellent result. Figure 24 shows silver adsorption isotherm on MC$-$[4]CH$_2^2$Py, together with ^{t}Bu[4]CH$_2^2$Py$-$IR. The maximum loading capacity for MC$-$[4]CH$_2^n$Py was 1.29 mol kg^{-1}, which corresponds to the [4]CH$_2^2$Py amount incorporated into MC$-$[4]CH$_2^n$Py, 1.21 mol kg^{-1}. The capacity is three times higher than that of ^{t}Bu[4]CH$_2^2$Py$-$IR.

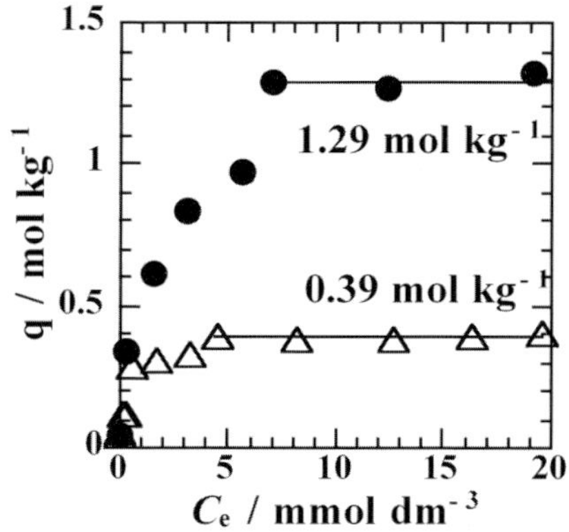

Fig. 24. Adsorption isotherm of silver on MC[4]CH$^{n}_{2}$Py(●) and ^{t}Bu[4]CH$^{2}_{2}$Py−IR (△), resin weight = 20 mg pH$_i$ = 3.00 at 30°C.

9.3 Silver Extraction with Tripodal Compounds

9.3.1 *Sulfur-containing tripodal compounds*

A tripodal compound such as alkyltrimethylol provides a narrower metal co-ordination site than that of calix[4]arene, whose co-ordination site is fit for the size of sodium and silver ions. In order to confirm the significance for the framework of a tripodal compound, sulfur-containing groups were selected as the typically soft functional groups introduced on the tripodal compound based on the HSAB theory [12]. Sulfur-containing tripodal compounds such as a thioester, thiol and dithioether and their related monopodal compounds were also employed [22, 23]. The structures of the compounds are shown in Fig. 25. The thioester-type tripodal compound, ^{8}None{3}CH$_2$SC(O)CH$_3$ was prepared from the reaction of the original triol compound, 1,1,1-tris(hydroxymethyl)-9-decene, ^{8}None{3}CH$_2$OH by a three-step reaction. That is, ^{8}None{3}CH$_2$OH was tosylated, brominated, then esterified by the reaction with potassium thioacetate. Thioether, ^{8}None{3}CH$_2$SH was prepared from ^{8}None{3}CH$_2$SC(O)CH$_3$ by the reduction. Dithioether, ^{8}None{3}CH$_2$SCH$_2$CH$_2$SMe was prepared from ^{8}None{3}CH$_2$SH. A series of reactions for tripodal compounds is shown in Fig. 26. The monopodal compounds were synthesized in a similar manner to the tripodal ones.

For the extraction method using tripodal and monopodal compounds, the monopodal concentration was adjusted to be three times higher than

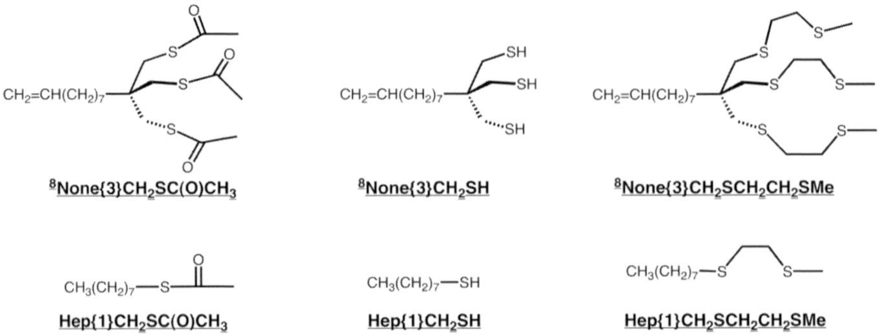

Fig. 25. Structures of the *S*-containing tripodal and its related monopodal compound.

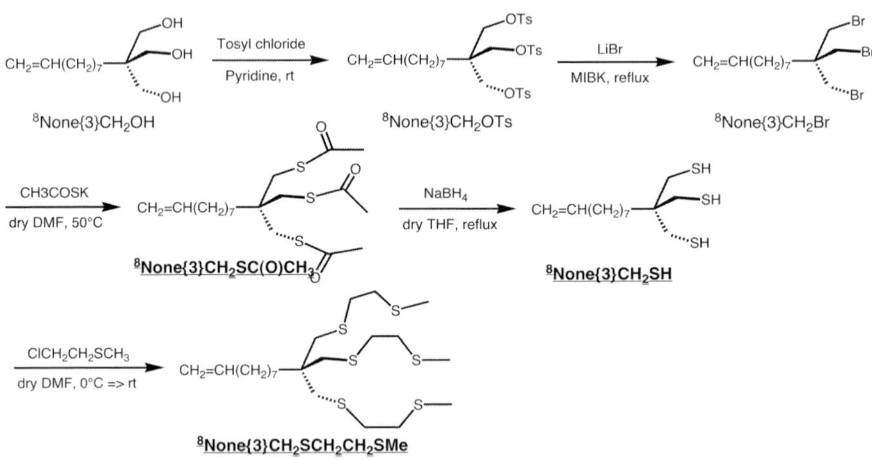

Fig. 26. Syntheses of sulfur-containing tripodal compounds.

the tripodal one so that the concentrations of functional groups in both compounds are the same.

Effects of initial pH on percentage extraction with sulfur-containing tripodal compounds together with its related monopodal compounds are shown in Figs. 27(a)–27(f). All sulfur-containing tripodal compounds quantitatively extracted silver and divalent palladium at pH 1, 3 and 5 examined, together with gold. The pH change before and after the extraction was not observed for both thiol compounds. It means that thiol compounds do not exchange ions with but co-ordinate to metal ions.

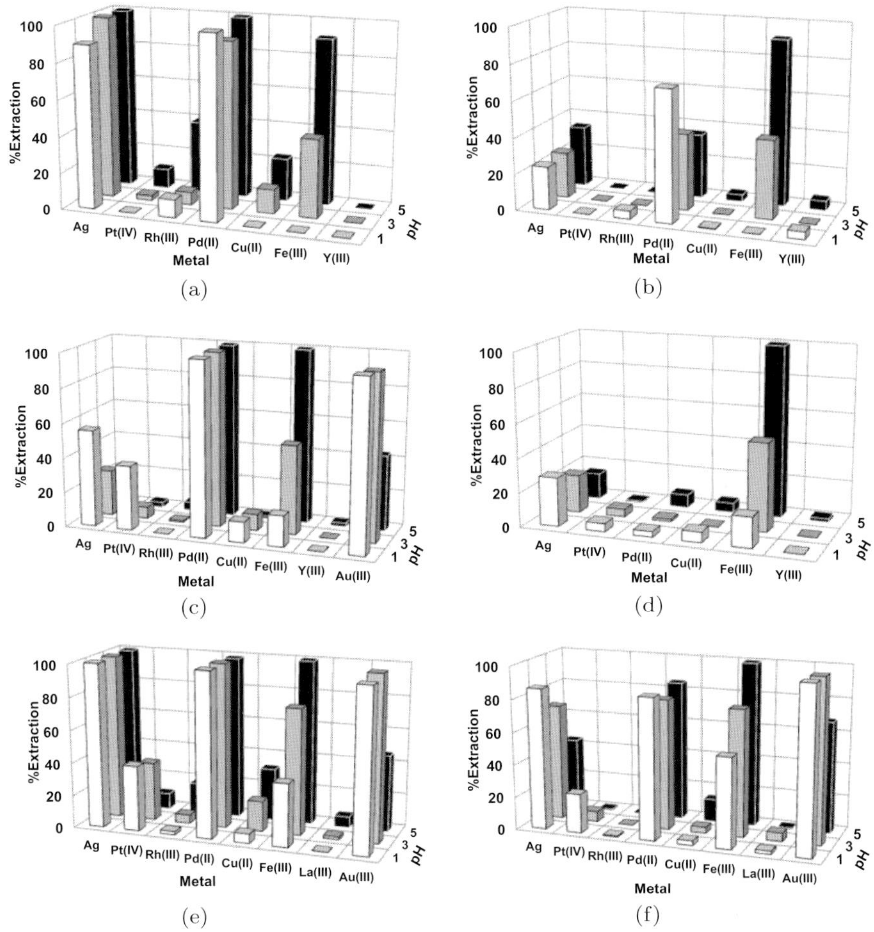

Fig. 27. Effect of initial pH on percentage extraction with (a) ^{8}None{3}$CH_2SC(O)CH_3$, (b) Hep{1}$CH_2SC(O)CH_3$, (c) ^{8}None{3}CH_2SH, (d) Hep{1}CH_2SH, (e) ^{8}None{3}$CH_2SCH_2CH_2SMe$, (f) Hep{1}$CH_2SCH_2CH_2SMe$ [tripodal compound] = 5 mM in chloroform [monopodal compound] = 15 mM in chloroform [metal] = 0.1 mM, pH adjustment = 0.1 M HNO_3 − 0.1 M HEPES, 30°C.

The difference in silver and divalent palladium extraction percentages between tripodal and monopodal compounds for dithioether compounds was smaller than that for thioester compounds. It was attributed to the chelating bidentate property of dithioether compounds. In particular, a large difference in silver extraction was observed. The high silver extraction

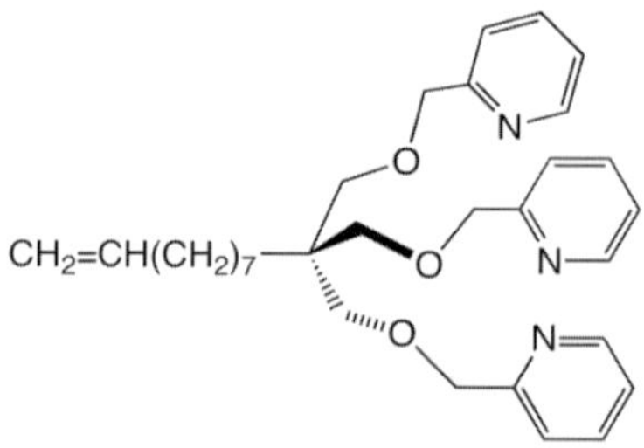

^{8}None{3}$CH_2OCH_2^2Py$

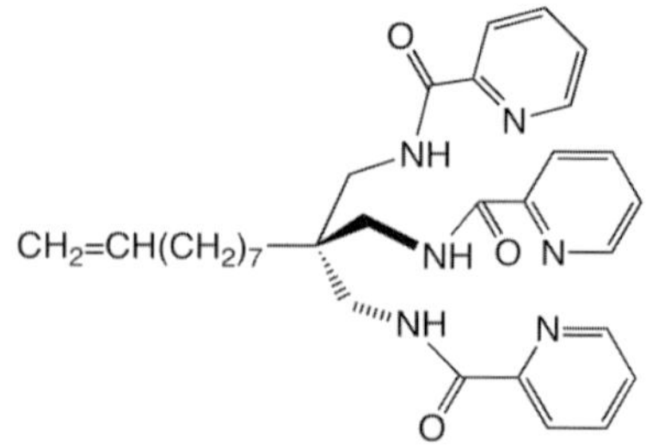

^{8}None{3}CH_2NHCO^2Py

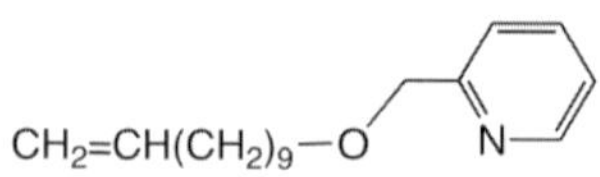

^{8}None{1}$CH_2OCH_2^2Py$

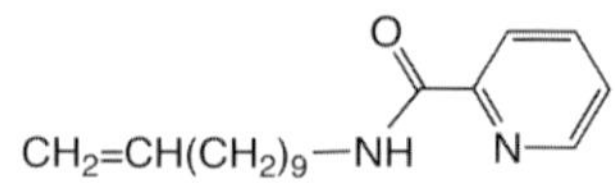

^{8}None{1}CH_2NHCO^2Py

Fig. 28. Structures of 2-pyridyl type of tripodal and monopodal compounds.

ability of tripodal compounds is attributed to the structural effect, such as the converging effect of functional groups and complimentary effect; consequently, the framework of a tripodal compound is significant for silver extraction.

9.3.2 *Pyridyl derivatives of tripodal compounds*

Next, pyridyl-type tripodal and the corresponding monopodal compounds were investigated. The structures are shown in Fig. 28. The tripodal ^{8}None{3}$CH_2OCH_2^2Py$ was prepared from the reaction of the original triol compound, ^{8}None{3}CH_2OH by a single-step reaction [10]. The tripodal ^{8}None{3}CH_2NHCO^2Py was prepared from the reaction of a triamino compound, ^{8}None{3}CH_2NH_2 with pyridine-2-carbonyl chloride hydrochloride salt. Both monopodal compounds were synthesized in a similar manner to the corresponding tripodal compounds.

Effect of pH on silver extraction percentage with ^{8}None{3}CH_2 OCH_2^2Py and ^{8}None{1}$CH_2OCH_2^2Py$ is shown in Fig. 29. Apparently the silver extraction ability of ^{8}None{3}$CH_2OCH_2^2Py$ is higher than ^{8}None{1}$CH_2OCH_2^2Py$ due to the structural effect of a tripodal framework.

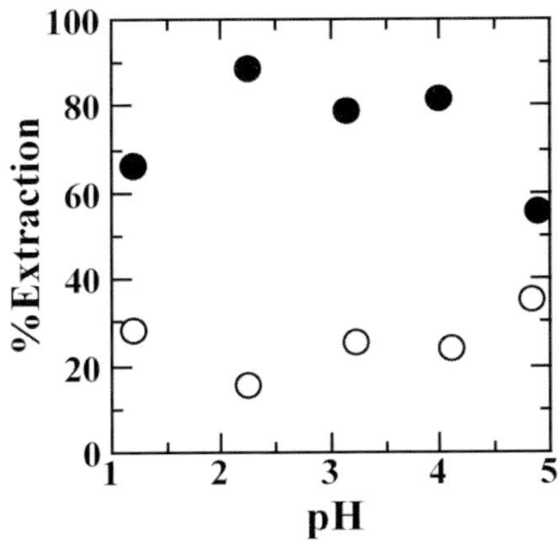

Fig. 29. Effect of pH on the extraction percentage of silver ion with ^{8}None{3}$CH_2OCH_2^2$Py (●, 5 mM), ^{8}None{1}$CH_2OCH_2^2$Py (○, 15 mM) in nitrate media. [silver ion] = 0.1 mM, 0.1 M HEPES at 30°C.

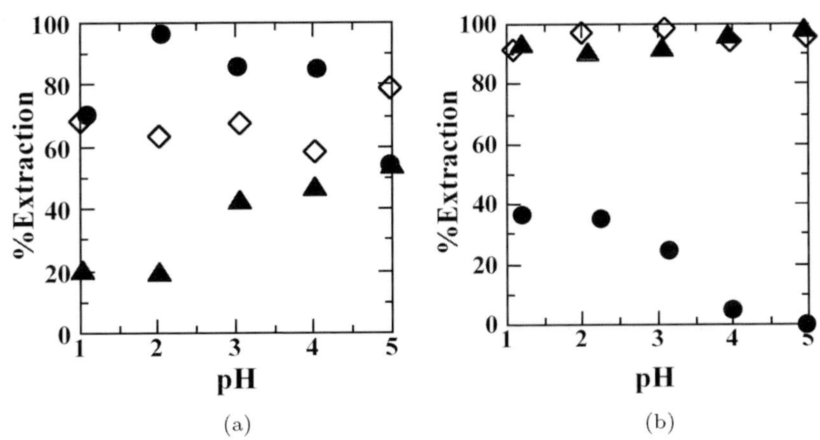

Fig. 30. Effect of pH on the extraction percentage of precious metals with (a) ^{8}None{3}CH_2NHCO^2Py (5 smM) and (b) ^{8}None{1}CH_2NHCO^2Py (15 mM) in a nitrate medium. Ag (●), Pd(II) (◇), Pt(IV) (▲), [metal ion] = 0.1 mM, 0.1 M HNO_3 − 0.1 M HEPES at 30°C.

This was supported by ^{1}H-NMR spectroscopy for ^{8}None{3}$CH_2OCH_2^2$Py spectra before and after silver loading.

Effects of pH on the extraction percentage of silver, divalent palladium and tetravalent platinum with ^{8}None{3}CH_2NHCO^2Py and ^{8}None{1}CH_2NHCO^2Py are shown in Figs. 30(a) and 30(b), respectively.

The extraction percentage of silver with ^{8}None{3}CH_2NHCO^2Py is also higher than that with ^{8}None{1}CH_2NHCO^2Py due to the structural effect, while those of palladium and platinum with ^{8}None{3}CH_2NHCO^2Py is smaller than those with ^{8}None{1}CH_2NHCO^2Py. It is caused by "negative structural effect". It means that the tripodal framework contributes to less extraction ability. Metal ions may be not surrounded by the tripodal framework but may be extracted outside of the metal co-ordination site of the pseudocyclic framework. And in general, the stoichiometry of ^{8}None{1}CH_2NHCO^2Py with a metal ion is lower than that of ^{8}None{3}CH_2NHCO^2Py, which may also contribute to the negative effect.

9.3.3 *Phenylamido derivative of tripodal compounds*

Finally, phenylamidotype tripodal and the corresponding monopodal compounds were investigated [24]. The structures of the tripodal and monopodal compounds are shown in Fig. 31. The tripodal ^{8}None{3}CH_2NHCOPh was prepared from the reaction of a triamino compound, ^{8}None{3}CH_2NH_2 with pyridine-2-carbonyl chloride hydrochloride salt. The monopodal compound was synthesized in a similar manner.

Effects of nitric acid concentration on the extraction percentage of various metal ions with ^{8}None{3}CH_2NHCOPh and ^{8}None{1}CH_2NHCOPh are shown in Figs. 32(a) and 32(b), respectively. Both compounds exhibited high silver and palladium selectivity over other precious and base metals. The extraction percentage of silver with ^{8}None{3}CH_2NHCOPh is also higher than that with ^{8}None{1}CH_2NHCO^2Ph due to the structural effect. However, for this extraction, remarkable spectral change of ^{8}None{3}CH_2NHCOPh by ^{1}H-NMR spectroscopy was not observed.

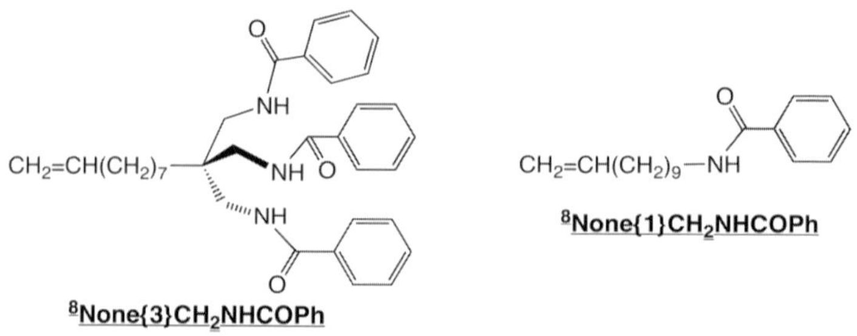

Fig. 31. Structures of phenylamido-type tripodal and monopodal compounds.

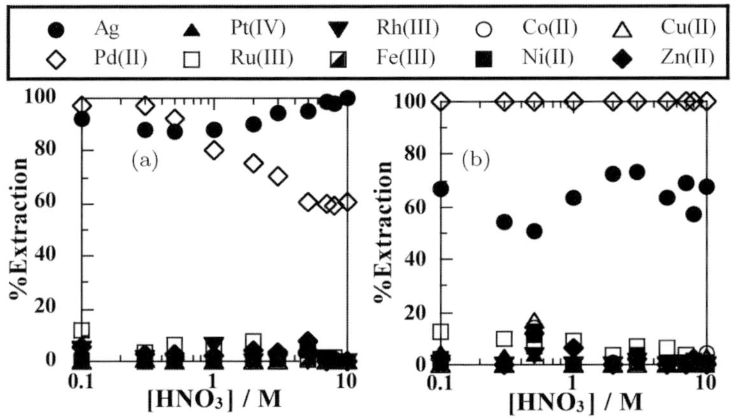

Fig. 32. Effects of nitric acid concentration on the extraction percentage of various metal ions with (a) ^{8}None{3}CH_2NHCOPh (5 mM) and (b)^{8}None{1}CH_2NHCOPh (15 mM). [Metal ions] = 0.1 mM at 30°C.

It means that silver was not surrounded and co-ordinated by carbonyl oxygen atoms in a tripodal framework, but may be outside of the framework Palladium was significantly extracted with both extraction reagents. The silver extraction percentage of ^{8}None{1}CH_2NHCO^2Ph was higher than that of ^{8}None{3}CH_2NHCOPh. Palladium may be also extracted outside of ^{8}None{3}CH_2NHCOPh due to the high functional group effect but not structural effect.

The ion-selective electrodes containing the above-mentioned reagents as a detecting device for silver ion were also prepared and investigated for potential response and showed silver selectivity over coexisting other metal ions [22, 25–27].

9.4 Conclusions

Various derivatives of macrocyclic calixarenes and pseudocyclic tripodal compounds were prepared for silver extraction. Both frameworks providing metal co-ordination site exhibited a remarkable structural effect consequently leading to the high silver extraction ability and selectivity. It was found that the selection of functional groups is also important for high silver selectivity over palladium and other metal ions. The extraction reagents based on calix[4]arene and nonenyl trimethylol with excellent extraction property were applied to use in the micro-reactor system and as ion-exchange resins and silver-ion selective electrodes.

References

[1] Cox M (2004) *Solvent Extraction Principles and Practice*, 2nd Ed., by Rydberg J, Cox M, Musikas C, Choppin GR (ed.) Marcel Dekker, Inc., New York, Solvent Extraction in Hydrometallurgy, pp. 455–505.

[2] Rice NM (1978) Recent developments and potential uses for carboxylic acid extractants — a review. *Hydrometallurgy*, **3**: 111.

[3] Preston JS (1985) Solvent extraction of metals by carboxylic acids. *Hydrometallurgy*, **14**: 171.

[4] Freiser H (1988) Solvent extraction of tervalent lanthanides as chelates -a systematic investigation of extraction equilibria. *Solvent Extraction and Ion Exchange*, **6**: 1093.

[5] Cox M (1988) "New Reagents." In *Developments in Solvent Extraction*, Chapter 8, S Alegret (ed.) (Ellis Horwood Ltd, Chichester) pp. 151–158.

[6] Otu EO, Westland AD (1990) Solvent extraction with organophosphonic mono-acidic esters. *Solvent Extraction and Ion Exchange*, **8**: 759.

[7] Ohto K (2010) Review of the extraction of metal cations with calixarene derivatives. *Solvent Extraction Development and Research, Japan*, **17**: 1.

[8] Ohto K (2014) Molecular design and metal extraction behavior of calixarene compounds as host extractants. *Ion Exchange and Solvent Extraction*, **21**: 81.

[9] Ohto K, Nakagawa H, Furutsuka H, Shinohara T, Nakamura T, Oshima T, Inoue K (2004) Solvent extraction of rare earths with a novel phosphonate extractant providing a narrow coordination site. *Solvent Extraction Research and Development, Japan*, **11**: 121.

[10] Yamaguma R, Yamashita, Kawakita H, Miyajima T, Takemura C, Ohto K, Iwachido N (2012) Selective extraction of precious metal ions with novel trident molecules with pyridyl groups. *Separation Science and Technology*, **47**: 1303.

[11] Ohto K, Murakami E, Shinohara T, Shiratsuchi K, Inoue K, Iwasaki M (1997) Selective extraction of silver (I) over palladium (II) with ketonic derivatives of calixarenes from highly concentrated nitric acid. *Analytica Chimica Acta*, **341**: 275.

[12] Pearson RG (1963) Hard and soft acids and bases. *Journal of the American Chemical Society*, **85**: 3533.

[13] Kim JY, Morisada S, Kawakita H, Ohto K, Kim Y (2015) Relationship between chemical structure and extraction efficiency toward palladium with

ketonic derivatives of p-tert-octylcalix[4]arene in nitric acid media. *Journal of Inclusion Phenomena and Macrocyclic Chemistry*, **82**: 25.

[14] Kim JY, Morisada S, Kawakita H, Ohto K (2014) Effect of interfacial activity on silver extraction rate with calix[4]arene derivatives in nitrate media. *Journal of Ion Exchange*, **25**: 136.

[15] Maeki M, Hatanaka Y, Yamashita K, Miyazaki M, Ohto K (2014) Solvent extraction behavior of metal ions with calixarene derivatives by using a microreactor. *Solvent Extraction Research and Development, Japan* **21**: 77.

[16] Ohto K, Kim JY, Morisada S, Maeki M, Yamashita K, Miyazaki M (2014) Microreactor extraction system with macrocyclic host compounds for rare metal recovery. *International Journal of the Society of Materials Engineering for Resources* **20**: 92.

[17] Ohto K, Yamaga H, Murakami E, Inoue K (1997) Specific extraction behavior of amide derivative of calix[4]arene for silver (I) and gold (III) ions from highly acidic chloride media. *Talanta*, **44**: 1123.

[18] Ohto K, Yano M, Inoue K, Yamamoto T, Goto M, Nakashio F, Shinkai S, Nagasaki T (1995) Solvent extraction of trivalent rare earth metal ions with carboxylate derivatives of calixarenes. *Analytical Sciences*, **11**: 893.

[19] Ohto K, Higuchi H, Inoue K (2001) Solvent extraction of silver ion with pyridino calix[4]arenes. *Solvent Extraction Research and Development, Japan*, **8**: 37.

[20] Hashiguchi N, Tanaka M, Ohto K, Kawakita H, Inoue K (2008) Solid phase extraction of silver ion by impregnated resins with *p*-tert-butylcalix[4]arene pyridyl derivatives. *Solvent Extraction Research and Development, Japan*, **15**: 99.

[21] Adhikari BB, Hashiguchi N, Ohto K, Kawakita H, Inoue K (2009) Preparation of calix[4]arene methylene crosslinked resins with 2-pyridyl pendant group and their adsorption behavior towards silver ion. *Journal of Inclusion Phenomena and Macrocyclic Chemistry*, **65**:121.

[22] Partially contained in Furugou H, Ohto K, Kawakita H, Harada H, Inoue K (2007) Preparation of thioester type of broom extractant with silver selectivity. *Ars Separatoria Acta*, **5**: 68.

[23] Ohto K, Furugou H, Yoshinaga T, Morisada S, Kawakita H, Inoue K (2017) Precious metal extraction with thiol and dithioether derivatives of a trident molecule. *Solvent Extraction Research and Development, Japan*, **24(2)**: 77.

[24] Ueda Y, Morisada S, Kawakita H, Ohto K (2013) Solvent extraction behavior of silver ion with secondary phenylamide derivative of trident molecule. *Solvent Extraction Research and Development, Japan*, **20**: 53.

[25] Shinohara T, Higuchi H, Senba Y, Ohto K, Yoshizuka K, Inoue K (2001) Silver ion selective electrode based on calix[4]arene methyl ketonic derivative. *Analytical Sciences*, **17**: 889.
[26] Higuchi H, Shinohara T, Oshima T, Ohto K, Inoue K (2002) Silver ion selective electrode based on (2-pyridylmethoxy)-p-t-octylcalix[4]arene. *Ars Separatoria Acta*, **1**: 99.
[27] Yamamoto C, Seto H, Ohto K, Kawakita H, Harada H (2011) Effect of π electrons on the detection of silver ions by ion-selective electrodes containing tripodal broom molecules as an ionophore. *Analytical Sciences*, **27**: 389.

Chapter 10
Environmental Impacts of Silver from Spent Nanosources

Marija Ljubojević, Mirta Milić and Ivana Vinković Vrček*

Institute for Medical Research and Occupational Health
Ksaverska cesta 2, 10000 Zagreb, Croatia
**ivinkovic@imi.hr*

10.1 Nanosilver Production

Due to its biocidal activity, colloidal silver has historically been used for more than century, but its use was significantly reduced after introduction of antibiotics in the 1940s. Silver has been listed as a biocidal material in the United States since 1954 [1]. After the 1990s, there has been a rise in its biocidal applications largely in alternative medicinal treatments, although the beneficial health effects of colloidal silver have not been scientifically supported [2, 3]. Colloidal silver contains silver in the form of particles in size range of 1–100 nm. Silver nanoparticles (AgNPs) are nowadays one of the mostly commercialized nanomaterials (NMs). More than 30% of all nanoproducts contain AgNP, as recorded by the Woodrow Wilson Database, but their production volume is still much lower than other NM types [4]. Thus, silver mass in the atmosphere is comparable to the levels found 30 years ago, while the silver content in sewage sludge is significantly lower than the levels measured during the golden ages of photography [5]. Besides biocidal activity, the main purposes of AgNP applications in consumer products are the reinforcement of physical properties of the material itself or its surface properties like scratch resistance, water repellency, reflectivity,

Table 1. Main sources of AgNPs in the environment.

Product type	AgNP function	Commercial items	AgNP release	References
Textile	Biocidal activity	Shirts, socks, underwear, sports clothing	Into water system via washing effluents	[6–9]
Food packaging	Biocidal activity	Fresh food bags, plastic storage bags, kitchenware	Migration into food	[10]
Medical devices	Biocidal activity	Wound dressing, dental hygiene, surgical meshes, vascular prosthesis, ventricular drainage catheters, orthopedics	Migration into tissues	[2, 11–13]
Electronic goods	Conductivity and electrical properties	Transparent conducting films, electrodes, tandem solar cells, etc.	Upon disposal	[14]
Others	Biocidal activity	Water treatment, washing machines, paints	Into water system via washing effluents	[1, 15]

and photo activity. Table 1 shows main sources of AgNPs in commercially available nanofunctionalized materials.

10.2 Release of Nanosilver into the Environment

The release of nanosilver into environmental compartments may occur intentionally or unintentionally. Nanosilver may be released in the form of individual NPs, weakly bound agglomerated NPs, or strongly bound aggregated NPs or even embedded in some other materials. In order to determine the production phase, that is, production, transport/storage, use, and disposal/recycling, and the technological process at which the release of AgNPs take place, the product life cycle should be systematically and comprehensively evaluated [16–18]. Figure 1 presents a schematic representation of the AgNP life cycle.

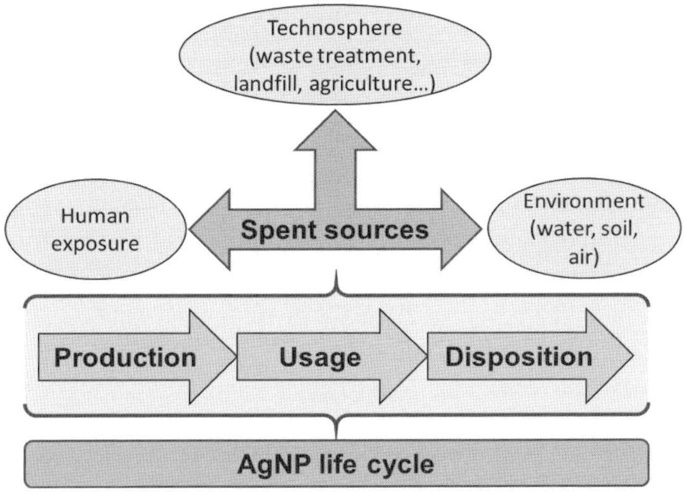

Fig. 1. Schematic representation of the AgNP life cycle.
Source: Adapted from Ref. [19].

Starting from crude resources material, products are manufactured for commercial use and finally disposed as waste material after use, leaving residues in our surroundings. A significant proportion of AgNPs and silver as waste sources comes from functionalized materials used mainly as biocides with anti-bacterial, anti-fungal, anti-viral, and anti-algae properties [20]. Also, various forms of AgNPs can be applied in the field of electronics as bio-sensors, anti-static materials, and superconductors, (transparent electrodes for flexible devices and conducting films, flexible thin film layer in solar cells, etc.) and end up as complex material of spent sources [19]. AgNP release, in turn, will affect certain environmental compartments, humans, or other organisms. Thus, during the unintentional release of AgNP from geotextiles, terrestrial organisms may be affected, whereas release from socks or underwear could affect humans when AgNPs come in contact with the skin, gastrointestinal tract, or even lungs. Both types of release may end up in the wastewater treatment facilities, while soils might become an endpoint of bio-solid applications. Another very important process in the life cycle of AgNP is incineration as a key waste treatment technology. Different scenarios for the access of AgNP to the incineration process can be assumed, including generation of municipal solid waste; wastes from nanotechnological research and development process; generation of hazardous, medical, and infectious wastes; and sewage sludge generation in wastewater treatment plants (WWTPs). The incineration process may significantly modify the physico-chemical properties and subsequent behavior of AgNP in an environment. The amount of waste that is incinerated worldwide is rising steadily [21]. Only in the United States, more than 30 million tons ($\sim$15%) of municipal waste is incinerated per year, whereas this quantity is even higher in the European Union (EU) where 20–50% of municipal waste is incinerated per year [22, 23]. Due to limited landfill space, Japan treats $\sim$80% of its municipal waste by incineration [21]. For hazardous waste, including research waste, and infectious waste, the incinerated amount is 3–44 million tons and almost 150,000 million tons, respectively, each year only in the United States [22, 24–26]. In addition, the United States and Europe incinerate 20–25% of WWTP sludge per year [27]. Release of nanosilver or ionic silver from WWTP to surface waters is predicted to be $\sim$60–100 kg per year [28]. The predicted environmental concentrations (PECs) of nanosilver are 0.088–10 ng/L in surface waters, 0.0164–17 μg/L in WWTP effluents, and 1.29–39 mg/kg in WWTP sludge [7, 27, 29, 30]. For the proper risk assessment of AgNPs, an estimation of the exposure concentrations is of vital importance. Estimation and risk characterization of released AgNPs into the environment was investigated in a few generalized researches in an attempt to establish the PECs, sometimes relative to predicted no-effect concentrations (PNECs), which are evaluated from the available toxicity data for relevant bio-available forms of silver. However, emissions from production, transport, and deposition processes of AgNP containing products and their waste in atmosphere (air), surface water (fresh, sea, and waste water) and soil is not yet united as a general model of AgNP and/or silver fate in the environment. Currently developed exposure models show that PEC values for AgNPs in different environments range from ng/L to mg/kg [31]. It should be stressed that exposure to AgNP exist even for several thousand years, since there are evidences about the presence of AgNP spontaneously produced as incidental nanomaterials and therefore the human contact with them can last an-aware for similar period as it has been demonstrated that metallic silver is capable to release large quantities of AgNP under specific conditions [32].

It has been estimated that release of silver, either in nano or ionic form, only from food packaging will result in mean value of 0.37 μg per year in drinking water systems [33]. Water quality criteria values for silver, as set by the United States Environmental Protection Agency (USEPA), are 1.9 μg/L and 3.4 μg/L in salt and fresh water, respectively. Although many authors predicted the removal of total Ag in WWTP was greater than 95%, the measured level of Ag in the WWTP effluents is higher by at least one order of magnitude than background levels [34]. As may be expected, after disposal of AgNPs from various sources, air, soil, and water are the environmental compartments from which silver and AgNPs can come into direct or indirect contact with all living creatures as "non-target organisms" [35]. Due to the ever-increasing production and use of AgNPs, there is an urgent demand to assess all consequences of their release into the environment and all consequences of subsequent exposure. Emissions from production, transport, and deposition processes of AgNP-containing products and their disposal in atmosphere (air), surface water (fresh, sea, and wastewater) and soil is not yet united as the general model of AgNP and/or silver fate in the environment. Silver presence and behavior in WWTPs and in bio-solids as its "products" are of huge importance because by effecting bacteria, which play a central role in the biological processes taking place in sewage treatment plants, changes essential to microbial communities in WWTP sand occurs, rendering the process of decomposition to be different and directly increasing nitrates in WWTP. Furthermore, all environmental areas are endangered with runoff (as rain washout material) from outdoor painted surfaces or down the façade coatings with incorporated AgNPs, together already resulting in a decrease in the diversity of natural freshwater bacteria as consequence of the occurrence of AgNPs in the environment [19, 20, 36].

In Europe, waste management is regulated by the Directive 2008/98/EC of the European Parliament and of the Council and Council Directive 1991/689. However, neither of them considers NPs [37, 38]. The primary challenge in regulation regards the question: do NMs pose an environmental risk in waste? The current framework of US Environmental Protection Agency (EPA) for regulation and management of hazardous waste is also limited by the difficulty in determining the toxicity of NPs via regular tests due to their unexpected transformations and interferences caused by their intrinsic and extrinsic properties. Australian regulation, including the Hazardous Waste Act and the Environment Protection Act, also fails to regulate waste containing NPs. The Japanese Ministry of the Environment released a guideline for the control and prevention of the environmental release of NPs [39]. It seems that there is no environmental agency worldwide which has developed a strategy for the safe management of nanowaste. NPs will adhere to soils and undergo transport in ground, surface, and drinking waters through completely different modes of action compared to bulk materials. An additional problem is the disposal of some nanowastes within the household waste, which does not come under hazardous waste regulations [40].

Currently, there is no specific international and distinct safety regulation for the production and use of nanoproducts [41]. The industry refers NPs under the same CAS number as their analogous chemical bulk counterparts, while their production and use are regulated in Europe by the EU chemical safety policy REACH (Registration, Evaluation, Authorisation and Restriction of Chemicals). According to REACH regulation, ecotoxicity must be evaluated for all chemicals that are produced annually at a volume of more than 1 ton and sold at the European market [42]. Producers or importers have to provide data on short-term (48 h) toxicity testing using crustaceans like *Daphnia magna* [43]; 72 h growth inhibition of aquatic plants, preferably algae [44]; and short-term (96 h) toxicity testing on fish [45].

10.3 Transformation of Nanosilver

In recent years, efforts to explain and predict the fate, transport, and toxicity of AgNPs on both human and environmental health have been significantly improved. However, proper analysis of human and/or environmental impacts of AgNPs is very difficult. Inappropriate characterization and dose metrics that are significantly affected by the conditions and components of media (environmental, cellular, tissue, etc.) are the main obstacles in the risk assessment of AgNPs [20]. One of the central questions in understanding AgNPs behavior in suspension is: do they persist under realistic environmental conditions? It is well accepted that detailed insight into transformation behavior of AgNPs under real environmental conditions represent the central question for their environmental risk assessment. However, the elucidation of AgNP reactivity in the environment with potential transformations encounter great analytical complexity of different processes such as photochemical transformation, oxidation and reduction, dissolution, precipitation, adsorption, desorption, combustion, bio-transformation, and abrasion, among other biogeochemically driven processes such as aging, surface reactions and changes of moieties, aggregation (strongly bound), and agglomeration (weakly bound) [46]. All these processes could greatly affect the pathway and extent of environmental release, influencing the bio-availability and bio-accumulation of AgNPs, as presented in Fig. 2. At the nanosize, the surface-to-volume ratios become huge, resulting in unexpected mechanical properties and catalytic activity of NPs [36, 47]. Also, the similarity in size between AgNPs and natural bio-molecules translates into an increased potential for NPs to interfere with biological systems/processes like cell membranes, bio-chemical pathways in cells, or even the genetic code

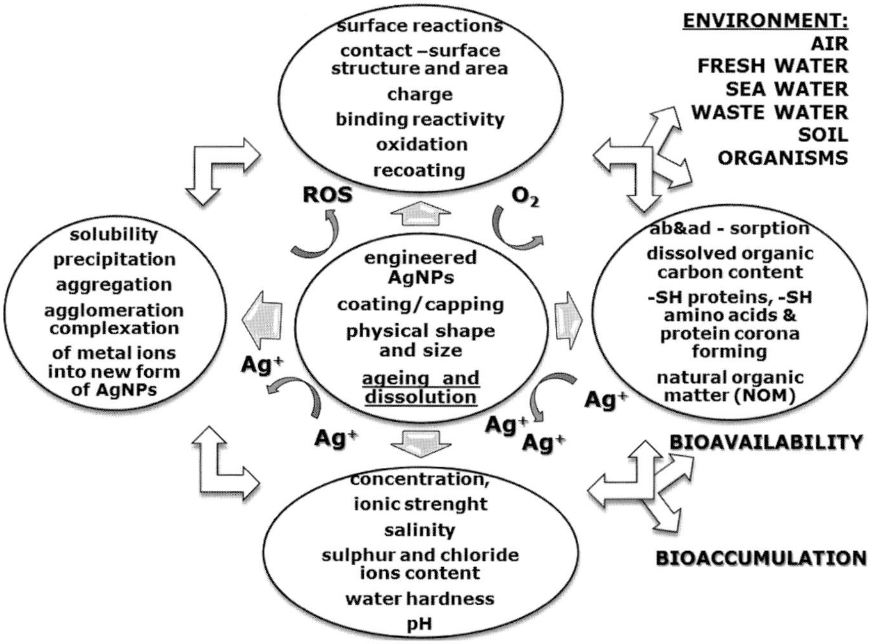

Fig. 2. Schematic view of the dynamic transformations that AgNPs potentially undergo in the organism and/or in the environment together with interplay between these changes (illustrated by two-end arrows).

Source: Schematic view inspired by [36].

itself [36, 47]. Thus, the dispersion of NPs in even the simplest bio-fluids such as blood plasma or in environmental fluids such as river water requires care and understanding in practice. Once discharged into the bio-system, uncoated or coated NPs will be subjected to alterations through interactions with naturally occurring bio-macromolecules, including proteins, polysaccharides, and fatty acids [36, 47]. It is likely that a layer of very-long-lived bio-molecular shell, called "corona", covers the surface of NPs [48]. The coronas result from the adsorption of a wide range of macromolecules and are dynamic in nature. However, many proteins and other macromolecules are effectively irreversibly adsorbed over relevant time scales, and therefore in part determine the "identity" of the NP (e.g., size, electrophoretic mobility, and surface composition). The nature, properties, and robustness of this macromolecular layer and interactions between the core and the corona are currently poorly understood. The adsorbed proteins may also offer a way of entry for NP into cells through the process of receptor-mediated endocytosis [48, 49]. When NPs are covered with adsorbed bio-molecules such as albumin, there is a possibility that a cell cannot recognize the NP as foreign particle. As a result, the particles are taken up into cells. This cellular uptake mechanism is called the "Trojan horse" effect [48, 49].

The most important factors affecting AgNP transformation and environmental fate are ionic strength, media composition, pH, dissolved organic matter, humidity of the environment, dissolved oxygen concentration, temperature, size, shape, and coating of AgNPs and their concentration [20]. Once released in the environment, AgNPs may stay as individual nanoscale particles or form clusters, dissolve, or chemically transform after contact with bio-molecules and/or organics [19]. The transformed AgNPs may be more heterogeneous than the AgNPs originally released into the environment, but a decrease in heterogeneity is also possible. Transformations and the properties of the transformed AgNPs are ultimately dependent on the characteristics and composition of the specific environment in which AgNPs are exposed.

The AgNP abundance decreases quickly after reaching the sewage water, undoubtedly due to particle dissolution and possible precipitation with sulfur and chloride ions, followed by aggregation or settling and ad/ab-sorption (e.g., to soil particles) or by uptake of micro-organisms. It has to be emphasized that AgNPs in the effluent from WWTP were not of original engineered source, and definitely there are a few reports in the literature of engineered AgNP dissolving and re-precipitating as new NPs, forming previously non-existing coating/capping with the surrounding organic or inorganic material [34]. Also, the appearance of bio-engineered, naturally occurring silver precipitates in form of NPs, including those in alfalfa plants (*Medicago sativa*, lucerne), and in bacteria has been described and characterized [34].

Simulations of AgNP alterations due to environmental conditions and aqueous medium are tried in the laboratory by changing the ionic strength that controls the speed and degree of aggregation of most AgNPs and by supplementing suspensions with natural organic matter (NOM). Consisting of high-molecular-weight humic and fulvic acids resulting from animal and plant material decay, NOM is an abundant, and weakly defined, constituent of environmental systems which instantly attaches onto the extremely reactive surface of NPs [36]. Generally, NOM binds to the reactive surface of AgNP, displaces a possible capping molecule, and establishes a heterogeneous, dynamic layer of fragments resembling the bio-molecular "corona" that has been the focus of a few analyses. Thus, in the presence of NOM, AgNPs may be stabilized, which may decrease their aggregation, leading to the higher percentage of single NPs [36]. Some other molecules, that is, polysaccharides secreted by bacteria, may cause an increase in the AgNP aggregation rate [36]. In addition to different organic molecules (NOM, proteins, carbohydrates), inorganic ions also may adsorb onto the AgNP surface. Study on the binding of molecules/atoms and conversion of the AgNP into other forms has an important role in the toxicity estimation, but to assess econanotoxicity, the dissolution process of AgNP into ions (from metallic Ag^0 to ionic Ag^+) has been demonstrated to be responsible for the biocidal character of AgNPs. However, researchers are aware of the complex interaction of the large reactive surface of AgNPs, where a huge percentage of constituent atoms are exposed to surface sorption with medium macromolecules, but are also prone to react with dissolved oxygen and sulfur. Similarly, as many naturally existing compounds have a predisposition to complex Ag^+ ions, large concentrations of free dissolved Ag^+ are unlikely to be present in any medium. Surface reactivity has an additional interesting characteristic that, by attracting the layers of opposite charged ions and molecules, or repulsion of the identically charged ions, it might possibly cause local proton depletion on the surface of bacteria, consequently leading to a disruption of the electrochemical gradient, with lethal effects. Based on the available data on the fate of AgNPs in WWTP, the transformation of AgNP into AgS_2 (of much lower toxicity) as well as into AgCl can be presumed [20].

All these possible scenarios emphasize a mutual endeavor to fill the knowledge gaps on the aggregation behavior for characterizing the transport of AgNPs through environmental compartments. However, research in this field may greatly suffer on oversimplification of both the experimental design and clarification of obtained results which certainly lead to invalid conclusions. Therefore, dynamic NP transformations and behavior of AgNPs necessitate careful, time-dependent characterization of AgNPs in environmentally relevant conditions that is critical for the subsequent elucidation of risks associated with the release of AgNPs in the environment. Estimating the potential for all possible processes to occur under different environmental conditions should be a research priority with benefits to the multiple stakeholder groups, including research communities, industry, and risk assessors.

10.4 Testing Methodology for Risk Assessment of Nanosilver

The European REACH regulation states that when chemicals are produced in a volume of more than one ton per year and sold at the European market, they must be characterized for their potential impact on the ecosystems [50]. For most NMs, including AgNPs, the scientific information on potential harmful effects is way behind the industrial production and wide applications [51]. This gap exists for three reasons: (i) the scientific research required to develop nanotechnology does not yield adequate data to assess the risks of those products; (ii) scientific research data are not robust and adequate enough for the regulatory agencies to conduct risk assessments; and (iii) research on the environmental health and safety of NMs receives less than 5% of the funding spent to develop new NMs [51]. To that end, ecotoxicologists have used standardized testing methodology and standard test organisms in the rush to understand the effects of AgNP exposure. In the recent years, studies of toxicity evaluation of AgNPs increased dramatically, and many more are in progress. However, these results are not always clear, while contradictory reports on AgNP toxicity have been frequently published [51]. The cause of these contradictory results can be categorically defined as insufficient physical and chemical characterization of the AgNPs. Engineered AgNPs differ largely from traditional chemicals although they share the same chemical composition. It is not fully understood yet what physical and chemical properties of AgNPs are related to their biological effects. Comprehensively, there is no standard material for AgNPs. Put simply, it is necessary to classify AgNPs not only by chemical composition but also by chemical and physical properties for evaluation of its biological activities. The factors and conditions that most influence the harmful effects of AgNP are still uncertain due to the lack of data on the physio-chemical properties of AgNP and their transformation under experimental and/or environmental conditions [51, 52]. If the risk evaluation of AgNP is based on information about their transformation and interaction in relevant environments, there may be very few contradictions (Fig. 3).

A rational approach to study AgNP's mode of action should begin with standardized protocols for the preparation of substantially characterized NPs. According to the recommendation of the EU NanoCommission [19], in order to achieve comparable studies about the effects and behavior of NPs, an assessment of the physico-chemical properties of the as-synthesized and/or purified NP suspensions should include:

- chemical composition, purity, impurities,
- particle size and particle size distribution,
- specific surface morphology (shape, crystal structure),
- surface chemistry/coating,
- extent of agglomeration/aggregation, or particle size under experimental conditions as appropriate (e.g., media with or without protein),
- water solubility (focusing on metastable particles).

There are many techniques available to measure the AgNP properties such as chemical composition, size, and charge, three categorizes representing the first steps in characterization of AgNP in all standardized protocols. However, different methods applied during AgNP characterization may lead to disagreement in the measured results, which is caused by different

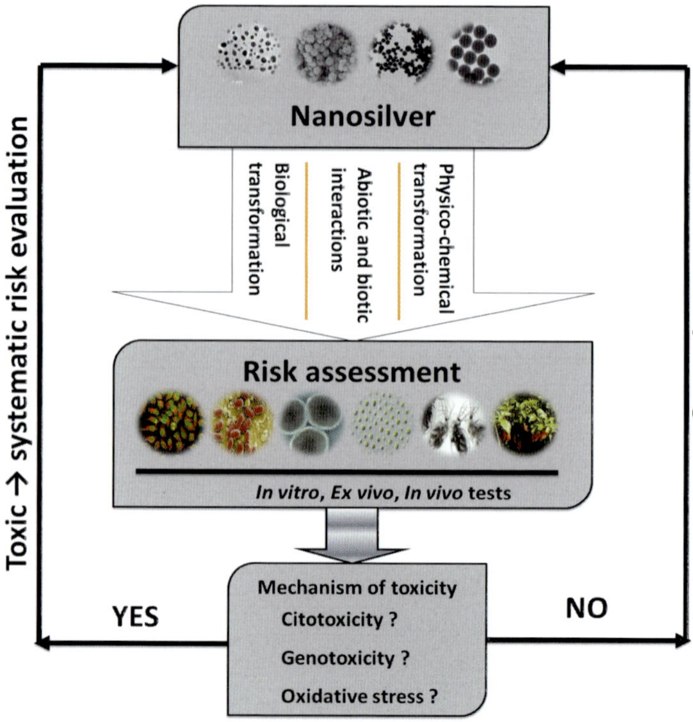

Fig. 3. A method of evaluating the safety of nanosilver.

principles applied for each technique involved. In addition, differences in measurement may arise from variations in sample preparation and instrument-operating procedures. Table 2 presents an overview of the different techniques available for the characterization of AgNPs.

The largest source of uncertainty when determining the environmental fate and transportation of AgNPs stems from the way in which AgNPs behave in different matrices (natural or used in the laboratory assays) and how the actual exposure of different organisms to AgNPs is quantified [36]. Current analytical methods yield limited insight into a subset of these transformations; development of sensitive, in situ characterization tools would significantly advance this field. Furthermore, the dissolution processes are of great interest in applications based on the biocide properties of AgNPs (paint, cleaning products, textiles). For instance, the anti-bacterial and anti-fungal activities of AgNPs depend on the release of

Table 2. Characterization techniques for NPs.

Technique	Parameters analyzed	Advantages and challenges
Dynamic light scattering	Size, size distribution, agglomeration	Measurements under relevant conditions; information pertaining to the stability of particles in different media with respect to time; polydispersity leads to a bias; provides average hydrodynamic size
Electrophoretic light scattering	Charge	
Nanoparticle tracking and analysis	Size, size distribution, agglomeration	Allows NPs to be visualized individually with simultaneous analysis of their Brownian motion; no intensity bias toward large particles
Brunauer–Emmett–Teller	Size, surface area	Measures size and surface area simultaneously; only provides average size, and not size distribution
Electron microscopy	Size, size distribution, shape, agglomeration, aspect ratio, elemental composition	Direct measurement of particle properties, including chemical composition (when combined with energy-dispersive X-ray spectroscopy); time-consuming, expensive; vacuum needed
Atomic force microscopy	Size, size distribution morphology, surface texture and roughness, agglomeration, aspect ratio	Visualization in three dimensions; information about multiple physical properties; greater statistical significance; in the absence of vacuum
Field flow fractionation (FFF)	Size, volume, elemental composition	Chromatography-like size-fractionating method; evaluation of particle size via diffusion coefficients; no stationary phase, usually combined with other techniques or modified to increase utility

(*Continued*)

Table 2. (*Continued*)

Technique	Parameters analyzed	Advantages and challenges
Inductively coupled plasma mass spectrometry	Elemental composition	High sensitivity; rapid analysis
Capillary electrophoresis	Size	Size determination with simultaneous analysis of absorbance properties; low sample volume; rapid; low operating cost

Ag^+ ions in solution due to the redox potential characterizing the AgNPs. Biocidal activity of such AgNPs is related to the binding of released metallic ions from the AgNP surface to electron-donor groups containing sulfur, oxygen, or nitrogen at the membrane of micro-organisms [36, 48]. Cellular uptake of AgNP may also lead to intracellular metal ion release, exogenous proteins uptake, and intracellular chemical reactions [48]. In cells, electronic transfers can induce an oxidation of proteins or the generation of an oxidative stress [48, 49]. Oxidative stress is a state of redox disequilibrium in which production of reactive oxygen species (ROS) overwhelms the antioxidant defense capacity of the cell, thereby leading to adverse biological consequences such as damage of macromolecules, lipids, DNA, or proteins, resulting in excess cell proliferation, apoptosis, lipid peroxidation, or mutagenesis [53]. For detailed ecotoxicological studies, an accurate comparison of the metal's biological activity in different forms (dissolved, NP, "bulk" particles) is crucial for defining an NP's mode of action and understanding the hazard issues. Dissolved metal speciation and bulk particle characteristics also need quantification. The time and resource requirements for characterization are usually under-appreciated, and there is a lack of co-located or collaborative expertise in both environmental/ecotoxicological and metrology/analysis, meaning that logistical challenges are not insignificant. As such, a multimethod approach is recommended, which should be as exhaustive as possible, that is, drawing upon a wide range of methodologies to define the important parameters for NPs such as size, shape, aggregate state, solubility, and binding capacities to bio-molecules and other organic material. AgNP properties should ideally be quantified over exposure periods in ecotoxicology experiments, in order to establish the causes and mode of toxicity of the AgNPs, and over relevant time periods for biological studies.

Quantifying the behavior and effect of AgNPs in organisms and environmental media has proven to be more difficult than previously experienced with soluble chemicals. Many of those methods are proving inadequate and do not adapt well to the unique properties of AgNPs. Ecotoxicologists usually treat AgNPs in the same way they have considered traditional chemicals [54]. They use the same bio-assays and endpoints and assume that AgNPs behave in the same way as soluble contaminants. The benefits of all conventional methods cannot be totally discarded for the evaluation of AgNPs, but there is large amount of evidence that many of the *in vitro* assays are not suitable for AgNPs [54]. The literature includes many examples where implementation of standard bio-assays leads to false-positive or false-negative results [48, 54]. Frequently, results from current toxicological tests are not comparable and do not provide the technical framework needed by stakeholders and policymakers as the basis for control and regulatory measures [48, 54]. Taken together, a large proportion of the published data may be inaccurate in the context of AgNPs and may lead to the drawing of scientifically invalid conclusions. Thus, a suitable testing platform for the safety evaluation of AgNPs is required. The risk assessment of AgNPs is brought up by examination both *in vivo* and *in vitro*. Although the application of AgNPs to model animals gives us reasonable toxicological information, compared to the *in vitro* system, biological reactions *in vivo* are very complex. *in vitro* evaluation is essential as the primary step in the elucidation of mechanisms of biological influences induced by AgNPs. The techniques that can be used to assess *in vitro* toxicity of AgNP include (i) cell viability/proliferation *in vitro* assays and mechanistic assays (ROS generation, apoptosis, necrosis, DNA damaging potential); (ii) microscopic evaluation of intracellular localization (SEM, TEM, AFM, fluorescence microscopy); (iii) gene expression analysis, high-throughput systems; (iv) *in vitro* hemolysis; and (v) genotoxicity [35, 47, 48, 54]. *in vitro* model systems provide a rapid and effective means to assess AgNPs for a number of toxicological endpoints. They also allow the development of mechanism-driven evaluations and provide refined information on how AgNPs interact with living cells in many ways [55, 56].

Such studies can be used to establish concentration–effect relationships and the effect-specific thresholds in cells. Therefore, in addition to the adaptation of standard tests, there is a need to develop better and rapid screening methods and to move into more predictive toxicology of AgNPs.

Since available information on nanotoxicology is scarce, a selection of methods and testing protocols are needed to better track the rate and extent of transformations of NPs under realistic conditions. Grand challenges of safety assessment of AgNPs include (i) functional-level toxicity assessment in complex ecosystems with relevant models and model-functions; (ii) detection of engineered AgNPs that are similar in chemical composition to natural particulate matter; and (iii) in situ characterization of AgNPs in complex, natural environments. The selection of methods that are reproducible, simple, inexpensive, and reliable are the basis for the protection of the environment relating to nanotechnology which, in turn, will be able to realize its full potential only through addressing the issues raised by toxicological studies. To accomplish the task of function-level toxicity assessment, there is a need for better multidisciplinary communication in order to determine the most relevant and important model systems and best practices in research data interpretation.

10.5 Econanotoxicology of Silver

Econanotoxicology investigates the environmental impacts of all engineered NPs, including AgNPs, following their fate from industrial or commercial use to waste products and their possible toxicity, combined with efforts toward the sustainable use of novel NMs. However, it is also accounted as process of research that should include various experts from multidisciplinary fields, including material engineers, chemists, biologist, and toxicologists. Studies of the AgNP toxicity observe its effects on a wide range of the ecosystem food chain/web constituents, including microorganisms, algae, plants, and other multicellular terrestrial and aquatic living organisms [36].

The following section presents a short review on data harvested so far on the toxicity effects of engineered AgNPs in experimental model organisms representing various groups in order to better grasp the comprehensive conclusions of the toxicological studies.

A large amount of data on the environmental consequences of AgNP release is collected to emphasize their effects on micro-organisms, including highly susceptible nitrifying bacteria *Nitrosomonas europaea* (oxidation of ammonia), denitrifying *Pseudomonas stutzeri* (release of nitrogen gas), and nitrogen-fixing bacteria *Azotobacter vinelandii* (nitrogen fixation). However, these effects on bacteria survival and/or on enzymes in the nitrification/denitrification cycle are mostly ascribed to Ag^+ ion release, without indicating nanospecific toxicity [35]. A highly informative critical review by the Anne Kahru group, where authors went through recent nanotoxicological studies containing relevant information about various NMs, encountered more than three quarters of results that considered silver, with ratio of AgNP versus Ag^+ ions being 2:1 [35]. A summary of this review on silver is presented in Table 3. The summarized results show the concentration sensitivity scale in a range of freshwater aquatic organisms representing the most vulnerable environmental organisms — from crustaceans and algae to fish. According to the EU classification [57], both Ag forms, ionic and nanoparticulate, should be classified as very toxic chemicals. Interestingly, all aquatic species are more sensitive to AgNPs and Ag^+ ions than bacteria [35].

Table 3. Median of minimal inhibitory concentrations (MIC) for bacteria and median of half-maximal lethal or effective concentration (L(E)C50) values for all other organisms exposed to AgNPs and ionic Ag form.

	Median L(E)C50 or MIC values (mg Ag/L)	
Group of organisms	AgNP	Ionic Ag
Crustaceans	0.01	0.00085
Algae	0.36	0.0076
Fish	1.36	0.058
Nematodes	3.34	4.8
Bacteria	7.10	3.3
Yeast	7.90	2.16
Mammalian cells (*in vitro*)	11.3	2
V. fischeria	32	5.7
Protozoa	38	1.5

10.5.1 *Micro-organism models*

Gram-positive and gram-negative bacteria are susceptible to silver ions. This is the main purpose of commercial interest for biocidal activity of AgNPs, especially when antibiotic resistance starts to be the main issue. However, resistance to Ag^{+} ions (and possible AgNPs) has already been seen in some strains [36]. Bacteria may develop resistance to ionic or nanoparticulate forms of silver by plasmid-carried horizontal transfer of genes possessing efflux pumps, metal scavengers, or some other inventive molecular mechanism [36]. Thus, it seems that resistance to silver could be even more widespread and a more serious problem in the future compared to different antibiotics. However, the promise of an additional mode of AgNP action toward bacteria is a potential way to decrease Ag concentration and increase the bactericidal properties of functionalized AgNPs during medical use.

Generally, it need not be emphasized that bacteria are the main target of intentional AgNP use, but they are also the constituents of the bottom of the food chain/web, and as ubiquitous members of ecosystems, they are of unique importance in global nutrient cycling. Ecological importance, together with the relative ease of maintaining monocultures, has made bacteria a primary model organism for econanotoxicity studies. Common microbial species, such as *Escherichia coli*, *Bacillus subtilis*, and *Pseudomonas aeruginosa* usually exclude the utilization of relevant species in the environment to evaluate the effects of AgNPs with various toxicity assays as presented in Table 4 [36]. All the above-mentioned factors, as exemplified in Table 5, increase the difficulty in making some generalized conclusions from experimental results on bacteria and applying it to the environmental reality (i.e., different conditions and analyses need to be applied).

Techniques in bacterial nanotoxicity have been used as for any xenobiotic giving results about bacterial viability and growth, with established methods as colony-forming unit (CFU) assays and optical density (OD) measurements to calculate the minimum inhibitory concentrations (MIC) or create growth curves. ROS generation using fluorescent probes and morphological changes, together with the accumulation of AgNPs, have been commonly observed with electron microscopy. Analyses have been performed on bacterial response to AgNP exposure as a function of

Table 4. Bacterial models used in AgNP nanotoxicity research and applied analytic methods for toxicity evaluation [36].

Bacterial model	AgNP dose, mg/L	Range assays
Escherichia coli	0.025–158	CFU, growth (OD), liquid-to-plate (colony area), viability (LIVE/DEAD Baclight and diameter of inhibition zone in disk diffusion), minimum inhibitory concentration (OD), β-galactosidase activity (MetPLATE bio-assay)
Bacillus subtilis	0.025–50	Growth (OD), CFU, liquid-to-plate (colony area), viability (disk diffusion, LIVE/DEAD BacLight)
Nitrosomonas europaea	0.2–20	Nitrification activity (nitrite production observed with NitriVer 3 Reagents), membrane integrity (LIVE/DEAD BacLight), select gene expression (qRT-PCR)
Shewanella oneidensis	2–7.5	Viability (diameter of inhibition zone in disk diffusion, MIC (OD), LIVE/DEAD BacLight)

Abbreviations: CFU (colony-forming unit); OD (optical density); ROS (reactive oxygen species); MIC (minimum inhibitory concentration); qRT-PCR (quantitative real time polymerase chain reaction).

Table 5. Bio-accumulation of silver in different aquatic organisms.

Organism	% of silver accumulated from food	% of transfer efficiency of silver from food
Snow crab	90	57
Marine oysters	44	11
Marine clams	20–40	5–7
Marine mussels	15–25	1–3
Sea urchin	34	2
Zooplankton	15–40	0.2
Fish predators	10	0.04

Source: Adopted from [68].

AgNP concentration, agglomeration state, coating with engineered capping, and/or natural transformations in the environmental conditions as UV-light exposure and interaction with NOM.

Simplified conclusions from literature may be condensed to a few informative points regarding the possible mechanisms of AgNP action.

(a) Dissolution and release of Ag^+ ions from AgNPs is considered the main mechanism of the anti-microbial activity. Liberation of ions, which can interfere with sulfur-containing proteins in the bacterial cell wall, may lead to all observed toxic effects.
(b) Contact inhibition of proliferation to both gram-positive and -negative bacteria may be due to the surface zeta potential of the AgNP, so anti-bacterial properties depend on strong adhesion/contact such that small-sized AgNPs with a large reactive surface area might establish the changing membrane properties of the bacterial surface.
(c) The above assumption has direct correlation with the observed changes in the proton electrochemical gradient an essential driving force for ATP synthesis in bacteria. Meaning, that if those processes are interrupted (by proton-exhausted regions due to micro-galvanic effect of AgNP, which causes proton depletion), essential energy for all subsequent dependent reactions cannot be supplied, which leads to bacterial death.
(d) Very small (i.e., just few nm in size) AgNPs showed the greatest biocidal effect and possibility that use (active) transport through the bacterial membrane, inducing ROS and resulting in oxidative stress and DNA damage.

10.5.2 *Bio-accumulation*

Since the bio-accumulation process has demonstrated even higher amount of bio-accumulation, it does not have to be the key. It has been found that phytoplankton bio-accumulate silver from marine waters in much higher concentration that it can be found in the water where these phytoplankton live. These concentrations can be higher, ranging from 10,000 up to 70,000 times [58]. The same effect has been seen in oceans, coastal waters, and estuaries due to chemical reactions that dominate silver speciation [56]. Therefore, small changes in exposure concentration can have a strong impact on the bio-accumulated level in the organisms. For example, marine invertebrates bio-accumulate silver more than zinc, cadmium, copper, cobalt, chromium, or selenium [59]. This high bio-accumulation effect means as longer the organisms lives, the more they will accumulate silver [60–66].

Digestion due to low pH and enzymes can generate more silver ions that can also combine with proteins and amino acids in the digestive tract or gut [67]. The gut membrane can then transport clusters of molecules called micelles into the cell either by intracellular digestion or through the cell membrane by endocytosis. Predators can accumulate chemicals at higher concentration due to the effective assimilation of silver in the gut, low rate of silver loss, and high potential to magnify the silver rate in the organisms or tissue in the rate much higher than the prey they have consumed [62, 63]. One example is the microscopic aquatic plants (phytoplankton) that cause clams (*Macoma petalum*) from the San Francisco Bay that consume them to take into their tissues 39–49% of ingested silver and accumulate it in concentrations five to seven times higher than in the phytoplankton itself [64]. As a result, exposure through diet explains 40–95% of the silver bio-accumulation by these animals [65]. We have taken previously gathered information from [56] to demonstrate the amount of silver deposited in some aquatic animals that are consumed by higher organisms, including humans (Table 5).

10.5.3 *Aquatic environment*

Algae are sometimes the target organisms of AgNP anti-algal products but also "non-target" collateral victims of other biocidal and technological waste. Moreover, crustaceans and algae, both invertebrate organisms, are the key constituent of food chain/web in freshwater aquatic ecosystems. Algae are critical for planktonic photosynthesis and are primary oxygen and food creators; they serve directly as a food source for zooplankton (with crustacean *Daphnia magna* as an example) that are subsequently consumed by other invertebrates and/or fish, transferring new composition of introduced and bio-transformed AgNPs into higher organisms. Changes in the composition and efficiency of the phyto- and zooplankton population may indirectly affect the ecosystem by affecting water quality and/or induce direct essential changes in the entire ecosystem [31, 35]. Consequently, freshwater crustaceans (*Daphnia* spp.) together with algae

(*Pseudokirchneriella subcapitata* and *Chlamydomonas reinhardtii*) and fish (*Danio rerio*, *Oryzias latipes*) are implied as aquatic model organisms for the key regulatory ecotoxicological assays suggested for the tests of toxicity according to the EU REACH regulation.

To each living organism, the routes of exposure to AgNPs depend on their habitat. For fish, gills, mouth, and the gut represent a means of entry of AgNPs that are deposed in the fresh or sea water. As already emphasized, AgNPs from spent sources could be drastically altered in the aquatic medium and through lower organisms, and their fate depends on the nature of the NP surface and all transformation than can be accounted. Japanese medaka (*Oryzias latipes*) and zebra fish (*Danio rerio*) are standard model organisms for vertebrate biology. There are numerous studies of the impact of AgNPs on the embryos of these fish, emphasizing that bio-availability, bio-accumulation, and resulting toxicity are specific to the properties of AgNPs exposed to the environment and animals. However, some conclusions about the main toxicity of free Ag^+ ions are not without doubts, as the shape of smaller AgNPs and contact of the NM with cell membranes might be an essential parameter of toxicity [31]. The general assumption for the primary mechanism of silver toxicity in fish resides in the direct impairment and reduction of Na^+ ions uptake by the gills. The mechanism of inhibition might be interference with function and primary activity of Na^+/K^+-ATP pump that provides electrochemical gradient for various secondary and tertiary active transports across the membrane [31]. This mechanism could be relevant for both water- and soil-residing invertebrates, but whether it is the direct interaction of Ag^+ ions and sulfates moieties in the pump has not yet been resolved [31].

10.5.4 *Multicellular terrestrial organisms*

Earthworms (*Eisenia fetida*) constitute a critical part of the ecosystem and are model organism of soil detrivores (organisms that decompose dead particulate organic material) in ecological studies [30–32]. They can sense and avoid, if possible, soil contaminated with ionic silver, but AgNPs can be sensed after a two-day period, again drawing attention to the dissolution of AgNPs. Accumulation of silver has been shown to be significantly dependent on the soil type: it is higher in the sandy soil in comparison to the artificial soil, with observed reproductive toxicity again ascribed to Ag^+ ions [30–32]. All relevant data for AgNP transformation, including nature of the microbial communities in the soil and measurement of the silver ions' possible release, are not defined in these studies.

Caenorhabditis elegans, as a model organism for nematode, was quite extensively used in the investigation of AgNP toxicity [30–32]. The possibility of using multicellular organisms with known genetic profiles and use of specific mutants (oxidative stress-sensitive and metal-sensitive) resulted in genetic (use of mutants), pharmacological (rescue with antioxidant), and physico-chemical (analysis of AgNP dissolution) complementary approaches. However, AgNP's toxicity in correlation with dissolved silver is still unresolved. One of the analyses of AgNP's behavior in suspension showed the dependence on the ionic strength of medium. Median effective concentration (EC50 values) and threshold lethal doses (LD50 values) for the same AgNPs were higher in lower and lower in a higher ionic strength medium [30–32]. The presence of ions in a solution of AgNP suspension significantly affects particle dissolution or agglomeration, thus influencing their bio-availability.

10.5.5 *Terrestrial plants*

The relation of plants with air, soil, and water possibly contaminated with AgNPs make them exceptionally significant in econanotoxicology, but entry through roots is a main concern. In addition, bio-accumulation of AgNPs by plants makes the particles more bio-available to all species in the food chain/web, making all contamination in short and long terms a relevant study for future.

The experimental design of plant seeds exposed to various AgNP suspensions and then monitored for germination and root elongation suggested that the AgNPs themselves (rather than Ag^+ ions produced during NP dissolution) impact the earliest stage of plant development and that different NMs showed varied effects. Exposure of the mature plant to AgNP agglomerates to follow uptake and genotoxicity showed an increased frequency of micronuclei induction and chromosomal aberrations (chromatid and isochromatid types of gaps, breaks, and fragments of chromosomes) in an AgNP dose-dependent manner [36]. However, after analyzing these studies, questions about whether the mode of action is through AgNP or Ag^+ ions are still open. As NPs enter and accumulate in the ecosystem through a variety of intentional and unintentional release pathways, the most likely way of plant exposure to NPs is through the soil or sand, and even if hydroponic exposure to NPs influence plant growth and health, that experimental system ignores many of the NP transformations that modify their bio-availability. Different settings of exposure (through the soil, air, and irrigation) and use of vastly different types of NPs at different exposure times and concentrations make cross-correlations difficult, so as with a previously described model, standard settings would help in possibly combining the results into general firm conclusions [36].

10.5.6 *Mesocosm models in econanotoxicology*

The understanding of the fate, cycling, and toxicity of engineered AgNPs has expanded in the last decade, in part due to the complex mesocosm models (*in situ*, outdoor experimental system that investigate the natural environment under controlled conditions) of aquatic environment (fresh-water/ocean) and soil (with/without plants). In that kind of system, over a very long period, the starting concentrations of AgNPs is brought into either the water or into the soil of the system that included many of species characteristic for one ecosystem. Clear conclusion from the studies was that the way of introducing AgNPs into the ecosystem matters, and confirmation is gained from the assumption that silver does react with environmental constituents, for example, it is oxidized to silver sulfide and other silver sulfhydryl compounds [36].

10.6 Human Exposure to AgNP

The exposure of humans to AgNPs from spent sources can be either due to occupational exposure (inhalation) or consumption of drinking water [69] and food contaminated with AgNPs (accumulated in the animals, vegetables, fish, meat, clamps or crops irrigated with clean waste water and applied with waste water). The toxic effects caused by this exposure could be classified as dermal toxicity, multiple organ toxicity, cytotoxicity, and genotoxicity. The methods used for the assessment should be the same as for the toxic assessment of the AgNP compound used for oral or dermal use [70]. Since there are no studies on nanosilver waste, we can only give some overview for the toxicity studies on animals' *in vivo* and human cells' *in vitro* studies [71].

It seems that in the case of exposure from spent sources, the amount of nanosilver should be lower, but then there is still the fact that the amount of nanosilver transferred into ionic silver is unknown, as also the fact of non-determined amount and types of complexes that nanosilver can create in the water or in the organisms that has accumulated it. There are no reports of human toxicity, but there are many reports of the toxicity of organisms [49], with no determined highest allowed amount of nanosilver in drinking water. There are also possible surface modifications caused by the light exposure (silver and photoreactivity) [71, 72].

There is also no report about the amount of NPs released into the atmosphere during the transportation of the sludge or during landfilling or incineration [73].

Still there are no defined criteria for the chronic environmental exposure amounts for the living being from the USEPA regarding silver toxicity. There are also no standards in hygiene that are applicable to NMs; for example, it is accepted nowadays that the maximum admissible concentration of the time-weighted average for silver nano-object fraction is 0.01 mg/m^3, and inhalable fraction 0.05 mg/m^3, with many scientist convinced that those values and explanations do not provide sufficient protection against the harmful effect of silver in the form of nano-objects [74]. A research on 25 studies has demonstrated median LCD values for mammalian cells at 11.3 mg/L, with toxic effects at 8–10 mg/L, and harmful from 10 mg/L onward [35]. The consumption of NPs is estimated to grow from 225,060 metric tons in 2014 up to 584,984 metric tons in 2019, with an average annual growth rate of approximately 21% [75] in modeling, with no exact data from the manufacturers and therefore no idea of the actual quantity released in the environment [76].

Since there are no studies on the effect of AgNP waste on human health, we can only assume that the methods for the testing should be equal as for the normal toxicity testing of chemicals that are not considered as waste, although in waste the concentration of the AgNPs or silver ions should be lesser than in the chemicals.

For the risk assessment, the dose and the exposure are both important. For the dose to be measured as the effect on the living organism or cell, silver first needs to enter into the cell or bind to the cell, and its bio-accumulation is the indicator of the dose to which organism has been exposed. Bio-accumulation is the term that sums the total amount of silver ingested, either through food, water, or inhalation or, in case of aquatic organisms, also from filtration of the sediment, water, or suspended material. AgNPs can perturb into silver ions, and there is no determined formula to say in what percentage this change occurs and whether the silver ions will again turn into AgNPs. This perturbation depends mostly on the stability of the AgNPs, usually coated with different coatings for stabilization, and depending on the surrounding medium, such as serum in the cell cultures [55, 77]. It has been demonstrated that, after entering into the cell, AgNPs can bind to the proteins, sulfur ions, or chloride ions, and they can also be oxidized into the sediment, for example, before they have been used (taken) from the aquatic animals.

A classical approach to understanding the absorption, distribution, metabolism, excretion (ADME) and toxicity properties of AgNP compounds is critical and has been tried to be taken into account and evaluated by a series of *in vivo* studies and various *in vitro* cell-specific approaches that select cells relevant to above-mentioned issues. Studies of distribution of AgNPs in animal models, regardless of the method absorption, concluded that liver, kidney, and pancreas could be considered the main sites of bio-accumulation of introduced AgNPs, and cells from these organs were attempted to be used in follow-up studies *in vitro* [71, 78]. However, those issues are not so sharply and clearly determined, because fibroblast cells or their progenitors are on front line of contact as a lot of products have the skin as a target. As AgNPs get into the human body directly via wound dressings, catheters, implants, etc., they come in contact with the human cells and tissues in their mostly unchanged, original, designed form [71, 78]. Moreover, lungs may be also considered as the target of AgNPs applied via sprays, or not so often from polluted air. The situation is similar when AgNPs come into contact with humans through cosmetic products directly applied on the skin or the mucous cavities, and is somewhat changed for AgNP released from textiles. However, in addition to direct contact from unchanged engineered coated AgNP consumer products, animals as well as humans can start with the uptake of engineered coated AgNP already modified with new bio-molecules by the influence of the environment and processes in the environment before the uptake. Thus, contact with AgNPs from spent sources that get into the environment and have undergone transformation and/or entered the food chain/web prior to uptake by humans or other organisms is entirely different and cannot be easily studied *in vivo* or *in vitro* except in some complex and composite ecotoxicological studies, such as the before-mentioned mesocosm models (that will also account for humans as part of the system) [71, 78].

Studies on rodents *in vivo* implicated that the liver is the main site of AgNP accumulation (or lungs, if it was administration of particles via inhalation), followed by kidney and pancreas, with some suggesting that silver has extended persistence in both testis and brain [71, 78]. However, the underlying mechanism that causes such a specific localization needs to be established. Potential accumulation is often attributed to macrophages, and the common observation is that small AgNPs do have a broader organ distribution.

From these *in vivo* studies focused on the tissue distribution of AgNPs, some relevant *in vitro* experimental designs are tried on relevant cell lines corresponding to the potential target sites of toxicity; for example, use of liver cell lines or a lung-derived cell line as the most relevant seems to be justifiable, together with primary fibroblasts from mouse and human mesenchymal stem cells used for the skin model of AgNP application [71, 78].

Dhawan and Sharma [78], in their comprehensive review, presented data on the overall toxicity of nanosilver under *in vitro* and *in vivo* conditions. Generally, nanosilver causes oxidative stress, cytotoxicity, chromosomal instability, apoptosis, intracellular calcium transients, cell cycle arrest, interference with DNA replication fidelity, and c-Jun N-terminal kinases (JNK) activation in mammalian cells. Also, *in vivo* studies caused free-radical–induced oxidative stress and alteration of gene expression, blood–brain barrier destruction and astrocyte swelling, neuronal degeneration, and in some instances, induced brain edema formation [78].

What silver is causing within the cell is generation of an excessive amount of ROS [79], which is usually generated during the normal cell cycle, but in amounts at which antioxidants produced in the cell can remove them with no harm. Excessive amounts of ROS render the work of antioxidant enzymes ineffective and their non-removal can cause cell membrane damage, lipid peroxidation, fragmentation of the DNA, and problems with transport within the cell compartments. *In vitro* investigations confirmed that cytotoxicity and genotoxicity were greater when cells were exposed to smaller compared to larger AgNPs for the same exposure time using comet, micro-nucleus, and cell viability tetrazole MTT

(3-(4,5-dimethylthiazol-2-yl)-2,5-diphenyltetrazolium bromide) assays. However, explicit interference with DNA replication and transcription processes was shown by directing AgNPs with a nuclear signal into the cell compartment and inducing cell cycle arrest and apoptosis, but the usual mode of action that causes overall toxicity, and specifically genotoxicity, is through high levels of ROS interfering with normal cellular homeostasis [78].

Since there are no studies on nanosilver waste, we can only give some samples for the toxicity studies on human cells *in vitro* or *in vivo* studies using laboratory animals. *In vitro* study on CHO-K1 and CHO-XRS5 cell lines demonstrated cytotoxicity and genotoxicity from the 100-nm AgNP exposure that was even greater than exposure to 10-nm AgNP after 24 h using comet assay, micronucleus, and cell viability XTT assay (XTT is 2,3-bis-(2-methoxy-4-nitro-5-sulfophenyl)-2H-tetrazolium-5-carboxanilide as replacement of has been MTT, yielding higher sensitivity and a higher dynamic range) [80]. In lymphocytes, AgNPs at 90–180 nm sizes cause genotoxicity at a concentration of 25 μg/mL, while DNA damage was observed in Swiss albino mice after single intraperitoneal administration of 10 mg AgNP/kg body weight [81]. An *in vivo* study on the Sprague Dawley rats treated for 12 weeks by inhalation of AgNP (14–15 nm diameter) at highest concentrations of 3.24×10^6 particles/cm^3 (381 μg/m^3) demonstrated significantly higher DNA damage in the lung cells assay [82].

In vitro exposure of HepG2 cell line to AgNPs of two distinct sizes (10 and 75 nm) and coatings (citrate and polyvinylpyrrolidone) showed higher oxidative stress, as measured by the level of change in glutathione, after the exposure to smaller (10-nm) citrate-coated AgNPs than polyvinylpyrrolidone-coated, bigger AgNPs [83]. The direct interaction of AgNPs with fish red blood cells resulted in hemolysis and membrane damage in a size- and dose-dependent manner, with greater power of smaller AgNPs (15 nm) than 50- and 100-nm sized AgNPs [84]. Cytotoxic responses of mammalian cells to AgNPs demonstrated that although similar in the response, there are some cell lines more susceptible to the damage when exposed to AgNPs, usually depending on the level of antioxidant defense and metallothionein expression. That has been demonstrated in six different cell lines, epithelial cells (Caco-2, NHBE, RLE-6TN, and

BEAS-2B) and macrophages (RAW 264.7 and THP-1) of human and rodent origin, exposed to 20 nm citrate- and polyvinylpyrrolidone-coated AgNPs [85].

Exposure to silver nanoclusters that could be found after exposure to nanosilver-functionalized consumer products demonstrated genotoxic and cytotoxic effects in concentrations smaller than expected on human peripheral blood mononuclear cells *in vitro*, also inducing intracellular ROS generation, DNA damage, and apoptosis after 3, 6, and 12 h in a dose-dependent response [86]. Cytotoxic and genotoxic effects with elevated ROS levels were seen in reproductive and pulmonary cells (CHO-9, 15P-1, and RAW264.7) with most toxic effects seen for nanosilver $\leq$ 10 nm [87]. HepG2 cell line and mice primary liver cells treated for 24 h with concentrations in the range of 1–10 mg/L or 1–400 mg/L (for primary cells) demonstrated a dependent decrease in cell viability in both cells, with IC50 values of 2.764 and 121.7 mg/L in HepG2 and primary liver cells, respectively [88]. It seems that the control of uptake of the Ag^+ ion differs from the uptake of AgNPs and their penetration within the cells. Usually the penetration occurs with different endocytotic mechanisms, depending on the size, charge, and coating, but there is still much work to be done to clarify the exact mechanisms. Questions about the pathway of entrance into the cell surely depends on the AgNP size and coating and also on the type of the cells in the experiment, but together with that question, the usual ones on dissolution, recoating in the medium with the formation of a protein corona, and the overall stability of applied AgNPs remain the main unresolved issues. Also, as already mentioned, the relevant applied concentration of AgNP would be paramount. Sometimes, the established correlation of localization with responses of cells is lacking in studies, and it would be great to have the kinetics response and direct measurements of, for example, enzyme inhibition. However, as already mentioned, a standard frame of questions that every research should answer needs to be established to make the results more comparable to present any conclusion on the mechanism of AgNP cell interactions and toxicity [89]. In that line, a few *in vitro* papers focused only on the transition studies of the physico-chemical nature of the AgNPs, which was measured ahead of, during, and after incubation with various solutions (mostly cell culture media). Studies of AgNP behavior in a culture medium showed that dissolution was influenced by pH, and varied by the presence of bovine serum albumin (BSA), which can protect the AgNPs from proceeding with the disintegration processes by forming a layer named protein corona. An interesting example is the artificial lysosomal fluid (ALF — low pH of 4.5), aimed to resolve questions about particle behavior on the endocytic-lysosomal pathway. The collected data gave the outcome that the polysaccharide or hydrocarbon coating of the AgNPs was degraded after integration into the lysosomes, and the AgNP seems to be decreasing in size with time because of the elimination of the coating and its dissolution to silver ions [78].

So, in general, AgNP causes cytotoxicity, chromosomal instability, oxidative stress, apoptosis, intracellular calcium transients, cell cycle arrest, interference with DNA replication fidelity, and JNK activation in mammalian cells *in vitro* [90–96], whereas free-radical–induced oxidative stress and alteration of gene expression, blood–brain barrier destruction and astrocyte swelling, neuronal degeneration, and induced brain edema formation have been demonstrated through *in vivo* studies [97–99].

It is crucial to also mention that traditional toxicology methods are not always applicable for nanosilver studies, since nanosilver can interfere with the results [100–102]; usually due to the high surface area and higher adsorption and catalytic capacity, there is interference with light absorption detection [101].

Fragmentation and diversity of experimental approaches opens even more questions that are regularly emphasized in reviews. The effect of AgNPs on mammalian cells produced has been evidenced in many published papers, but with only few important points as a consequences that generally may be associated with AgNP exposure, such as impairment of the mitochondrial repiratory system resulting in oxidative stress, followed by inflammatory response, leading to observed cytotoxic and genotoxic effects [89]. However, there is still lack of data on detailed mechanism of AgNP action in living system following its exposure.

Diseases caused by silver are connected with the intake of an excessive amount of silver on the skin or in the body. Silver is not thought to be toxic to the cardiovascular, nervous, or reproductive systems of humans or a cancer-causing agent [103]. Silver can be absorbed into the body through the lungs, gastrointestinal tract, mucous membranes of the urinogenital tract, and the skin [3]. If silver is ingested, the efficiency with which it is absorbed is thought to be low ($\sim$10%) [103]. The most commonly reported response of humans to prolonged silver exposure is argyria (pigmentation or discoloration of the skin and nails) or argyrosis (silver in internal organs), which is a non-reversible and incurable disease, with no obvious long-term health effects found to be associated with either disorders [103], and in both conditions, the skin color changes to gray, gray-blue, or even black. Toxicity in humans occurs due to bio-available forms of silver and at very high doses. The acute symptoms of overexposure to silver nitrate in massive doses are decreased blood pressure, diarrhea, stomach irritation, and decreased respiration. Prolonged intake of low doses of silver salts can cause ulcers [104], fatty degeneration of the liver and kidneys, and changes in blood cells [103], whereas prolonged exposure of fibroblast and endothelial cells involved in wound healing caused cell loss in concentrations from 0.4 to 8.2 mg/L [105].

10.6.1 *Detoxification patterns for AgNPs*

Almost all organisms have detoxifying mechanisms to tackle metal exposure. In humans, it occurs as a precipitation of silver salts, usually as silver chloride, silver phosphate, or silver sulfide. These granules are deposited in the connective tissue outside the cell, called the basement membrane. High concentrations of silver can also occur in the basement membrane of the digestive tract, which has a strong ability to accumulate, retain, and eliminate the metal [106]. Silver penetration into the cell can be also stopped by macrophages [106] or lyzosomes, but before lyzosomes start to act, silver should be bonded to the protein with a large proportion of sulfhydryl groups, usually called metal-specific binding proteins, and then this complex aggregate is encapsuled into lyzosomes. If the concentration is too high, the lyzosome will break and start to leak. Silver deposits can also be found close to peripheral nerves and the blood–brain barrier, with no influence on the nervous tissue [107]. The principal detoxifying organ is the liver, with its detoxifying proteins called metallothioneins, and the liver is one of the target organs where most of silver accumulation happens.

10.7 Further Steps

The conclusion obtained from all the above-mentioned studies confirms the transformation of AgNPs by their environment, which considerably influences their chemical composition and configuration, bio-availability, and consequently, bio-accumulation and (eco)toxicity. A particular concern is how those transformations affect the impact of AgNPs on humans. The impact of AgNPs on human cells and tissues is significantly different, depending on the route of uptake and whether the AgNPs were unchanged or changed by environment. Experiments with AgNPs and other NMs has opened up a new level of complexity that must be encountered in all studies, beginning with the characterization of AgNPs not only in water but also in the exposure medium. Interference with the current applied methods to study AgNPs cannot be ignored. Information not only about the reactions of cells or organisms exposed to NPs suspended in solution, but also on the interactions of AgNPs in the surrounding medium seems to be crucial for understanding what actually interacts with cells/organisms, and this kind of approach should be performed regularly under agreed standard conditions for any given model that needs to be observed for response to AgNP exposure in order to avoid environmental disasters and losing species on a daily basis because of not only NM but also general pollution.

A good but still uncertain step in the risk assessment of AgNP is the modeling approach, which may investigate the whole life cycle of AgNPs in order to first minimize the release and then a possible contact with and contamination in the environment [108–110]. The release of the nanosilver into the environment can be controlled by size, surface modification, oxidant availability, composition of the medium, structured released materials, or release device and locality [111]. Therefore, it is necessary to establish and perform a risk assessment for products containing AgNPs, and the most important information will be the environmental mass loading [68]. Further toxicity testing should focus on realistic exposures and conditions addressing long-term, low-dose exposure (meaning the concentration exposure ranges in ng/L) and not perform only short-term acute toxicity testing. We must not forget the fact that ionic silver can be toxic at the levels of 50 ng/L and above, while AgNPs demonstrated toxicity at very low concentrations and caused changes in the development of aquatic organisms (e.g., fish). Also, bio-availability studies should also be included, since it seems that bio-availability and accumulation of AgNPs in the organisms exceeds all expected ratios, depending on the complexes which AgNPs can form in the water, sediment, or organism. The bio-availability of these materials will also depend on their uptake when ingested by organisms. Although AgNP may not be active or available when it binds, for example, with sulfides, cellular membranes are full of sulfides, thus providing a way for the tight attachment of AgNPs to the cell membranes, and although AgNP is not more acting as the active pollutant in the water, and by itself would not enter into the cell, binding to the membrane through sulfides is enabling the *Trojan horse* entering into the cell in any case. To quote Luoma, "silver is an old problem, with nanosilver as a new challenge" [68].

References

[1] Nowack B, Krug HF, Height M (2010) 120 years of nanosilver history: Implications for policy makers. *Environment Science Technology*, **45**: 1177–1183.

[2] Wong KKY, Liu X (2010) Silver nanoparticles-The real-silver bullet in clinical medicine? *Medicinal Chemistry Communications*, **1**: 125–131.

[3] Lansdown ABG (2004) A review of the use of silver in wound care: Facts and fallacies. *British Journal of Nursing*, **13**: 6–19.

[4] Project on emerging nanotechnologies. Available at: http://http://www.nanotechproject.org/cpi/. Accessed 3 April 2017.

[5] Jonas PR (1979) Environmental impacts of artificial ice nucleating agents. *Quarterly Journal of the Royal Meteorological Society*, **105**: 728–728.

[6] Walser T, Demou E, Lang DJ, Hellweg S (2011) Prospective environmental life cycle assessment of nanosilver t-shirts. *Environment Science & Technology*, **45**: 4570–4578.

[7] Blaser SA, Scheringer M, MacLeod M, Hungerbühler K (2008) Estimation of cumulative aquatic exposure and risk due to silver: Contribution of nano-functionalized plastics and textiles. *Science* of the *Total Environment*, **390**: 396–409.

[8] Benn TM, Westerhoff P (2008) Nanoparticle silver released into water from commercially available sock fabrics. *Environment Science & Technology*, **42**: 4133–4139.

[9] Geranio L, Heuberger M, Nowack B (2009) The behavior of silver nanotextiles during washing. *Environment Science & Technology*, **43**: 8113–8118.

[10] Cushen M, Kerry J, Morris M, Cruz-Romero M, Cummins E (2012) Nanotechnologies in the food industry — recent developments, risks and regulation. *Trends in Food Science and Technology*, **24**: 30–46.

[11] Lansdown ABG (2004) A review of the use of silver in wound care: Facts and fallacies. *British Journal of Nursing*, **13**: 6–19.

[12] Cao H, Liu X, Meng F, Chu PK (2010) Silver nanoparticles-modified films versus biomedical device-associated infections. *Wiley Interdisciplinary Review on Nanomedicine and Nanobiotechnology*, **2**: 670–684.

[13] Austin LA, Kang B, Yen CW, El-Sayed MA (2011) Nuclear targeted silver nanospheres perturb the cancer cell cycle differently than those of nanogold. *Bioconjugate Chemistry*, **22**: 2324–2331.

[14] Zeng XY, Zhang QK, Yu RM, Lu CZ (2010) A new transparent conductor: Silver nanowire film buried at the surface of a transparent polymer. *Advanced Materials*, **22**: 4484–4488.

[15] Li Q, Mahendra S, Lyon DY, Brunet L, Liga MV, Li D Alvarez PJJ (2008) Antimicrobial nanomaterials for water disinfection and microbial control: Potential applications and implications. *Water Resources*, **42**: 4591–4602.

[16] Davis JM (2007) How to assess the risks of nanotechnology: learning from past experience. *Journal of Nanoscience and Nanotechnology*, **7**: 402–409.

[17] Koehler A, Som C, Helland A, Gottschalk F (2008) Studying the potential release of carbon nanotubes throughout the application life cycle. *Journal of Cleaner Production*, **16**: 927–937.

[18] Som C, Berges M, Chaudhry Q, Dusinska M, Fernandes TF, Olsen SI, Nowack B (2010) The importance of life cycle concepts for the development of safe nanoproducts. *Toxicology*, **269**: 160–169.

[19] Som C, Wick P, Krug H, Nowack B (2011) Environmental and health effects of nanomaterials in nanotextiles and façade coatings. *Environment International*, **37**(6):1131–1142.

[20] Reidy B, Haase A, Luch A, Dawson KA, Lynch I (2013) Mechanisms of silver nanoparticle release, transformation and toxicity: A critical review of current knowledge and recommendations for future studies and applications. *Materials*, **6**(6): 2295–2350.

[21] Holder AL, Vejerano EP, Zhou X, Marr LC (2013) Nanomaterial disposal by incineration. *Environmental Science: Processes & Impacts*, **15**: 1652.

[22] U.S. Environmental Protection Agency (2008) Office of Solid Waste, Municipal Solid Waste in the United States: 2007 Facts and Figures, EPA530-R-08-010, Washington, DC.

[23] Quina MJ, Bordado JCM, Quinta-Ferreira RM (2011). Air pollution control in municipal solid waste incinerators, the impact of air pollution on health, economy, environment and agricultural sources.

MK Khallaf (ed.) (InTech, Rijeka, Croatia) DOI: 10.5772/17650. Available from: https://www.intechopen.com/books/the-impact-of-air-pollution-on-health-economy-environment-and-agricultural-sources/air-pollution-control-in-municipal-solid-waste-incinerators.

[24] National RCRA Hazardous Waste Biennial Report (2008), EPA530-R-08-012, Washington, DC.

[25] U.S. Environmental Protection Agency (2006) National Biennial RCRA Hazardous Waste Report. Available at: http://www.epa.gov/waste/inforesources/data/br05/national05.pdf. Accessed 3 April 2017.

[26] U.S. Environmental Protection Agency (2012) Proposed Amended Federal Plan Hospital/Medical/Infectious Waste Incinerators (HMIWI) Inventory, EPA-HQ-OAR-2011-0405-0002, in rulemaking docket: Federal Plan Requirements for Hospital/Medical/Infectious Waste Incinerators Constructed on or Before December 1, 2008. Available at: http://www.regulations.gov. Accessed 3 April 2017.

[27] Gottschalk F, Sonderer T, Scholz RW, Nowack B (2009) Modeled environmental concentrations of engineered nanomaterials (TiO(2), ZnO, Ag, CNT, Fullerenes) for different regions. *Environmental Science and Technology*, **43**: 9216–9222.

[28] Adams NWH, Kramer JR (1999) Silver speciation in wastewater effluent, surface waters, and pore waters. *Environmental Science and Technology*, **18**: 2667–2673.

[29] Mueller NC, Nowack B (2008) Exposure modeling of engineered nanoparticles in the environment. *Environmental Science and Technology*, **42**: 4447–4453.

[30] Klaine SJ, Alvarez PJJ, Batley GE, Fernandes TF, Handy RD, Lyon DY, Mahendra S, McLaughlin MJ, Lead JR (2008) Nanomaterials in the environment: Behavior, fate, bioavailability, and effects. *Environmental Toxicology and Chemistry*, **27**: 1825–1851.

[31] Fabrega J, Luoma SN, Tyler CR, Galloway TS, Lead JR (2011) Silver nanoparticles: Behaviour and effects in the aquatic environment. *Environment International*, **37**: 517–531.

[32] Glover RD, Miller JM, Hutchison JE (2011) Generation of metal nanoparticles from silver and copper objects: Nanoparticle dynamics on surfaces and potential sources of nanoparticles in the environment. *ACS Nano*, **5**: 8950–8957.

[33] O'Brien N, Cummins E (2010) Nano-scale pollutants: Fate in Irish surface and drinking water regulatory systems. *Human Ecological Risk Assessment*, **16**: 847–872.

[34] Reidy B, Haase A, Luch A, Dawson KA, Lynch I (2013) Mechanisms of silver nanoparticle release, transformation and toxicity: A critical review of current knowledge and recommendations for future studies and applications. *Materials*, **6**: 2295–2350. doi:10.3390/ma6062295.

[35] Bondarenko O, Juganson K, Ivask A, Kasemets K, Mortimer M, Kahru A (2013) Toxicity of Ag, CuO and ZnO nanoparticles to selected environmentally relevant test organisms and mammalian cells in vitro: A critical review. *Archives of Toxicology*, **87**(7): 1181–1200.

[36] Maurer-Jones MA, Gunsolus IL, Murphy CJ, Haynes CL (2013) Toxicity of engineered nanoparticles in the environment, *Analytical Chemistry*, **85**(6): 3036–3049.

[37] Franco A, Hansen SF, Olsen SI, Butti L (2007) Limits and prospects of the "incremental approach" and the European legislation on the management of risks related to nanomaterials. *Regulatory Toxicology and Pharmacology*, **48**: 171–183.

[38] Hansen SF, Baun A (2012) European regulation affecting nanomaterials — review of limitations and future recommendations. *Dose-Response*, **10**: 364–383.

[39] Guideline for control and prevent the environmental release of NPs (2009) (Japan Ministry of the Environment, Guideline for Preventive Environmental Impact from Industrial Nanomaterials, Tokyo, 2009.). Available at: http://http://www.env.go.jp/en/index.html. Accessed 3 April 2017.

[40] Breggin LK, Pendergrass J (2007) Where does the nano go? New report on end-of-life regulation of nanotechnologies, Woodrow Wilson International Center for Scholars, PEN, Washington, DC. Available at: http://www.nanotechproject.org/filedownload/files/NanoEnd-of-LifePen10.pdf.

[41] EC (2008) Follow-up to the 6th Meeting of the REACH Competent Authorities for the implementation of Regulation (EC) 1907/2006; (REACH). European Commission, Brussels, Belgium. Available at: shttp://ec.europa.eu/environment/chemicals/reach/pdf/nanomaterials.pdf.

[42] European Parliament and European Council (2006) Directive 2006/121/EC. *Official Journal of the European Union*, **561**(L396): 850.

[43] OECD (2004) Guidelines for the testing of chemicals. Section 2: Effects on Biotic Systems. Test No. 202: Daphnia sp. Acute immobilisation test. Organisation for Economic Co-operation and Development, Paris, France.

[44] OECD (2011) Guidelines for the testing of chemicals. Section 2: Effects on Biotic Systems. Test No. 201: Freshwater Alga and Cyanobacteria, growth inhibition test. Organisation for Economic Co-operation and Development, Paris, France.

[45] OECD (1992) Guidelines for the testing of chemicals. Section 2: Effects on Biotic Systems. Test No. 203: Fish, acute toxicity test. Organisation for Economic Co-operation and Development, Paris, France.

[46] Nowack B, Ranville JF, Diamond S, Gallego-Urrea JA, Metcalfe C, Rose J, Horne N, Koelmans AA, Klaine SJ (2012) *Environmental Toxicology and Chemistry*, **31**(1): 50–59. doi: 10.1002/etc.726.

[47] Auffan M, Rose J, Wiesner MR, Bottero JY (2009) Chemical stability of metallic nanoparticles: A parameter controlling their potential cellular toxicity in vitro. *Environmental Pollution*, **157**: 1127–1133.

[48] Horie M, Kato H, Fujita K, Endoh S, Iwahashi H (2012). In vitro evaluation of cellular response induced by manufactured nanoparticles. *Chemical Research in Toxicology*, **25**: 605–619.

[49] Levard C, Hotze EM, Lowry GV, Brown Jr.GE (2012) Environmental transformation of silver nanoparticles: impact on stability and toxicity. *Environmental Science and Technology*, **46**: 6900–6914.

[50] European Commission (2013) Communication from the Commission to the European Parliament, the Council and the European Economic and Social Committee. Second Regulatory Review on Nanomaterials. Brussels, October 3, 2012, COM (2012) 572 final.

[51] Klaine S, Koelmans AA, Horne N, Carley S, Handy RD, Kapustka L, Nowack B, von der Kammer F (2012) Paradigms to assess the environmental impact of manufactured nanomaterials. *Environmental Toxicology and Chemistry*, **31**: 3–14.

[52] Kahru A, Savolainen K (2010) Potential hazard of nanoparticles: From properties to biological and environmental effects. *Toxicology*, **269**: 89–91.

[53] von der Kammer F, Ferguson PL, Holden PA, Masion A, Rogers K, Klaine SJ, Koelmans AA, Horne N, Unrine JM (2012) Analysis of nanomaterials in complex matrices (environment and biota): General considerations and conceptual case studies. *Environmental Toxicology and Chemistry*, **31**: 32–49.

[54] Vinković Vrček I, Pavičić I, Crnković T, Jurašin D, Babič M, Horák D, Lovrić M, Ferhatović L, Ćurlin M, Gajović S (2015) Does surface coating of metallic nanoparticles modulate their interference with in vitro assays? *RSC Advances*, **5**: 70787–70807.

[55] Milić M, Leitinger G, Pavičić I, Zebić Avdičević M, Dobrović S, Goessler W, Vinković Vrček I (2015) Cellular uptake and toxicity effects of silver nanoparticles in mammalian kidney cells. *Journal of Applied Toxicology*, **35**(6): 581–592.

[56] Luoma SN, Ho YB, Bryan GW (1995) Fate, bioavailability and toxicity of silver in estuarine environments. *Marine Pollution Bulletin*, **31**: 44–54.
[57] CEC (1996) CEC (Commission of the European Communities) Technical Guidance Document in Support of Commission Directive 93/67/EEC on Risk Assessment for New Notified Substances. Part II, Environmental Risk Assessment. Office for Official Publications of the European Communities, Luxembourg.
[58] Fisher NS, BoheM, Teyssie JL (1994) Accumulation and toxicity of Cd, Zn, Ag, and Hg in four marine phytoplankters. *Marine Ecology Progress Series*, **18**: 201–213.
[59] Wang WX (2001) Comparison of metal uptake rate and absorption efficiency in marine bivalves. *Environmental Toxicology and Chemistry*, **20**: 1367–1373.
[60] Wang WX, Fisher NS, Luoma SN (1996) Kinetic determinations of trace element bioaccumulation in the mussel Mytilus edulis. *Marine Ecology Progress Series*, **140**: 91–113.
[61] Wang WX (2002) Interactions of trace metals and different marine food chains. *Marine Ecology Progress Series*, **243**: 295–309.
[62] Reinfelder JR, Wang WX, Luoma SN, Fisher NS (1997) Assimilation efficiencies and turnover rates of trace elements in marine bivalves: A comparison of oysters, clams, and mussels. *Marine Biology*, **129**: 443–452.
[63] Wang WX, Fisher NS (1999) Assimilation efficiencies of chemical contaminants in aquatic invertebrates: a synthesis. *Environmental Toxicology and Chemistry*, **18**: 2034–2045.
[64] Reinfelder JR, Fisher NS, Luoma SN, Nichols JW, Wang WX (1998) Trace element trophic transfer in aquatic organisms: A critique of the kinetic model approach. *Science of the Total Environment*, **219**: 117–135.
[65] Griscom SB, Fisher NS, Luoma SN (2002) Kinetic modeling of Ag, Cd and Co bioaccumulation in the clam Macoma balthica: Quantifying dietary and dissolved sources. *Marine Ecology Progress Series*, **240**: 127–141.
[66] Cheung MS, Wang WX (2005) Influences of subcellular compartmentalization in different prey on the transfer of metals to a predatory gastropod from different prey. *Marine Ecology Progress Series*, **286**: 155–166.
[67] Luoma SN (1989) Can we determine the biological availability of sediment-bound trace elements? *Hydrobiologia*, **176/177**: 379–396.
[68] Luoma SN (2008) Silver Nanotechnologies and the Environment Old Problems or New Challenges? PEN 15. Project on Emerging Nanotechnologies, Woodrow Wilson International Center for Scholars, Washington, DC.

[69] Panyala N R, Pena-Méndez E M, Havel J (2008) Silver or silver nanoparticles: a hazardous threat to the environment and human health. *Journal of Applied Biomedicine*, **6**(3): 117–129.

[70] Canas JE, Shresta B, Li S (2013) Environmental and human health effects of nanomaterials used in water and waste water treatments. In *Nanotechnology for Water and Wastewater Treatment*, PNL Lens *et al.* (eds.) (IWA Publishing, London, UK), pp. 27–43.

[71] Turco RF, Bischoff M, Tong ZH, Nies L (2011) Environmental implications of nanomaterials: Are we studying the right thing? *Current Opinion in Biotechnology*, **22**(4): 527–532.

[72] Aiken GR, Hsu-Kim H, Ryan JN (2011) Influence of dissolved organic matter on the environmental fate of metals, nanoparticles, and colloids. *Environmental Science and Technology*, **45**(8): 3196–3201.

[73] Laera G, Lens PNL (2013) Nanotechnology for water and wastewater treatment: Potentials and limitations. In *Nanotechnology for Water and Wastewater Treatment*, PNL Lens *et al.* (eds.) (IWA Publishing, London, UK), pp. 1–26.

[74] Świdwińska-Gajewska AM, Czerczak S (2015) Nanosilver–occupational exposure limits. *Medycyna Pracy*, **66**(3): 429–442.

[75] Research 2015: Global Markets for Nanocomposites, Nanoparticles, 2 Nanoclays, and Nanotubes. BCC Research, published in May 2014, Analyst — Andrew McWilliams, Code — NAN021F, Available at: https://www.bccresearch.com/market-research/nanotechnology/nanocomposites-market-nan021f.html. Accessed 3 April 2017.

[76] Piccinno F, Gottschalk F, Seeger S, Nowack B (2012) Industrial production quantities and uses of ten engineered nanomaterials in Europe and the world. *Journal of Nanoparticle Research*, **14** (9): 1e11.

[77] Vrček IV, Žuntar I, Petlevski R, Pavičić I, Dutour Sikirić M, Ćurlin M, Goessler W (2016) Comparison of in vitro toxicity of silver ions and silver nanoparticles on human hepatoma cells. *Environmental Toxicology*, **31**(6): 679–692.

[78] Dhawan A, Sharma V (2010) Toxicity assessment of nanomaterials: methods and challenges. *Analytical and Bioanalytical Chemistry*, **398**(2): 589–605.

[79] Manke A, Wang L, Rojanasakul Y (2013) Mechanisms of nanoparticle-induced oxidative stress and toxicity. *Biomedical Research International*, **2013**: 942916.

[80] Souza TA, Franchi LP, Rosa LR, da Veiga MA, Takahashi CS (2016) Cytotoxicity and genotoxicity of silver nanoparticles of different sizes

in CHO-K1 and CHO-XRS5 cell lines. *Mutation Research — Genetic Toxicology* and *Environmental Mutagenesis*, **795**: 70–83.

[81] Ghosh M, Sinha S, Chakraborty A, Mallick SK, Bandyopadhyay M, Mukherjee A (2012) In vitro and in vivo genotoxicity of silver nanoparticles. *Mutation Research*, **749**(1–2): 60–69.

[82] Cho HS, Sung JH, Song KS, Kim JS, Ji JH, Lee JH, Ryu HR, Ahn K, Yu IJ (2013) Genotoxicity of silver nanoparticles in lung cells of Sprague Dawley rats after 12 weeks of inhalation exposure. *Toxics*, **1**: 36–45.

[83] Prasad RY, McGee JK, Killius MG, Suarez DA, Blackman CF, DeMarini DM, Simmons SO (2013) Investigating oxidative stress and inflammatory responses elicited by silver nanoparticles using high-throughput reporter genes in HepG2 cells: Effect of size, surface coating, and intracellular uptake. *Toxicology In Vitro*, **27**(6): 2013–2021.

[84] Chen LQ, Fang L, Ling J, Ding CZ, Kang B, Huang CZ (2015) Nanotoxicity of silver nanoparticles to red blood cells: size dependent adsorption, uptake, and hemolytic activity. *Chemical Research in Toxicology*, **28**(3): 501–509.

[85] Zhang H, Wang X, Wang M, Li L, Chang CH, Ji Z, Xia, Nel AE (2015) Mammalian cells exhibit a range of sensitivities to silver nanoparticles that are partially explicable by variations in antioxidant defense and metallothionein expression. *Small*, **11**(31): 3797–3805.

[86] Orta-García ST, Plascencia-Villa G, Ochoa-Martínez AC, Ruiz-Vera T, Pérez-Vázquez FJ, Velázquez-Salazar JJ, Yacamán MJ, Navarro-Contreras HR, Pérez-Maldonado IN (2015) Analysis of cytotoxic effects of silver nanoclusters on human peripheral blood mononuclear cells 'in vitro'. *Journal of Applied Toxicology*, **35**(10): 1189–1199.

[87] Zapor L (2016) Effects of silver nanoparticles of different sizes on cytotoxicity and oxygen metabolism disorders in both reproductive and respiratory system cells. *Archives of Environmental Protection*, **42**(4): 32–47.

[88] Faedmaleki F, H Shirazi F, Salarian AA, Ahmadi Ashtiani H, Rastegar H (2014) Toxicity effect of silver nanoparticles on mice liver primary cell culture and HepG2 cell line. *Iran Journal of Pharmaceutical Research*, **13**(1): 235–242.

[89] de Lima R, Seabra A B, Durán N (2012) Silver nanoparticles: A brief review of cytotoxicity and genotoxicity of chemically and biogenically synthesized nanoparticles, *Journal of Applied Toxicology*, **32**(11): 867–879.

[90] Asharani PV, Hande MP, Valiyaveettil S (2009) Antiproliferative activity of silver nanoparticles. *BMC Cell Biology*, **10**: 65.

[91] AshaRani PV, Low Kah Mun G, Hande MP, Valiyaveettil S (2009) Cytotoxicity and genotoxicity of silver nanoparticles in human cells. *ACS Nano*, **3**(2): 279–290.

[92] Foldbjerg R, Dang DA, Autrup H (2010) Cytotoxicity and genotoxicity of silver nanoparticles in the human lung cancer cell line, A549. *Archives in Toxicology*, **85**(7): 743–750.

[93] Hsin YH, Chen CF, Huang S, Shih TS, Lai PS, Chueh PJ (2008) The apoptotic effect of nanosilver is mediated by a ROS- and JNK-dependent mechanism involving the mitochondrial pathway in NIH3T3 cells. *Toxicology Letters*, **179**(3): 130–139.

[94] Kawata K, Osawa M, Okabe S (2009) In vitro toxicity of silver nanoparticles at noncytotoxic doses to HepG2 human hepatoma cells. *Environmental Science and Technology*, **43**(15): 6046–6051.

[95] Miura N, Shinohara Y (2009) Cytotoxic effect and apoptosis induction by silver nanoparticles in HeLa cells. *Biochemical and Biophysical Research Communications*, **390**(3): 733–737.

[96] Yang W, Shen C, Ji Q, An H, Wang J, Liu Q, Zhang Z (2009) Food storage material silver nanoparticles interfere with DNA replication fidelity and bind with DNA. *Nanotechnology*, **20**(8): 085102.

[97] Tang J, Xiong L, Wang S, Wang J, Liu L, Li J, Yuan F, Xi T (2009) Distribution, translocation and accumulation of silver nanoparticles in rats. *Journal of Nanoscience and Nanotechnology*, **9**(8): 4924–4932.

[98] Rahman MF, Wang J, Patterson TA, Saini UT, Robinson BL, Newport GD, Murdock RC, Schlager JJ, Hussain SM, Ali SF (2009) Expression of genes related to oxidative stress in the mouse brain after exposure to silver-25 nanoparticles. *Toxicology Letters*, **187**(1): 15–21.

[99] Sharma HS, Hussain S, Schlager J, Ali SF, Sharma A (2010) Influence of nanoparticles on blood–brain barrier permeability and brain edema formation in rats. *Acta Neurochirurgica Supplement*, **106**: 359–364.

[100] Doak SH, Griffiths SM, Manshian B, Singh N, Williams PM, Brown AP, Jenkins GJ (2009) Confounding experimental considerations in nanogenotoxicology. *Mutagenesis*, **24**(4): 285–293.

[101] Kroll A, Pillukat MH, Hahn D, Schnekenburger J (2009) Current in vitro methods in nanoparticle risk assessment: limitations and challenges. *European Journal of Pharmaceutics and Biopharmaceutics*, **72**(2): 370–377.

[102] Monteiro-Riviere NA, Inman AO, Zhang LW (2009) Limitations and relative utility of screening assays to assess engineered nanoparticle toxicity in a human cell line. *Toxicology and Applied Pharmacology*, **234**(2): 222–235.

[103] Drake PL, Hazelwood KJ (2005). Exposure related health effects of silver and silver compounds: A review. *Annals of Occupational Hygiene*, **49**: 575–589.

[104] Wadhera A, Fung M (2005) Systemic argyria associated with ingestion of colloidal silver. *Dermatology Online Journal*, **11**(1): 12. Available at: http://dermatology.cdlib.org/111/case_reports/argyria/wadhera.html, Accessed 13 March 2017.

[105] Hidalgo E, Domínguez C (1998) Study of cytotoxicity mechanisms of silver nitrate in human dermal fibroblasts. *Toxicology Letters*, **98**: 169–179.

[106] Baudin JP, Garnier-Laplace J, Lambrechts A (1994). Uptake from water, depuration and tissue distribution of 110m Ag in a freshwater fish Cyprinus carpio. *Water, Air, and Soil Pollution*, **72**: 129–141.

[107] Lansdown ABG (2007) Critical observations on the neurotoxicity of silver. *Critical Reviews in Toxicology*, **37**: 237–250.

[108] Gutowski TG (2002) Environmentally benign manufacturing and ecomaterials; product induced material flows. *Materials Transactions*, **43**(3): 359–363.

[109] Kobayashi H (2005) strategic evolution of eco-products: a product life cycle planning methodology. *Research in Engineering Design*, **16**(1–2): 1–16.

[110] Serban D, Man E, Ionescu N, Roche T (2004) A TRIZ approach to design for environment. In *Product Engineering*, D Talab, T Roche (eds.) (Springer, Netherlands), pp. 89–100.

[111] Lem KW, Hsu SH, Lee DS, Iqbal Z, Sund S, Curran S, Brumlik C, Choudhury A, Hu DSG, Chiu N, Lem RC, Haw JR (2012) Waste minimization for the safe use of nanosilver in consumer products — Its impact on the eco-product design for public health, public health — Methodology, environmental and systems issues. In *Public Health — Methodology, Environmental and Systems Issues*, J Maddock (ed.) (InTech, Rijeka, Croatia), pp. 217–248. Available at: http://www.intechopen.com/books/public-health-methodology-environmental-and-systems-issues/waste-minimization-for-the-safe-use-of-nanosilver-in-consumer-products-its-impact-on-the-eco-pro. Accessed 13 March 2017.

Index

B

C

R

S

X

Y

Z